XII Maghreb Days of Material Sciences

IOP Conference Series: Materials Science and Engineering
Volume 186

Fez, Morocco
19 – 21 November 2015

Editors:

Ali Ahaitouf
Ben Mohammadi

ISBN: 978-1-5108-3773-7
ISSN: 1757-8981

Printed from e-media with permission by:

Curran Associates, Inc.
57 Morehouse Lane
Red Hook, NY 12571

Some format issues inherent in the e-media version may also appear in this print version.

Printed by Curran Associates, Inc. (2017)

For permission requests, please contact the Institute of Physics
at the address below.

Institute of Physics
Dirac House, Temple Back
Bristol BS1 6BE UK

Phone: 44 1 17 929 7481
Fax: 44 1 17 920 0979

techtracking@iop.org

Additional copies of this publication are available from:

Curran Associates, Inc.
57 Morehouse Lane
Red Hook, NY 12571 USA
Phone: 845-758-0400
Fax: 845-758-2633
Email: curran@proceedings.com
Web: www.proceedings.com

Table of contents

Volume 186

XII Maghreb Days of Material Sciences
19–21 November 2015, Fez, Morocco

Accepted papers received: 1 March 2017
Published online: 13 March 2017

Preface

Nanotechnologies and new-generation materials

Building materials & Heritage

Energy

Environment

IOP Conf. Series: Materials Science and Engineering **186** (2017) 011001 doi:10.1088/1757-899X/186/1/011001

Preface

XIIth Maghreb Days of Material Sciences JMSM'2015

Overview

The JMSM (the XIIth Maghreb Days of Material Sciences) are organized each two years in a Maghreb country, the 2015 edition was held between **19-21 November 2015** on the campus of **Faculty of Sciences and Technology Sidi Mohamed Ben Abdellah University of Fez, the imperial city of Morocco.**

The JMSM event was an opened exchange discussion forum, scientific debate and interchange ideas and knowledge between the Maghreb communities and their national and international colleagues. The conference focused on materials science, from the synthesis step, processes development, to industrial use and exploitation, through modeling and observation of the physicochemical phenomena that govern their functioning. It was also a good opportunity for researchers to establish cooperation relationship, enhance mutual exchanges, and promote industrial research and development, which ensure much better socio-economic integration of the university and strengthening its role as a nucleus of development and pioneer through research and innovation.

The JMSM'2015 program included many topics related to Material Sciences namely: Nanotechnologies and new-materials generation, Biomedical and Biology, Photonics, Electronics, Building materials &Heritage, Energy, Environment, Security and Transport.

JMSM'2015 has attracted 250 submissions, after the review process, 155 papers have been accepted between oral and posters communications and 40 papers were selected for publication.

The conference was well attended by participants from many different organizations including researchers, PhD students, industrials and academics from different countries.

All our thanks and acknowledgments go to our sponsors who supported JMSM'2015 Fès, and all those who contributed to the success of this conference. Especially, we would like to acknowledge ERSI Laboratory members, the organizing and scientific committees for their valuable help, time and effort in the peer-reviewing and the organization.

Editors

IOP Conf. Series: Materials Science and Engineering **186** (2017) 011001 doi:10.1088/1757-899X/186/1/011001

Introduction

Laboratory of Renewable Energies and Intelligent Systems (LERSI)

Organizer lab

Laboratory of Renewable Energies and Intelligent Systems

Electronics and renewable energies are among the strategic development axis, sustained by the high level authorities of the Morocco kingdom. Both of these fields are substantial deposition of direct and indirect jobs.

The ERSI laboratory is coming to support this work. Our university and our institutions have long initiated the training of young graduates in the fields of electronics microelectronics (Master FST), embedded systems (engineers cursus ENSA) and telecommunications systems (engineers cursus FST and ENSA). With their expertise in these areas and given the recent funded research projects in the field of renewable energy as intelligent systems, three teams from both institution (FST) and (ENSA - National School of Applied Sciences) have decided to pool their efforts and establish a research laboratory specializing in renewable energy and intelligent systems.

This laboratory is formed by three teams:

-Systems of renewable energies and energy efficiency (SEREE): this team is formed by researcher coming from the FST, headed by Prof. MECHAQRANE Abdellah, and has a long history and facilities that allow it to qualify solar and wind fields in the Fez Boulmane region.

-Telecommunication network and software engineering (MSE), this team is formed by the ENSA researcher, headed by Prof. BENNANI DOSSE Saad. This team has a long research experience in the fields of telecommunications, antennas and related items. Also some of involved researchers work on software engineering to develop some tools for internet of things, data processing and mailing.

-The third team is Microelectronic and embedded systems. This team is a mixed, as it involves researcher from the two establishments (FST and ENSA). It is headed by Prof. A. AHAITOUF and works on microelectronics component, integrated analogue and digital circuits design, VLSI design, embedded systems, video processing etc…

Motivation:

Our skills are varied, diverse and complementary. They include electronics, telecommunications, processing of the signal applied mathematics, numerical methods, and hardware programming languages. We chose to combine renewable energy and intelligent systems already at the name of the laboratory for better visibility and to show the consistency that does exist between these two themes and that sometimes goes unnoticed. Indeed any modern system of renewable energy incorporate an electronic system for command, sun tracking and/ or energy management and optimization. The electronic systems are today incorporating advanced software programs, communications systems ranging from classic wire-connected antennas and networks to the more modern, functioning with all types of waves or wireless systems such Wi-Fi, Zig-Bee, Bluetooth ... etc.

Of course their design responds to a drastic set of specifications both in terms of response time (software design) the space occupation (hardware design). These constraints show how the balance between these various skills is desirable for successful smart and futuristic systems. In fact an intelligent system is not only a system controlled by computer, but it integrates some additional elements including temperature, pressure, humidity and radiation sensors, etc ... However, it can be qualified as smart that when it is used to control and manage the multitude of sensors, data and treatments and making the good decision in the convenient time.

IOP Conf. Series: Materials Science and Engineering **186** (2017) 011001 doi:10.1088/1757-899X/186/1/011001

Moreover, alongside the renewable energy systems, the development of embedded systems remains a priority for our laboratory, these miniaturized and intelligent systems can make our teams competitive again large international groups in these high technology areas and high added values.

We have initiated innovative training at national level and hope to move forward in the areas research and develop useful intelligent systems in territorial surveillance, biomedical, video surveillance etc...

It is as natural as any design or invention that will be innovative when it fits perfectly into its environment, this is done in most cases by the communication and exchange of information between a command emitter (master) and a command receiver (slave) and this can only be implemented by a true knowledge of standards and standard telecommunications. The huge potential provided by the RTL team of the ENSA of Fez.

Also in this laboratory we will open our field of investigations on all other issues relating to the progress of our specific research and ranging from optimization algorithms, numerical methods through Web Services, the driven engineering models, and design and implementation of intelligent systems enhancing the concept ABC (Always Best connected) for the next generation of connected vehicles.

Given the complementarily of our skills, given the desire of each member and given the developed research at the university, we decided the common creation and accreditation of this laboratory, to enlarge and diversify the offer of our university and to fix us a new challenging project.

Partnership

National Partners

- Moroccan Foundation for Advanced Science, Innovation and Research (MASCIR)
- National Center for Scientific and Technical Research (CNRST)
- Research Institute for Solar Energy and New Energies (IRESEN)
- International University of Rabat
- Cadi Ayyad University Marrakech
- Faculty of Sciences and Technology Mohammedia
- National School of Applied Sciences Fez, Morocco (ENSAF)
- ENS Casablanca
- Zodiac Aerospace
- AIC Metallurgie

International Partners

- Lorraine University LMOPS Laboratory
- Georgiatech Lorraine UMI2958GT-CNRS
- Sherbrooke University Canada
- Lille University 1 IEMN
- INSA Lyon
- INSA EuroMed
- ESEO Angers
- ESIEE Paris
- Bordeaux University
- ENIS de Sfax -Tunisia
- Belfort University
- Limoges University
- Sophia Antipolis Nice LEAT Laboratory
- University Beijing, China
- Iasé Faculty, Romania University

Editors

IOP Conf. Series: Materials Science and Engineering **186** (2017) 011001 doi:10.1088/1757-899X/186/1/011001

XIIth Maghreb Days of Material Sciences JMSM'2015
Photographs
19-21 November 2015 on the campus of Faculty of Sciences and Technology
Laboratory of Renewable Energies and Intelligent Systems (LERSI)
Sidi Mohamed Ben Abdellah University Fez Morocco

IOP Conf. Series: Materials Science and Engineering **186** (2017) 011001 doi:10.1088/1757-899X/186/1/011001

IOP Conf. Series: Materials Science and Engineering **186** (2017) 011001 doi:10.1088/1757-899X/186/1/011001

IOP Conf. Series: Materials Science and Engineering **186** (2017) 011001 doi:10.1088/1757-899X/186/1/011001

IOP Conf. Series: Materials Science and Engineering **186** (2017) 011001 doi:10.1088/1757-899X/186/1/011001

List of Committee: JMSM'2015

Honour Committee

Prof. Omar ASSOBHEI

President of Sidi Mohamed Ben Abdellah University
Fez, Morocco

Prof. Mustapha IJJAALI

Dean of Faculty of Sciences and Technology
Fez, Morocco

Prof. El Mestafa EL HADRAMI

Vice Dean, in charge of Scientific Research and Cooperation
FST Fez, Morocco

Prof. Taoufiq ACHIBAT

Vice Dean, in charge of pedagogical and student affairs
FST Fez, Morocco

Chairman and Editors

Ali AHAITOUF

Director of Laboratory Renewable Energies and Intelligent Systems (LERSI)
Electrical Department, Faculty of Sciences and Technology
Sidi Mohamed Ben Abdellah University Fez, Morocco
ali.ahaitouf@usmba.ac.ma

Zahra BEN MOHAMMADI

Laboratory of Renewable Energies and Intelligent Systems
FST, Sidi Mohamed Ben Abdellah University Fez, Morocco
zahra.benmohammadi@usmba.ac.ma

Local Organizing Committee

A Ahaitouf
(FST -Fès)

Az Ahaitouf
(FP-Taza)

L Bejjit
(EST -Meknès)

S Bennani Dosse
(ENSA -Fès)

F Errahimi
(FST -Fès)

N ES-Sbai
(FST -Fès)

M Haddad
(FS -Meknès)

S Haouache
(FST -Fès)

IOP Conf. Series: Materials Science and Engineering **186** (2017) 011001 doi:10.1088/1757-899X/186/1/011001

Jorio M.
(FST -Fès)

Y Lakhrissi
(ENSA -Fès)

A Mansouri
(ENSA -Fès)

A Mechaqrane
(FST -Fès)

K Moughamir
(FST -Fès)

M Razi
(FST -Fès)

Organizing Committee (PhD. Students)

Z Ben Mohammadi
(FST -Fès)

M Bououd
(FST -Fès)

S El Himer
(FST -Fès)

S El-Yahyaoui
(FST -Fès)

Website and Online submission

Ben Mohammadi Z (FST -Fès)
Bououd M. (FST -Fès)

IOP Conf. Series: Materials Science and Engineering **186** (2017) 011001 doi:10.1088/1757-899X/186/1/011001

Registration and payment

Amor S.
(FST -Fès)

Ben Mohammadi Z.
(FST -Fès)

Bououd M.
(FST -Fès)

Cheddadi Y.
(FST -Fès)

Diouri O.
(FST -Fès)

El Ansari A.
(FST -Fès)

S El-Yahyaoui
(FST -Fès)

Tabakh I.
(FST- Fès)

Scientific Committee Members

Pr. Abarkan M. (FP -Taza)
Pr. Abd-Lefdil M. (FS -Rabat)
Pr. Ahaitouf A. (FST -Fès)
Pr. Ahaitouf Az. (FP -Taza)
Pr. Aroui H. (FS- Tunisie)
Pr. Atmani M. (FST -Fès)
Pr. Baitoul M. (FSDM -Fès)
Pr. Barhdadi A. (ENS -Rabat)
Pr. Bejjit L. (EST -Meknès)
Pr. Bennani Dosse S. (ENSA -Fès)
Pr. Benyoussef A. (FS -Rabat)
Pr. Bouabid K. (FS -Agadir)
Pr. Boukir A. (FST -Fès)
Pr. El-Barkany A. (FST -Fès)
Pr. El-Khalfi A.(FST -Fès)
Pr. Ennadi A. (FST -Fès)
Pr. Errahimi F. (FST -Fès)
Pr. ES-Sbai N. (FST -Fès)

IOP Conf. Series: Materials Science and Engineering **186** (2017) 011001 doi:10.1088/1757-899X/186/1/011001

Pr. Haddad M. (FS -Meknès)
Pr. Hamedi L. (FST -Fès)
Pr. Hazm J.-E. (FST -Fès)
Pr. Hourmatallah A. (ENS -Fès)
Pr. Idrissi Kandri N. (FST -Fès)
Pr. Ihlal A.(FS -Agadir)
Pr. Ijjaali M. (FST -Fès)
Pr. Jabrane R. (FST -Fès)
Pr. Jorio A. (FSDM -Fès)
Pr. Jorio M. (FST -Fès)
Pr. Kerouad M. (FS -Meknès)
Pr. Lhassani A. (FST -Fès)
Pr. Mansouri A. (ENSA -Fès)
Pr. Mechaqrane A. (FST -Fès)
Pr. Moughamir K. (FST -Fès)
Pr. Ouazzani-Jamil M. (UP -Fès)
Pr. Ougazzaden A. (GT -Lorraine)
Pr. Soltani A. (IEMN -Lille)
Pr. Zazoui M. (FST -Mohammédia)
Pr. Zorkani I. (FSDM -Fès)

Sponsoring organisations/funding acknowledgements

Research Institute for Solar Energy and New Energies (IRESEN)

National Center for Scientific and Technical Research (CNRST)

Urban Community of Fez

Fès SAISS

Banque Populaire Morocco (BP)

Le Groupement d'Assurance du Supérieur (GASUP)

Organizers

Sidi Mohamed Ben Abdellah University Fez, Morocco (USMBA)

Faculty of Sciences and Technology Fez, Morocco (FSTF)

National School of Applied Sciences Fez, Morocco (ENSAF)

Laboratory of Renewable Energies and Intelligent Systems (LERSI)

IOP Conf. Series: Materials Science and Engineering **186** (2017) 011001 doi:10.1088/1757-899X/186/1/011001

Keynote & Invited Speakers (Plenary conference)

Prof. Hassan MAHER
Sherbrooke University, Canada
Plenary conference title: **Gallium nitride microelectronic components for high power applications**

Prof. Abdelkader SOUIFI
INSA Institut des Nanotechnologies Lyon, France
Plenary conference title: **The advances in nanotechnology for the energy, demographic and diversification challenges of functions on nanoelectronic chips on silicon**

Prof. Abdelilah BENYOUSSEF
MAScIR-AH2ST- Mohammed V University Rabat, Maroc
Plenary conference title: **Design of new materials and functional nanomaterials by the functional theory of density**

Prof. Mimoun EL MARSSI
Université de Picardie Jules Verne, France
Plenary conference title: **The contribution of Raman spectroscopy to the study of ferroelectric oxides**

Prof. Ali SOLTANI
IEMN Université des Sciences et Technologies de Lille, France
Plenary conference title: **Potentials and applications of the GaN. What possible future for GaN on silicon substrate for power applications in microwave and millimeter wave?**

Prof. Izeddin ZORKANI
FSDM, Sidi Mohamed Ben Abdellah University Fez, Maroc
Plenary conference title: **Nanomaterials, optical properties and application opportunities for photovoltaics**

IOP Conf. Series: Materials Science and Engineering **186** (2017) 011001 doi:10.1088/1757-899X/186/1/011001

List of Participants

Title
Participant's name

Mécanismes de transport dans les diodes Schottky réalisées sur GaN
AMOR Sarrah

Propriétés diélectriques et magnétiques des Composites CoFeNiBSiMo, CoFeBSiCr et CoMnSiB dans une large bande de fréquence.
MyChrif ELBOUBAKRAOUI

Contribution to the thermal and mechanical behavior of the two materials at the base of clay reinforced by fibers alfa and of straw fibers.
Yassine EL HAMDOUNI

Numerical study of a trapezoidal cavity filled by nanofluids and heated from below.
Abdelmajid EL HASSNAOUI

BehaviouralElectrothermal Model of Silicon and Silicon Carbide Power Devices
Abderrazak LAKRIM

Improvement of Direct Torque Control by using a space vector modulation control of three level inverter
Amina ACHALHI

Etude de l'influence du dopage sur les nano composites conducteurs à base de Polypyrrole (Cl)/TiO2
Nacer-Eddine DJELALI

Electron transfer and photoluminescence quenching investigation of Reduced graphene oxide- Meso-tetrakis(4-phenylsulfonica-acid)porphyrin composite
Omar BAJJOU

Mécanique Statistique des Matériaux Biologiques Confinés
Radouane EL KINANI

The improvement of mechanical properties of the PbSbAl and PbSbAlSn alloys for energy storage for the new generation.
Kanza MARBOUH

Study of structural and electronic properties of [NH3-(CH2) n-NH3] ZnCl4, n=8, 10 using DFT.
Rajae EL MRABET

Etude du confinement des membranes lipidiques de topologie toroïdale et l'influence de l'adhésion au container
Lila BOUZAR

Synthèse d'une apatite silicatée à partir du carbonate de calcium et de l'acide phosphorique dans un milieu eau-éthanol et son évaluation biologique in vivo
Houda LABJAR

Thermal stability of a modified sol-gel derved hydroxyapatite nanopoders
Smaiel HERRADI

Etude biologique et anti-bactérienne de bioverres élaborés par voie sol-gel dans les systèmes SiO2-CaO, SiO2-CaO-P2O5 et SiO2-CaO-P2O5-Ag2O
Sarra BOUAHZMA

Monte Carlo simulation of a spin-3/2 Blume- Capel nanoparticle
Noura ZAIM

Monte Carlo study of a ferromagnetic nanoparticle system with disordered shell.
Noura ZAIM

IOP Conf. Series: Materials Science and Engineering **186** (2017) 011001 doi:10.1088/1757-899X/186/1/011001

Élaboration d'un pigment de structure spinelle par voie sol gel
Youssef EL JABBAR

Etude comparative de quatre algorithme de commande MPPT pour un générateur PV
Sarah ELHIMER

The Effect of ZrO2 and Al2O3 addition on the mechanical and physical properties of HAP.
Mohammed ES-SADDIK

Pressure Drop Analysis of Packed Bed Thermal Energy Storage Systems.
Zineb ERREGUERAGUI

Sliding mode controller for a photovoltaic pumping system.
Sabah MIQOI

The influence of calcium content on the performance of geopilymeric-based materials
Fouzia ALLALI

The contribution of EDXRF, 13C NMR/CP-MAS, XRD and FTIR spectroscopy in art conservation study of Historical Moroccan manuscripts from the 16th, 17th, 18th and 19th centuries exposed to accelerated ageing.
Latifa HAJJI

Structural and Morphological Characterization of Moroccan Wooden Artifacts Dating to 21th, 17th, 12th Centuries Using ATR-FTIR And Raman Spectroscopy, SEM And X-Ray Diffraction.
Somia FELLAK

Structural Assessment of the Natural Degradation Effects in Works on Wood Based Materials: A Case Using SEM, X-Ray Diffraction, FTIR and Raman Spectroscopy.
Somia FELLAK

Simulation numérique de l'encrassement dans un échangeur de chaleur à plaques lors d'une phase de traitement thermique du lait par un code Fluent en présence d'un milieu poreux
Mahdi YOUCEF

Structure and evolution of the phase space of the Paul trap system
Jaouad KHARBACH

Quantum Monte Carlo study of the electric properties of a ferroelectric super lattice
Abdellah FERAOUN

Structural and optical properties of vanadium oxide based nanocomposites
Issam DERKAOUI

Influence of dipolar interactions on the super paramagnetic relaxation time of g-Fe2O3
Asmae LABZOUR

Conversion treatment of chromiumlayersdeposited on high carbonsteelsubstrates
Younes BENARIOUA

Étude d'électrosorption du Zinc sur charbon préparé à base d'écorces de banane activées chimiquement.
Houda BENAKOUCHE

Conception et réalisation d'un système solaire autonome.
Ahmed GAGA

Une nouvelle approche pour l'adsorption des ions ammoniums des eaux usées.
Yassmina ANGAR

Bio-elmination d'un colorant acide (ETL) par des cellules sèches de streptiomyces.
Rahma BERRACHED

L'influence des paramètres physico-chimiques sur le comportement des armatures métalliques utilisées dans les constructions du génie civil dans des solutions synthétiques du béton carbonaté
Driss BENMESSAOUD LEFT

Mécanisme de dégradation des bâtiments anciens
Ibtissam KOURDOU

Caractéristiques spectrométriques de marbres du Maroc
Salam Khrissi

NIEL Calculations for Estimating the Displacement Damage Introduced in GaAs Irradiated with Charged Particles
El Mehdi El ALLAM

Growth Mechanism, Structural and Optical Properties Hydrothermally Synthesized Vanadium Oxides Nanobelts
IssamDerkaoui

L'effet d'un champ magnétique aléatoire sur la relaxation super paramagnétique
Abdelkhalek HOUSNI

Elimination d'un colorant de textile par adsorption sur charbons actifs obtenus à partir de noyaux de jujubes. Etude de l'influence de l'adjuvant.
Mounir DAOUD

Study of the corrosive effect of derivatives of 5-bromo-1H-indole-2,3-dione on carbon steel in a 1M hydrochloric acid solution.
Yassine KHARBACH

Localisation des ondes électromagnétiques dans les cristaux photoniques 2D périodiques et quasi-périodiques
Rihab ASMI

Middle infrared omnidirectional mirror based on deformed one dimensional photonic crystal
Tayssir GAHEF

The influence of the doping element on the calcinations temperatures, the grain size and the crystal phase the lead titanate.
Lamiae MRHARRAB

Effect of magnesium substitution on the magnetic properties of Ni-Mg ferrites with compositions Ni1-xMgxFe2O4
Najib BENZAKOUR

Physical and optical properties of nanostructured Na and Sb doped CuInS2 thin films produced by Glancing Angle Deposition (GLAD)
CHAFFAR AKKARI Férid

Synthèse et caractérisation de nouveaux catalyseurs pour piles à combustible microbiennes pour le traitement des eaux usées à base de matériaux ferroélectriques de formules Li1-xTa1-xWxO3.
Abdellah BENZAOAUK

Influence de la nature ferroélectrique de niobate/tantalate de lithium sur le processus photocatalytique pour la réduction de l'oxygène dans la pile à combustible microbienne (MFCs).
Nour-Eddine TOUACH

Performances of reverse osmosis and Nano filtration membranes for desalination: characterization and modeling.
Youssef Amine BOUSSOUGA

L'utilité des filtres polychromatiques a base des cristaux photoniques symétriquement déformée (Bg5/Cu3/Bg5).
Zina BARAKET

Structural, vibrational and luminescent properties of Sr(1-x)PbxWO4 solid solution
Abdelali HALLAOUI

Filtres interférentiels poly-chromoteurs à base de cristaux photoniques unidimensionnels construits selon la distribution Cantor généralisée
Osswa SOLTANI

Posters

Effect of charge polydispersity and charge residence time on the dynamics of a micellar system
Lamiae TALHA

IOP Conf. Series: Materials Science and Engineering **186** (2017) 011001 doi:10.1088/1757-899X/186/1/011001

Key micromechanics issues in integrated material design
Mounia BENNOURA

Experimental and Dielectric Properties of Pb1-xLaxTi1-x/4O3 Ceramic Compounds
Mohammed ZOUHAIRI

Structural Study of Emulsions Stabilized by Charged Nanoparticles
Soukaina EL MOUDNY

Brownian Diffusion of Particles on Liquid-Liquid Interfaces of Pickering Emulsions
Soukaina EL MOUDNY

Dirac-Equation for Graphene with an Arbitrary Potential: Exact Analytical Results and General Proof of Bloch's Theorem
Youssef KHALED

Ab initio investigation of Structural, electronic and elastic properties of YMnO3
Abdel-Hakim CHADLI

Efficacité Énergétique Et Récupération De La Chaleur Perdue Dans Les Systèmes Industriels.
Soukaina MOUNKID

Etude numérique de la convection naturelle dans Une cavité tridimensionnelle remplie de Nanofluides.
Mohamed SANNAD

Etude des propriétés physico-chimiques de nanocomposite Oxyde de graphène-Fe3O4
RhizlaneHATEL

Numerical study of liquid film absorption in vertical canal
SaraARMOU

Modélisation par data mining d'un four de galvanization.
BadreddineIFRAH

The Effects of temperature and size on the Quantum dots p-i-n solar cell's performance
Amine ABDELAZIZ

Indium free and flexible organic solar cells
KheireddineBENCHOUK

Conception d'un régulateur solaire MPPT commandé par mode de glissement
Youssef CHEDDADI

Compensation of DC/DC Buck-Boost converter four switches with PID controller for mobiles devices applications
Abdessamad BENLAFKIH

Caractérisation non destructive des mortiers par mesure de largeur de résonance
Hassan BITA

Études théoriques des propriétés physico-chimiques de matériaux inorganiques au sein de la théorie de la fonctionnelle de la densité DFT
BouhalouaneAMRANI

Form effects on the Diamagnetic Susceptibility of a magneto-donor in GaAs/Ga1-xAlxAs homogenius Quantum Dot
SanaeJANATI EDRISSI

Study of low temperature conduction in 2D p-GaAs
Asmaa CHAKHMANE

The effect of a strong magnetic field on the binding energy of an exciton in a double quantum ring.
Ahmed SALI

The combined effect of pressure and temperature on the binding energy of a donor impurity in a spherical quantum dot using the finite element method.
Ahmed SALI

Synthesis and characterization of Vanadium catalysts supported on copper orthophosphates
Mbarka OUCHABI

Effet de l'augmentation du taux de Mn sur les propriétés structuraux de titanate de plomb dopé au manganese
Lamiae MRHARRAB

THE PHYSICO-MECHANICAL CHARACTERISTICS OF THE CLAYS IN AND AFTER DOPING WITH METAL OXIDE AL2O3 IN THE REGION OF FEZ (MOROCCO).
Iatimad AKHRIF

Polarizability and binding energy of shallow donors in spherical Quantum dots Quantum well.
Safae M'ZERD

Etude comparatif entre le concentrateur parabolique composé (CPC) et le CPC à section rectangulaire
Sara EL-YAHYAOUI

Light Dependent Resistor based solar tracking for Concentrator photovoltaic
Zahra BEN MOHAMMADI

Conception et modélisation d'un concentrateur solaire parabolique
Mahmoud BOUOUD

Étude et modélisation des transferts thermiques dans un échangeur de chaleur fluide-air
Mahmoud BOUOUD

Electrical properties of $LiNbO_3$ ceramics.
Abdelhadi EL BACHIRI

Organic materials based on oxathiazol and tiophene for photovoltaic devices. Correlation structure/electronicproperties.
Fatima AGDA

Quantum chemical investigations study of the Bis-Dipolar Diphenylamino-EndcappedOligoarylfluorenes: Correlation Structure – Properties and Optoelectronic Applications.
Tayeb ABRAM

Electrostatic Properties Study of 4-Methyl-N-[(5- Nitrothiophen-2-Ylmethylidene)] aniline Compound
NourdineBOUKBCHA

BIOSORPTION DU COLORANT BLEU BEZEMA SBL EN SOLUTION AQUEUSE PAR L'ALGUE VERTE ULVA LACTUCA
Narimane KHEDDAM

Synthesis and study of the corrosion activity derivatives chloroisatin "5-Chloroisatin-1H-indole-2, 3-dione.
Zineb TRIBAK

Evaluation de la capacité de dioxyde de titane P25 à l'élimination du Delthamétrine
Nadia EL MOURAILLE

Etat de conservation du patrimoine moderne de la ville de Casablanca
IbtissamKOURDOU

Elementary and structural non-invasive analyses of coloring materials in ancient Moroccan manuscripts
AbdelmajidEL BAKKALI

Early Fault Prediction of DFIG Wind Energy system Based on External Environmental Factors
IdrissiImane

Synthesis, crystal structure and characterization of 2D-hybrid Cobalt hypophosphite, potential material for catalysis.
ATIPO ITOUA NGOPOH Fernand

The lithium effect on structural, morphological, Optical, and nonlinear optical properties of Sn-doped zinc oxide thin films
RHERARI AMAL

Impedance complex, dielectric relaxation and electrical conductivity studies of $Ba_{1-x}Sr_xTiO_3$ ceramics
Abdelhalim ELBASSET

Application of the high temperature series expansions method to the spinels systems: $AxA'_{1-x}B_2X_4$
Khalid BOUSLYKHANE

Spin and orbital magnetism of NiFe: a density functional theory study, mean field and Monte Carlo simulation
Rachid MASROUR

Electronic and magnetic structures of $CoCr_2O_4$ ferrimagnetic investigated by first principle, mean field and series expansions calculations
Rachid MASROUR

propriétés optiques du dioxyde de vanadium
Abderrahim Ben Chaib

Etude théorique par calculs quantique de quelques propriétés moléculaires de quinoxalin-2 (H)-one, 3-méthyl quinoxalin-2(1H)-one et 3-benzylquinoxalin-2(1H)-one
Rachid JDAA

Electrical and optical properties of Graphene derivatives/Poly vinyl alcohol composites
Meryem GOUMRI

IOP Conf. Series: Materials Science and Engineering **186** (2017) 011001 doi:10.1088/1757-899X/186/1/011001

Intrinsic defect concentrations dependence on Mass, charge and electrostatic interactions in lithium niobate
Mohamed BELBAKARA

Etude du pouvoir inhibiteur des derives de pyrazole de la corrosion de l'acier en milieu acide et leur adsorption à la surface metallique
Siham EL ARROUJI

Radiative forces on a dielectric nanosphere By MQBG beams
LoubnaOUAHID

Vortex hollow sinh-Gaussian beams through Fractional Fourier Transform opticalSystem
Ahmed Abdulrab Ali EBRAHIM

A theoretical investigation of the axial intensity distribution of generalized Laguerre-Bessel-Gaussian in turbulent atmosphere
Fatiha, BOUFALAH

Propagation of hollow double half-Gaussain beams and double half inverse Gaussain hollow beams through an ABCD optical system with Spiral Phase
Saad FAROQ

Propagation of Humbert-Gaussian beam Through an ABCD optical system
Naima NOSSIR

Axial intensity of double half-Gaussian hollow beams in a turbulent atmosphere
Fatima KHANNOUS

Focal shift in radially polarized Bessel-like beam
El Mostafa El HALBA

Spectroscopic Analysis of Polyfuran and Theoretical Investigation of Electronic Properties of Oligofurans destined for the solar cell applications.
Mustapha ABARKAN

Intrinsic defect concentrations dependence on Mass, charge and electrostatic interactions in lithium niobate
Smahane BOUAZZA

The physico-mechanical characteristics of the clays in and after doping withmetal oxide al2o3 in the region of Fez (Morocco)
Laila MESRAR

Propriétés structurales de nouvelles phases de matériaux de type M-P-O : TiP2O7 et ZrxTi1-xP2O7
Ismaïl MAAROUF

Synthese Et Analyse Structurale D'une Nouvelle Famille De Phosphates : M2+Fe23+(P2O7)2 avec M = Co2+, Mg2+, Zn2+)
Hafsae LAMSAF

IOP Conf. Series: Materials Science and Engineering **186** (2017) 011002 doi:10.1088/1757-899X/186/1/011002

Peer review statement

All papers published in this volume of *IOP Conference Series: Materials Science and Engineering* have been peer reviewed through processes administered by the proceedings Editors. Reviews were conducted by expert referees to the professional and scientific standards expected of a proceedings journal published by IOP Publishing.

Transport mechanisms in Schottky diodes realized on GaN

AMOR Sarrah[1,3,4], **AHAITOUF Ali**[1], **AHAITOUF Abdelaziz**[2], **SALVESTRINI Jean Paul**[3], **OUGAZZADEN Abdellah**[4]

[1]USMBA, FST, LERSI, B.P. 2202 Fez Morocco
[2]USMBA, FP, LSI, Taza Morocco
[3]UL, Supelec, LMOPS, 2 Rue E. Belin, 57070 Metz France
[4]GTL, UMI 2958, 2 Rue Marconi, 57070 Metz France

E-mail: amor.sarrah@usmba.ac.ma

Abstract.
This work is focused on the conducted transport mechanisms involved on devices based in gallium nitride GaN and its alloys. With considering all conduction mechanisms of current, its possible to understanded these transport phenomena. Thanks to this methodology the current-voltage characteristics of structures with unusual behaviour are further understood and explain. Actually, the barrier height (SBH) is a complex problem since it depends on several parameters like the quality of the metal-semiconductor interface. This study is particularly interesting as solar cells are made on this material and their qualification is closely linked to their transport properties.

1. Introduction

Gallium nitride (GaN) and its ternary and quaternary alloys are attracting more and more interest in the scientific and industrial communities for their potential for devices such as photodetectors, solar cells of new generations, power electronics etc ... Even if the industrial components based on GaN have evolved [1, 2, 3], a number of differences still persist about the physical mechanisms that manage the electronic transport in these components and a wide diversity of interpretations can be found in the literature. The study presented in this work aims to understand and also to control the transport phenomena which are the basis of operation of the performed structures.

This study was conducted through I-V and C-V measurements in function of temperature. After a detailed analysis of different power equations, the important parameters of a Schottky diode such as barrier height, the ideality factor and the doping density are determined. Finally, some physical mechanisms, including the role of deep traps will be presented to explain the different observed behaviours.

2. Experimental procedure

For this study, the BGaN/GaN samples are unintentionally doped, with a 400 nm thick of BGaN. Before metallization, the samples were cleaned with acetone and etched in hydrochloric acid HCl for 5 min to remove the native oxide. For the Ohmic contact, a metallic deposit Au/Ti/Al/Ti of respective thicknesses 200/15/200/15 nm were deposited by thermal evaporation under vacuum to 10^{-6} Torr. To form a Schottky contact layer of Pt (150 nm) was deposited using an electron beam evaporation. All these procedures were performed in a clean room. An example of the produced structures is given in Figure 1.

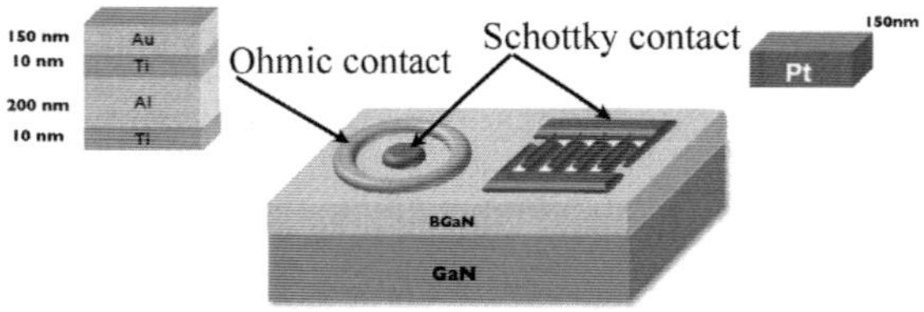

Figure 1. Example of structure showing ohmic and Schottky contacts.

3. Results and discussion

In Schottky diodes, four major mechanisms may contribute to the current in the two directions of polarization. The current thermionic emission I_{te}, the recombination-generation current I_{gr}, tunnel assisted traps current I_{TAT} and direct tunneling I_{TUN}. If the four processes coexist in a structure, the total current can be written:

$$I = I_{te} + I_{gr} + I_{tun} + I_{TAT} \tag{1}$$

In this expression, the thermionic current is given by [4] :

$$I_{te} = I_{te0}[exp(qV/kT) - 1] \tag{2}$$

Where

$$I_{te0} = A^* AT^2 exp((q\phi_b)/kT) \tag{3}$$

Where A^* is the Richardson constant modified ($\approx 26.4 \times 10^4 A/M^2K^2$) ϕ_b is the effective height of the Schottky barrier, q the electron charge, k and T are the Boltzmann constant and the absolute temperature respectively, and A is the area of the diode. The current generation-recombination is expressed as [4, 5]:

$$I_{gr} = I_{gr0} exp(qV/2kT) \tag{4}$$

With

$$I_{gr0} = \frac{qAW(V)n_i}{(V_{bi} - V)\tau} \tag{5}$$

Where τ is the lifetime of minority carriers taken equal to 7×10^{-9}s and W (V) is the depletion region width, which depends on the bias voltage V, and the barrier potential V_{bi}:

$$W(V) = \sqrt{\frac{2\epsilon_0\epsilon_{sc}|(V - V_{bi})|}{qN_D}} \tag{6}$$

Where n_i is the intrinsic carrier density and N_D is the doping concentration ($N_D = 10^{24}m^{-3}$).

The observed traps in the depletion region associated to the electric field enable the tunnelling electron from the valence band to the traps and the traps to the conduction band using the same way. This is called the tunnel assisted traps current and its expressed by the following equation [3, 6]:

$$I_{TAT} = \frac{\pi^2 q^2 V m_e M^2 N_T}{h^3(E_g - E_T)} \times exp\{-\frac{8\pi(2m_e)^{1/2}(E_g - E_T)^{(3/2)}}{3qE_{max}h}\} \tag{7}$$

With m_e the effective electron mass, h is Planck's constant, E_T is the position of the considered trap level, N_T the concentration of traps, E_{max} is the maximum electric field and M is the matrix element associated to traps ($\approx 10^{-23} eV^2 cm^3$) [6]. The fourth type of current is the tunnel current, which depends mainly on the value of the effective masses of the gap and the electric field in the junction. This current results from the crossing of carriers through the formed potential barrier formed [4, 5, 6].

$$I_{tun} = I_{tun0} exp(qV/E_0) \tag{8}$$

$$I_{tun0} = A^* A \frac{T\sqrt{\pi E_{00} q(\phi_b - V - \chi_i)}}{kcoth((qE_{00})/kT)} exp(-\frac{q\chi_i}{kT}) exp\{-\frac{q(\phi_b - \chi_i)}{E_0}\} \tag{9}$$

Where χ_i is the Fermi level energy of the semiconductor relative to the minimum of the conduction band. E_{00} is the tunnelling factor given by the following expression:

$$E_{00} = \frac{qh}{4\pi}\sqrt{\frac{N_D}{\epsilon_0 \epsilon_{sc} m^*}} \tag{10}$$

and

$$E_0 = E_{00} coth(\frac{qE_{00}}{kT}) \tag{11}$$

Figure 2 shows the curves of the various currents calculated separately. Note that generation-recombination the current is very small compared to others. The tunnel assisted traps current seems dominant between 0 and 0.5V. Above 0.5V we see that the tunnel current and thermionic current dominate.

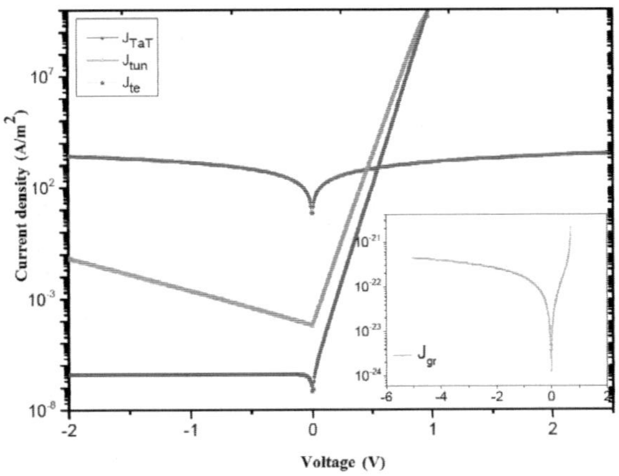

Figure 2. Representation of different currents given in equation (1)

Figure 3 shows an example of comparison between the result of different current calculated and the total current of a Schottky diode at 300K.

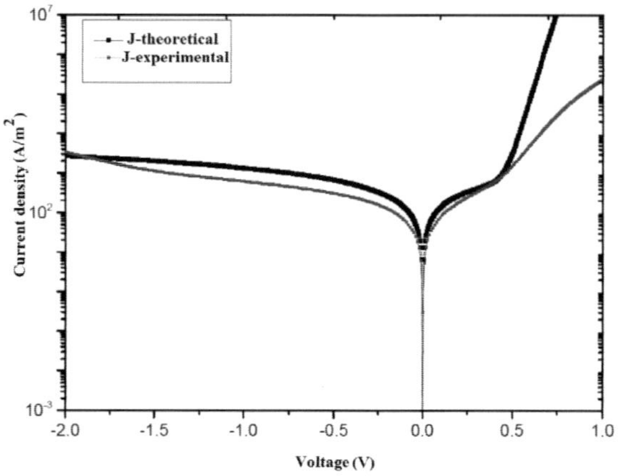

Figure 3. Comparison of theoretical dark current and the measured current in a Schottky diode at 300K

An adjustment between the experimental and the theoretical curve was performed by varying the lifetime of minority carriers τ and the level trapsE_T which are involved respectively the recombination-generation mechanisms and in the conduction mechanism by tunnel assisted trap. The analysis of these curves shows that:

- In the reverse bias the current is dominated by the I_{TAT} component, which confirms the existence of a high density of defects in the bulk of the structure.

- For forward bias, at lower bias voltage ($V < 0.5V$), the theoretical curve and the experimental measurements fit well, while at high bias ($V > 0.5V$) the current is dominated by thermionic emission despite a considerable disagreement between experimental measurements and theoretical calculation. This is probably due to a high value of the series resistance in our structures and a large ideality factor which can be, depending on the applied voltage [2, 7].

To verify our assumptions we have tried to determine the values of the series resistance and the ideality factor considering only the thermionic emission, the resulting values are respectively 107.07Ω and 4.06. These rather large values may explain in part the disagreement observed between measurements and theoretical curves.

4. Conclusion

The study was made at room temperature on samples of different concentrations of boron. Analysis of the current-voltage characteristics of Schottky diodes on BGaN allowed us to demonstrate the main conduction mechanisms observed in our samples and the contribution of different mechanisms: thermonic emission, generation-recombination, tunnel and tunnel assisted traps . The work is currently in progress to consider the density of defects, the true value of n and better describe our experimental results.

References

[1] Srour H, Salvestrini J P, Ahaitouf A, Gautier S, Moudakir T, Assouar B, Abarkan M, Hamady S and Ougazzaden A 2011 *Applied Physics Letters* **99** 221101

[2] Ahaitouf A, Srour H, Hamady S O S, Fressengeas N, Ougazzaden A and Salvestrini J 2012 *Thin Solid Films* **522** 345 – 351 ISSN 0040-6090

[3] Gopal V, Plis E, Rodriguez J B, Jones C E, Faraone L and Krishna S 2008 *Journal of Applied Physics* **104** 124506
[4] Sze S and Ng K 2006 *Physics of Semiconductor Devices* (Wiley) ISBN 9780470068304
[5] Mohammad S and Morkoc H 1996 *Progress in Quantum Electronics* **20** 361 – 525 ISSN 0079-6727
[6] Yang Q K, Fuchs F, Schmitz J and Pletschen W 2002 *Applied Physics Letters* **81** 4757–4759
[7] Ahaitouf A, Ahaitouf A, Salvestrini J P and Srour H 2011 *Journal of Semiconductors* **32** 104002

Effect of lithium on linear and nonlinear optical properties of Sn-doped zinc oxide prepared by spray pyrolysis

A Rherari[1], M Addou[1, 2], Z Sofiani[1], M El jouad[1], M Jbilou[2] and M Diani[2]

[1]Department of Physics, Laboratory of Optoelectronic and Physical Chemistry of Materials Ibn Tofail University, Kenitra 14000, Morocco
[2]Laboratory of Materials and valorization of Natural Resources, University Abdelmalek Essaadi- FST Tangier, Morocco

Email: amalrherari@gmail.com

Abstract. Li and Sn codoped ZnO (LTZO) thin films have been successfully deposited on heated glass substrates at 450°C using the spray pyrolysis technique, the effect of lithium of Sn-doped zinc oxide on the structural, morphological, optical and nonlinear optical properties was investigated using X-ray diffraction, transmission, the RMS average surface roughness, and third harmonic generation (THG). The value of optical band gap Eg was found to be decreased from 3.24 eV to 3.16 eV when the concentration of Li from 0 to 7%, while the concentration of Sn is fixed at 2%.The doping of ZnO films improves the nonlinear response and the highest susceptibility value $\chi^{(3)} = 13.422 \times 10^{-12}$ (esu) is found at concentration of lithium 7%.

1. Introduction

Zinc oxide (ZnO) thin films have been extensively studied and used for different domains such asmicroelectronic and optoelectronic applications [1], with direct band gap of 3.37 eV, high exciton binding energy of 60 meV at room temperature, is attractive for optoelectronic application in the blue and UV region, especially the data storage , which is the main work of the year , the blue emission generated to produce the new light by physics Nobel laureates [2], it can be used also at transparent conductive oxide thin films such as for application in solar cells [3]. Furthermore, the effect of ZnO doping on its nonlinear optical properties have been studied for several decades[4, 5, 6].

It was shown that the Sn doping of ZnO improves its electrical and structural properties [7],and the doping by Li was shown to be attractive for modifying the microstructure , ferroelectricity and optoelectronic properties [8] and enhance the rate of the crystal growth due to an increase of its diffusivity [9]. Furthermore, Yousefi et al [10] have reported that the crystallite and particle sizes can be increased by doping with group- I elements, reduce the oxygen vacancies in ZnO Lattice, Behera et al [11] reported that Li doping has always been in controversy, since Li ion can take up at interstitial sites, thus acting as shallow donor and compensate the acceptor action (self-compensation). Due to unequal ionic radius of lithium (0,68 Å) and zinc (0,74 Å) atoms, lithium is also expected to occupy substitution sites and thus can influencelately the structural and optical properties [11].Li-Mg [12],Li-Al [13], Li-N [14], Cu-Li [8] co-doping of ZnO has also been investigated.

In this work, Li and Sn co-doped ZnO (LTZO) thin films were prepared using a spray pyrolysis technique, where the deposition temperature, the used precursors, concentrationsthe spray rate most be controlled [15],and its influence on the linear and non-linear optical properties is studied and discussed.

2. Experimental detail

LTZO thin films were grown on glass substrates and heated at 450°C by spray pyrolysis, experimental setup has been described elsewhere [5]. The spraying solution was 0.05M zinc chloride. The doping was achieved by adding $SnCl_2$ in a concentration is fixed at 2%. Different concentration of Li were used for co-doping the ZnO by adding $LiClO_4$ in a concentration of 3, 5 and 7 %. The spray time was 3 min with solution flow rate of 5 ml $min^{-1}$and the distance between the nozzle and substrate distance was fixed to be 40cm. The structural properties of thin films were investigated using X-ray diffractometry (X'Pert³ Powder PANalytical diffractometer) with Cu (Kα) radiation (λ=1054Å). The optical transmission measurements were performed with Jasco V-570 UV/Vis/NIR spectrophotometer. Nano-surf Flex FM digital instruments operated in the tapping mode was used for the AFM imagesand the Third order nonlinear susceptibility was performed using third harmonic generation (THG) method [6].

3. Results and discussion

Figure 1 shows the XRD patterns of LTZO films with an atomic concentration of 2% Sn and different concentrations of Li (0, 3, 5,7%). From the results, we can see that all films are polycrystalline with a hexagonal wurtzite structure[16].

The calculatedlattice parameters a and c are summarized in the table 1. The full-with at a half maximum (FWHM) of the (002)peak are 0.13°, 0.12°, 0.11° and 0.10° and using the debay Scherrer formula ,the calculated grains size are 63.142, 72.593, 76.414 and 85.404 nm ,when the concentration of Li is increased from 0,3,5 and 7% respectively .

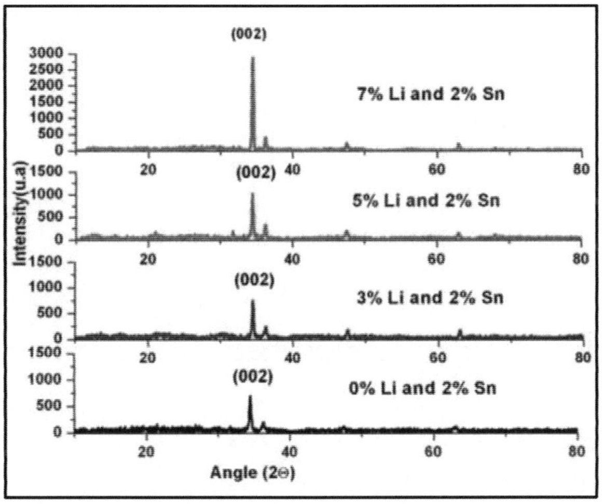

Figure 1.XRD patterns of LTZO films with various dopingconcentration. All films were grown at 450 [16]

Table 1. Lattice parameters, FWHMs and average crystal size obtained from XRD patterns

Films	a (Å)	c (Å)	FWHM β(deg)	Average crystal size (D)
0% Li and 2% Sn	3.2491	5.2005	0.13	63.142
3% Li and 2% Sn	3.2517	5.1984	0.12	72.593
5% Li and 2% Sn	3.2491	5.2000	0.11	76.414
7% Li and 2% Sn	3.2529	5.2003	0.10	85.404

Figure 2(a). Depicts the optical transmission spectra of LTZO thin films. All films exhibit an average optical transparency exceeding 85% in the visible region, the average transmittance was increased after Li/Sn codoping, reaching the maximum for Li and Sn concentration of 7% and 2% respectively. In effect Fellahi et al [17] reported that, the light scattering decreases with the improvement of the crystallinity of the ITO films. In fact the crystallinity probably affects the light scattering on the transparency of the LTZO films.

Figure 2(b) illustrates a plot of $(\alpha h\nu)^2$ vs. photon energy hν for the LTZO films. The extrapolation of the slopes of the curves, gives the optical band gap Eg. The values of optical band gaps decreases from 3.24 to 3.16 eV with increasing Li concentration from 0 to 7% .It has been reported that the increased concentration of Li in a Li/Al co-doped ZnO can induce a resonance structure in the density of state, which slipt off theband [13].

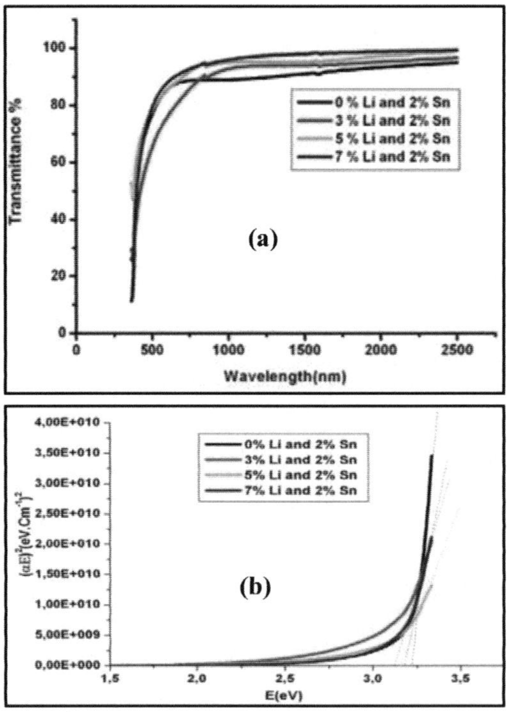

Figure 2. Optical properties of LTZO films. (a) Transmittance spectra and (b) Plots of $((\alpha h\nu)^2$ vs. hν).

The transparency of ZnO at 1064 nm, which is the laser excitation wavelength, makes it a good candidate for nonlinear optical applications [18]. Figure 3 shows the THG spectra of LTZO films. The third-order optical nonlinearities of Li-Sn codoped ZnO (LTZO) films were investigated by THG technique in order to study the effect of lithium on nonlinear optical properties of Sn-doped zinc oxide thin films, Figure 3 exhibit Maker–fringes obtained by rotating the samples through the range from ± 60° to the normal.

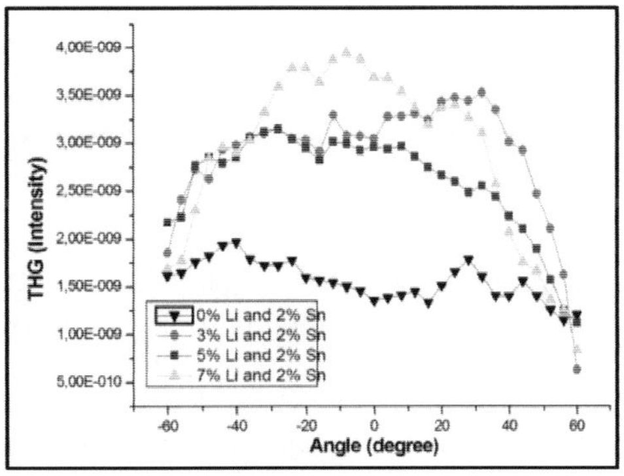

Figure3. The THG spectra of LTZO films deposited on glass substrates

The $\chi^{(3)}$ can be expressed as [19]

$$\chi^{(3)} = \chi_s^{(3)} \left(\frac{2}{\pi}\right) \left(\frac{L_{c,s)}}{l}\right) (\frac{I_{3\omega}}{I_{3\omega,s}})^{1/2} \quad (1)$$

Where the$I_{3\omega}$ is the THG intensity, the suffix s refers to the reference material, as silica. Lc,s is the coherence length of microscope slide which is around 18μm. The value of the nonlinear susceptibility is about: $\chi_s^{(3)}$= 5. 7841×10^{-12}(esu).

The values of $\chi^{(3)}$calculated using the formula (1) are summarized in the table (2) it can be seen that the Li-Sn co-doping of zinc oxide thin films improves the nonlinear response, and the highest susceptibility value $\chi^{(3)}$ =13.422 ×10^{-12} (esu) was found at concentration of lithium (7%),which can be explained by the fact that the surface smoothened by the Li doping as shown in the table (2),so the light diffusion is reduced, this agree with the increase of the transmission of the films when the Li concentration is increased as shown in the figure 2.a.Moreover,as we showed in previous work [16]the scattering of light is strongly affected by the roughness and the structural properties of the films, While can be deducted that the smoothening of the surface reduce the scattering of light and generates the highest nonlinear optical response .Same behaviour was reported by Addou et al [20].

Table 2. Third order nonlinear susceptibility $\chi^{(3)}$, the RMS average surface roughness of LTZO films with different concentrationsof Li .

Films	RMS(nm)	($\chi^{(3)}$)×10^{-12} (esu)
0% Li and 2% Sn	26.950	10.916
3% Li and 2% Sn	25.558	12.772
5% Li and 2% Sn	19.587	11.424
7% Li and 2% Sn	13.678	13.422

4. Conclusion

LTZO thin films have been deposited by spray pyrolysis technique. the films were polycrystalline with an hexagonal wurtzite structure, with an average optical transparency exceeding 85%in the visible range .The optical band gap Eg was found to decrease from 3,24 to 3,16 eV with increasing Li concentration from 0 to 7% .It was found that the codoping of ZnO thin films improves the nonlinear response .The highest susceptibility value $\chi^{(3)}$ =13,422 $\times 10^{-12}$ (esu) was found for the highest concentration of lithium (7%),which indicates the correlation between the structural, optical and morphological properties of co-doped ZnO thin films with the nonlinear optical properties .

5. Acknowledgment

This work is supported by Hassan II Academy of Science and Technology, ERANETMED- NInFEE and PPR 2015/9 Morocco

References

[1] B. Kulyk, B. Sahraoui, V. Figà, B. Turko, V. Rudyk, and V. Kapustianyk 2009. "Influence of Ag, Cu dopants on the second and third harmonic response of ZnO films," *J. Alloys Compd.*, **481** 819–825.

[2] A. & N. Heber, Joerg. "Nobel Prize 2014: Akasaki 2014, "for the invention of efficient blue light-emitting diodes which has enabled bright and energy-saving white light sources," *Nat. Phys.*, **10.11** 791–791,

[3] D. Fang, P. Yao, and H. Li2014, "Influence of annealing temperature on the structural and optical properties of Mg-Al co-doped ZnO thin films prepared via sol-gel method," *Ceram. Int.*,**40**5873–5880.

[4] K. Bahedi, M. Addou, M. El Jouad, Z. Sofiani, H. El Oauzzani, and B. Sahraoui2011, "Influence of strain/stress on the nonlinear-optical properties of sprayed deposited ZnO:Al thin films," *Appl. Surf. Sci.*, **257** 8003–8005.

[5] M. El Jouad, M. Alaoui Lamrani, Z. Sofiani, M. Addou, T. El Habbani, N. Fellahi, K. Bahedi, L. Dghoughi, a. Monteil, B. Sahraoui, S. Dabos, and N. Gaumer**2009**, "Roughness effect on photoluminescence of cerium doped zinc oxide thin films," *Opt. Mater. (Amst).*, **31**1357–1361.

[6] Z. Sofiani, B. Sahraoui, M. Addou, and R. Adhiri 2007, "Third harmonic generation in undoped and X doped ZnO films (X: Ce, F, Er, Al, Sn) deposited by spray pyrolysis," *J. Appl. Phys.*,**101** 063104.

[7] A. Bougrine, M. Addou, A. Kachouane, J. C. C. Bérnède, and M. Morsli2005, "Effect of tin incorporation on physicochemical properties of ZnO films prepared by spray pyrolysis," *Mater. Chem. Phys.*,**91** 247–252.

[8] J. C. Li, Q. Cao, X. Y. Hou, B. F. Wang, and D. C. Ba 2013, "Microstructure and optoelectronic properties of Cu–Li codoped ZnO film: Role of CuZn and Lii defects," *Superlattices Microstruct.*,**59** 169–177.

[9] S. Fujihara, C. Sasaki, and T. Kimura2001, "Crystallization behavior and origin of c-axis orientation in sol-gel-derived ZnO:Li thin films on glass substrates," *Appl. Surf. Sci.*, **180**341–350.

[10] R. Yousefi, A. K. Zak, and F. Jamali-Sheini2013, "The effect of group-I elements on the structural and optical properties of ZnO nanoparticles," *Ceram. Int.*, **39**1371–1377..

[11] D. Behera and B. S. Acharya 2010, "Study of the microstructural and photoluminescence properties of Li-doped ZnO thin films prepared by spray pyrolysis," *Ionics (Kiel).*, **16**543–548.

[12] S. Aksoy, Y. Caglar, S. Ilican, and M. Caglar 2012, "Sol-gel derived Li-Mg co-doped ZnO films: Preparation and characterization via XRD, XPS, FESEM," *J. Alloys Compd.*, **512**171–178.

[13] H. Nian, S. H. Hahn, K. K. Koo, J. S. Kim, S. Kim, E. W. Shin, and E. J. Kim2010., "Preparation and characterization of sol-gel Li and Al codoped ZnO thin films," *Mater. Lett.*, **64** 157–160.

[14] C. Ravichandran, G. Srinivasan, C. Lennon, S. Sivanathan, and J. Kumar2010., "Investigations on the structural and optical properties of Li, N and (Li, N) co-doped ZnO thin films prepared by sol–gel technique," *Mater. Sci. Semicond. Process.*, **13** 46–50.

[15] S. Bayoud, M. Addou, K. Bahedi, M. El Jouad, Z. Sofiani, M. A. Lamrani, J. C. Bernéde, and J. Ebothé2013, "The lithium effect on the blue and red emissions of Er-doped zinc oxide thin films," *Phys. Scr.*, **157** 014045,.

[16] A. Rherari, M. Addou, Z. Sofiani, M. El, M. Jbilou, M. Diani, and A. Chahboun 2016, "The lithium effect on roughness , structural , morphological properties of Sn doped zinc oxide thin," *J. Mater. Environ. Sci.***7** 554-557.

[17] N. Fellahi, M. Addou, Z. Sofiani, M. El jouad, K. Bahedi, S. Bayoud, M. Haouti, B. Sahraoui and J.C. bernede2010, "Effect of fluorine incorporation on the third-order nonlinear-optical properties of indium oxide thin films prepared by spray pyrolysis," *J. Optoelectron. Adv. Mater.*, **12** 1087–1091.

[18] O. G. Morales-Saavedra and L. Casta??eda 2007, "Second harmonic generation of fluorine-doped zinc oxide thin films grown on soda-lime glass substrates by a chemical spray technique," *Opt. Commun.*, **269**370–377.

[19] X. H. Wang, D. P. West, N. B. McKeown, and T. A. King 1998, "Determining the cubic susceptibility (3) of films or glasses by the Maker fringe method: a representative study of spin-coated films of copper phthalocyanine derivation," *J. Opt. Soc. Am. B*, **15** 1895–1903,.

[20] M. Addou, M. El Jouad, A. M. Lamrani, Z. Sofian, B. Sahraoui, A. Monteil, N. Gaumer, S. Dabos, T. El Habbani, N. Fellahi, K. Bahedi, and L. Dghoughi *2007*, "Effect of roughness on third-order nonlinear-optical susceptibility of rare earth doped zinc oxide thin films," *ICTON-MW'07 - Int. Conf. Transparent Opt. Networks "Mediterranean Winter" - Conf. Proc.*, 2–6.

IOP Conf. Series: Materials Science and Engineering **186** (2017) 012003 doi:10.1088/1757-899X/186/1/012003

Vibrational and optical properties of Meso-tetrakis(4-phenylsulfonica-acid) porphyrin decorated with graphene oxide.

O Bajjou[1,2,3], M Khenfouch [2,3], M Baïtoul[1] , B Mothudi[2], M Dhlamini[2] , E Faulques[6] and M Maaza[4,5]

[1] University Sidi Mohamed Ben Abdellah, Faculty of Sciences Dhar el Mahraz, Laboratory of Solid state Physics, Group of Polymers and nanomaterials, PO Box 1796 Atlas Fez 30 000, Morocco.
[2] Physics department, College of Science, Engineering and technology Science Campus, Corner of Christiaan de Wet Road & Pioneer Avenue, Florida, 1709, Johannesburg, South Africa
[3] Africa Graphene Center, 2 Boekenhout Street, Florida, Johannesburg. 1709. South Africa
[4] iThemba LABS-National Research Foundation of South Africa, Old Faure Road, PO Box 722, Somerset West 7129, Western Cape Province, South Africa.
[5] UNISA Africa Chair in Nanosciences-Nanotechnology, College of Graduate Studies, University of South Africa, Muckleneuk ridge, PO Box 392, Pretoria, South Africa.
[6] Institut des Matériaux Jean Rouxel, Nantes, 2 rue de la Houssinière, BP32229, 44322 Nantes, Cedex3, France.

Email: bajjou.omar@gmail.com

Abstract. Functionalized graphene oxide have received much attention in a broad range of applications from light harvesting.The interactions between graphene and Meso-tetrakis(4-phenylsulfonica-acid)porphyrin ($H_4TPPS_4^{2-}$), especially π–π interactions have also been extensively investigated. The composite material formed stable dispersions was characterized by X-ray diffraction, Transmission electronic microscopies, FTIR, Raman. The effect pH and concentration was study by UV-visible absorption.

1. Introduction

Graphene with a superelectron conjugation system is an ideal ultra-broadband and fast saturable absorber that has ultrafast carrier dynamics, and large absorption range. Graphene oxide (GO), holding some characteristics of graphene due to the presence of pristine graphitic nanoislands, exhibits some heterogeneous optical transition and nonlinear dynamics because smallsp2 carbon nanoislands are isolated by the sp3matrix[1].Furthermore, graphene's strong interactions with photons and electrons, and chemical function ability could add more functions to photoactive composites[2, 3].

Porphyrins and related tetrapyrroles have been extensively studied because of their importance in biological processes and they are often used in the development of artificial light-harvesting, energy conversion[4, 5], sensors systems and many others[6]. They are also interesting from a fundamental point of view, as systems to study the nature and dynamics of excited states in molecular assemblies of reduced dimensionality, as well as models for understanding the photophysics of natural light-harvesting complexes. The extraordinary optical properties of these molecular aggregates are to a large extent governed by exciton delocalization, transport, and relaxation phenomena[7]. To develop these properties, composites based on those structures show their potentials for optical and optoelectronic applications by the association of porphyrin properties to carbon nanotubes, polymers and other molecular materials[8-10].

In this paper, we study of Meso-tetrakis(4-phenylsulfonica-acid)porphyrin and graphene oxide composites, by means of X-ray diffraction, FTIR and Raman spectroscopy. Furthermore we conduct a studyon the pH effect and the GO concentration on the optical properties that were characterized by UV-vis absoption.

2. Materials setup

The graphene oxide (GO) was synthesized from graphite using Hummer's method[11]. The well dispersed GO sheets were prepared using ($KMnO_4/H_2SO_4$), and then the precipitate was washed many times with de ionized water. After preparation The pH of the solution was adjusted with hydrochloric acid HCL and NaOH solution, and the products were labeled as GO (neutral, pH=7), GO (acidic, pH2, 3, 4), and GO (basic, pH8, 10, 11). The final solution was diluted and sonicated for a major exfoliation of the oxidized sheets. The solution of the GO (2 ml) was mixed with$H_4TPPS_4^{2-}$(2 ml).

3. General characterization

X-ray diffraction (XRD) patterns of GO, $H_4TPPS_4^{2-}$ and the composite were performed with Bruker's D8 advanced X-ray diffractometer using CuKα radiation (λ=1.5418° Å). Fourier transform infrared (FT-IR) spectra were recorded on a PE Spectrum One spectrometer with KBr pellets in the 4000–450 cm^{-1} region. Ramanspectra were carried out using Bruker spectrometer SENTERRAR a man Stokes and spectral range of 90-3200cm-1 with an excitation wavelength λ =785nmandforoursamplesthe laser power is adjusted to a low power 10 mW UV–vis spectra were measured on a PerkinElmer Lambda 35 spectrometer.

4. Results & discussion

4.1 X-ray diffraction

The XRD was carried out to investigate the structure of the GO–$H_4TPPS_4^{2-}$hybrid. Fig. 1 shows the XRD patterns of graphite, GO, $H_4TPPS_4^{2-}$and GO–$H_4TPPS_4^{2-}$. While the characteristic (001) diffraction peak of GO appears at about 12.18°attributed to the interlayer d-spacing of 0.71 nm[12, 13]. The XRD pattern of GO– $H_4TPPS_4^{2-}$ displays a diffraction peak that has a relatively low intensity at about 11.82° with the d-spacing of 0.75 nm, indicating a further expansion of the (001) inter-planar spacing of GO owing to the incorporation of the H4TPPS42- molecules. The increase of d-spacing of GO with respect to that of GO–$H_4TPPS_4^{2-}$ can be attributed to the introduction of oxygenated functional groups. In addition, one additional broad weak peak at around 23.9° is also observed, which is probably due to partial stacking of $H_4TPPS_4^{2-}$, with itself or with the GO sheets during the introduction of $H_4TPPS_4^{2-}$ molecules onto GO sheets[14].

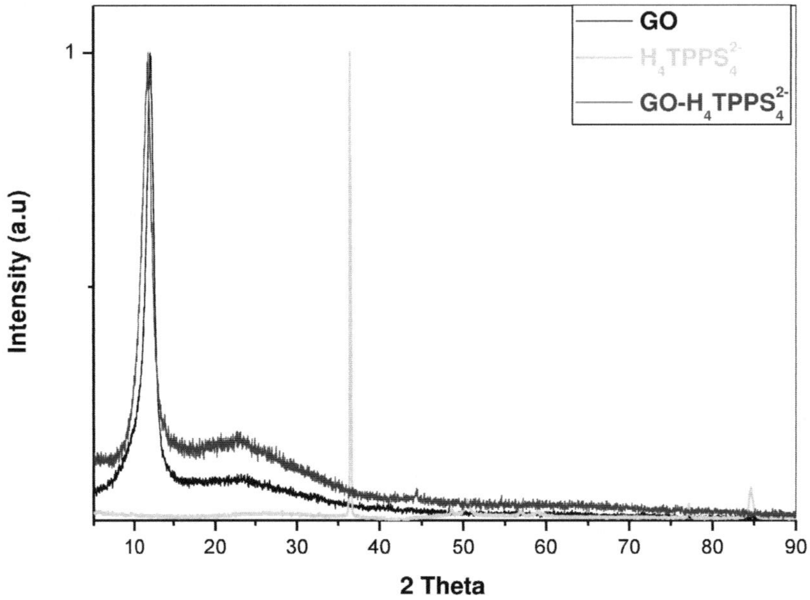

Figure 1. XRD of GO, $H_4TPPS_4^{2-}$ and GO- $H_4TPPS_4^{2-}$

4.2 Raman measurement

Raman spectroscopy is a widely used tool for the characterization of carbon materials; it is particularly informative on the structure of graphene oxide, as well as the existence of defects and the extent of functional. The Raman spectrum of GO exhibit the G-band and D-band around 1590 and 1340 cm^{-1}, respectively. The G-band is associated with the E2g vibration mode of sp^2 lattice of the graphitic materials. The D-band is due to the breathing modes of six-atom rings, which are associated with the structural defects in sp^2 lattice. [15, 16]. The height intensity of the D band directly depends on the number of the sp^3 carbon atoms of graphene surface, and thus on the number of defects in the graphene nanosheets. The spectrum of $H_4TPPS_4^{2-}$shows a characteristic band appearing around 1390 cm^{-1} which is assigned to the asymmetric stretching Cβ- Cα- NH bonds and a few bands assigned to the stretching C-C vibrations at 1550, 1604, and 1877 cm^{-1},respectively[9, 17]. The Raman spectrum of GO-$H_4TPPS_4^{2-}$shows two bands at 1390 and 1592 cm^{-1} corresponding to D and G respectively.

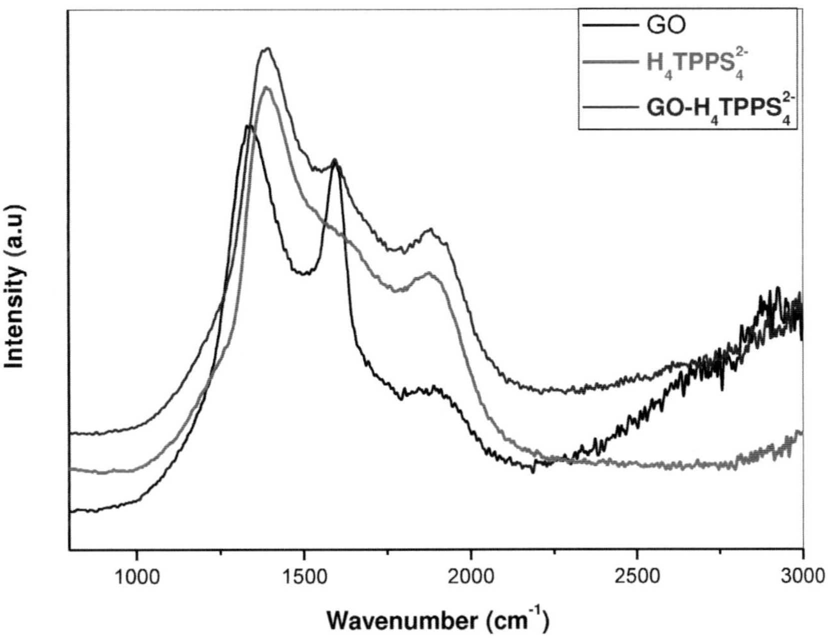

Figure 2. Raman spectra ofGO, $H_4TPPS_4^{2-}$ and GO- $H_4TPPS_4^{2-}$

Using the D-band and G-band peak intensities ratio (ID/IG) of Raman Spectra, often affords information about the structural changes and covalent modification. When ID/IG decreases it means that disorder in graphene decreases, in our case the calculated intensity ratio of D to G band shows a slight decrease from 0,87 for GO-$H_4TPPS_4^{2-}$to 0,85 for the GO film, which is consistent with the reduction of defects on the GO surface and/or an enhanced order in the GO layers. Moreover, the shift of the D-band indicates a strong interaction between the $H_4TPPS_4^{2-}$and the graphitic sheets that can be assigned to a charge transfer.

4.3 Infrared measurement

FT-IR spectroscopy is recognized as an important tool to study different types of functional groups. From the FT-IR characterization, the GO spectrum shown in Figure.3, exhibits several peaks described as follows. Firstly, the peak at 1408 cm^{-1} is assigned to a C–O vibrational mode. Secondly, the peak centered at 1,616 cm^{-1}is attributed to aromatic C=C bonds. Another peak at 1715 cm^{-1} corresponds to the vibrational mode of the ketone groups C=O[14, 18]. Finally, the peak at 3200 cm^{-1} denotes C–OH stretching. For the characteristic bands of $H_4TPPS_4^{2-}$, we note the presence of a broad intense band centered at 3400 cm^{-1} which is attributed to the N-H bond stretching, and the band situated at 2914 cm^{-1} is related to the stretching vibration of aromatic C-H bonds. In addition, a band was observed at around 1629 cm^{-1} and corresponds to the stretching bond in the phenyl nuclei. Moreover, the band at 1540 cm^{-1} is attributed to Cβ-Cβ' stretching which may be due to symmetrical deformation of the pyrrole rings and the bending strain Cα- N- Cα'. The band at 1415 cm^{-1} is attributed to the symmetric stretching Cα- N- Cα' and the Cβ-Cβ ' bond stretching in pyrrole nuclei, there is also a band which appears at 1318 cm^{-1} corresponding to the asymmetric HN-Cα-Cβ stretching bond. The bands at 1261 cm^{-1} and 1298 cm^{-1} are assigned to the asymmetric Cα-N-Cα' stretching[19-21]. In the spectrum GO-

$H_4TPPS_4^{2-}$ the intensity of 1411 cm^{-1} band greatly decreases and a new band appears at 1531 cm^{-1} corresponding to Cβ -Cβ 'vibrations of porphyrins. Furthermore, we find that the intensities of these bands decrease significantly in the case of composite GO- $H_4TPPS_4^{2-}$. The bands in the region 2850-2960 cm^{-1}, attributed to the C-H stretching vibrations of the aromatic rings, further confirm the existence of the porphyrin on GO. These changes clearly indicate a strong interaction between GO and $H_4TPPS_4^{2-}$.

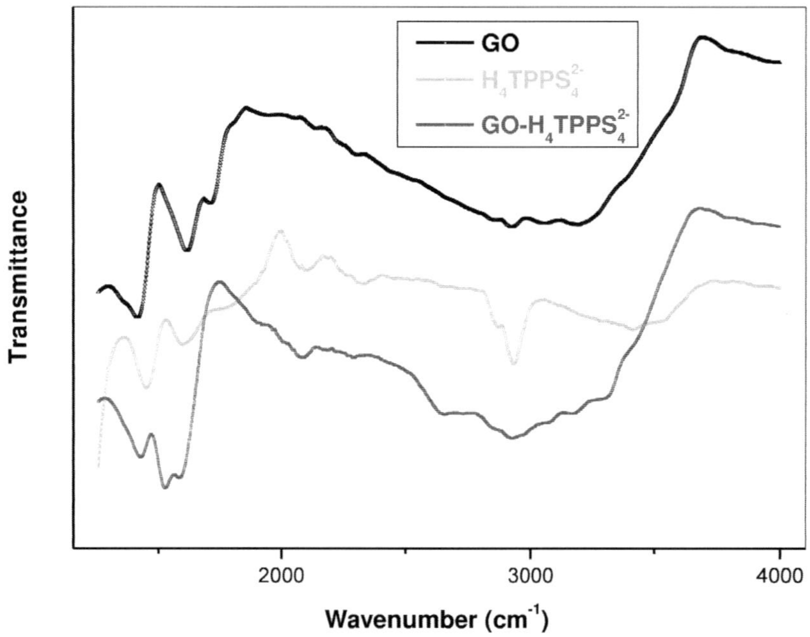

Figure 3.FTIR spectra ofGO, $H_4TPPS_4^{2-}$ and GO- $H_4TPPS_4^{2-}$

4.4 UV-Vis absorption

The Figure.4 shown UV–vis absorption spectra of GO and GO–$H_4TPPS_4^{2-}$composites. Good dispersion is of particular importance for graphene process ability and applications because most of their attractive properties are only associated with individual graphene sheets. Solution-phase UV–vis spectra has been reported to demonstrate a linear relationship between the absorbance and the relative concentrations of various graphene oxide hybrids, which obeys Beer's law at low concentrations, and has been used to determine the solubility of the hybrids. The absorption spectrum of GO displays a strong broad absorption at around 230nm, attributed to π-π* transitions of aromatic C=C bonds[11, 22, 23]. The absorption spectra exhibit typical characteristic of $H_4TPPS_4^{2-}$B-band located at 433 nm in the visible region, a number of absorptions known as the Q bands are present, two major absorption bands situated at 596 nm and 646 nm attributed Q(0.0) and Q(1.0) respectively. The lower energy band, Q(0,0), is attributed to excitation to the first excited state. The higher energy band, Q(1,0), is a result of vibrational transitions interacting with the electronic transitions of the porphyrin, and has been referred to as a vibronic overtone of the Q(0,0) band[24-26].Following the attachment of GO with $H_4TPPS_4^{2-}$, the broad absorption peak at 230 nm change the intensity when passing from concentration

0.1%to1%and2%.Thenwidening the bands with the increase in concentration can be understood in terms of size layer graphene increase.
Moreover, GO-$H_4TPPS_4^{2-}$we observe a change in the intensity in the composite when passing fromPH2 toPH11 can be explained by the fact that intramolecular donor-acceptor structure usually allows charge-transfer interaction or the highest occupied molecular orbital (HOMO)-lowest unoccupied molecular orbital (LUMO) transitions to occur. The change of GO absorption peak position may be explained by a change of its band gap structure and $H_4TPPS_4^{2-}$- peaks show a dramatic modification of intensity except the peak situated at 433 nm. Also composite a weak peak at 433 nm might arise from the splitting of the Soret band of $H_4TPPS_4^{2-}$due to the conformational changes of the local stacking structure or the formation of porphyrin H-aggregates to examine this effect on the optical properties of the nanocomposite. According to exciton theory, when molecules lie in a head-to-tail arrangement (J-aggregation), the allowed state is lower in energy, hence producing a red-shift to the monomer, and since the Soret band of $H_4TPPS_4^{2-}$ is red-shifted upon aggregation (J-aggregate), it implies the planar porphyrin molecules are stacked in an arrangement where individual porphyrin units are not positioned directly one on top of the other. Furthermore, the Q bands situated at 596 and 646 nm of the $H_4TPPS_4^{2-}$becomes broader low when de passing pH=7 and the relative intensity of the UV absorption band decreases clearly. This is due to the markedly increased absorption of GO, as well as the formation of GO–$H_4TPPS_4^{2-}$. Therefore, we can figure out that a π–π interaction exists between the two components of the GO- H4TPPS42-Composite.

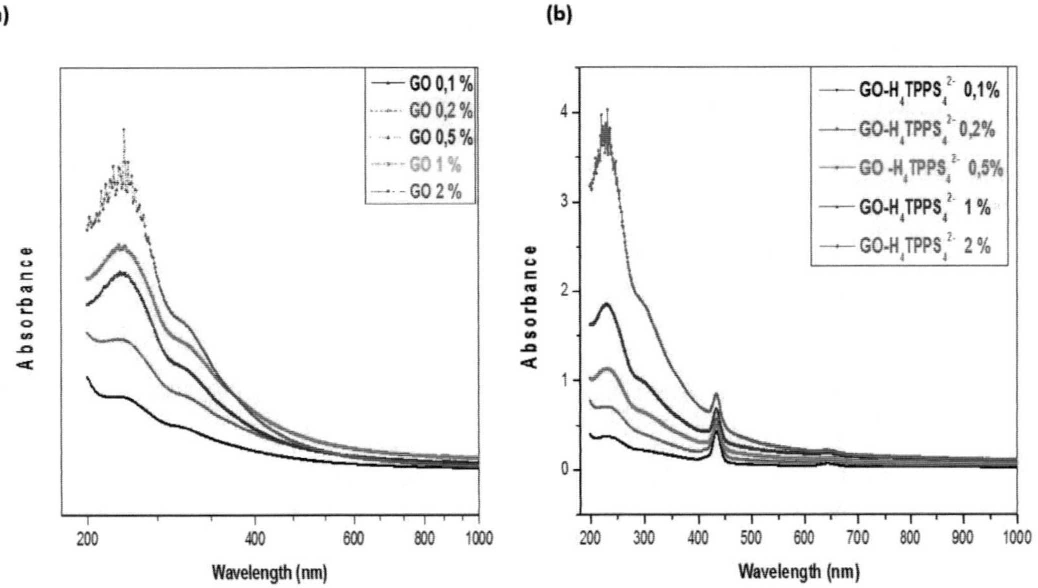

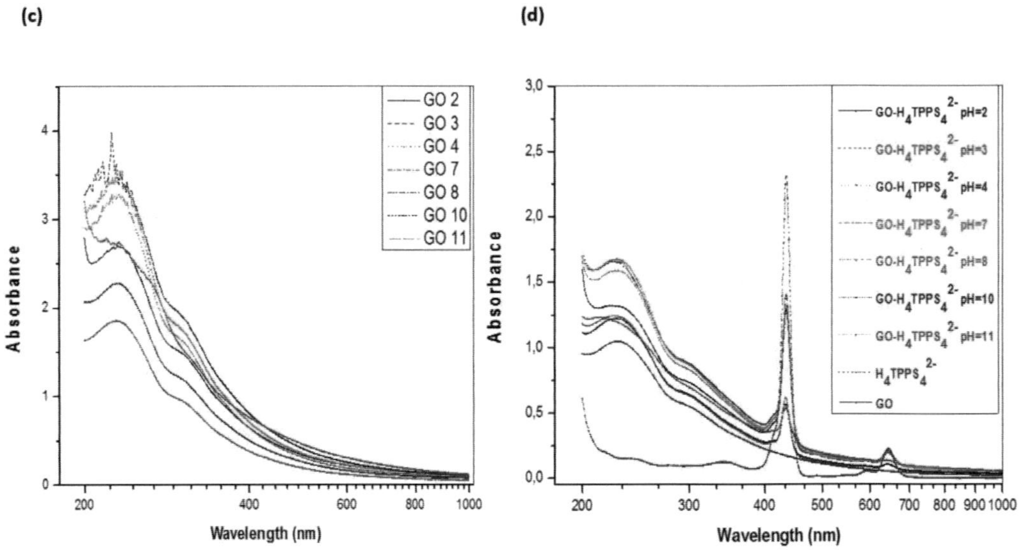

Figure 4. Optical absorption of GO, $H_4TPPS_4^{2-}$ and GO- $H_4TPPS_4^{2-}$ (a,b: concentration effect, c and d pH effect

5. Conclusion

In conclusion, we have presented in this work the synthesis of GO- $H_4TPPS_4^{2-}$film composite and investigated the structural and optical properties. Raman, FTIR and XRD studies confirm the enhancement of the structure of the composite after that the $H_4TPPS_4^{2-}$molecules functionalize GO via π–π stacking interactions without having a great effect on the electronic properties of the GO. We demonstrated, that generally a higher pH and concentration leads to an enhancement of the optical absorption in the GO-$H_4TPPS_4^{2-}$composites.

References

[1] Jianfeng Shen N , Min Shi, Yizhe Hu, Mingxin Ye, Covalent synthesis of organophilic chemically functionalized graphene sheets, 2010 Journal of Colloid and Interface Science, **348** 1-7.

[2] H. Chang, H. Wu, Graphene-based nanocomposites: preparation, functionalization, and energy and environmental applications, 2013 Energy & Environmental Science, **6** 3483-3507.

[3] Ferrari C, Raman spectroscopy of graphene and graphite: Disorder, electron–phonon coupling, doping and nonadiabatic effects, 2007 J. Solid State Communications, **143** 1-11.

[4] Panda P , Ladomenou K , Coutsolelos A, Porphyrins in bio-inspired transformations: Light-harvesting to solar cell, 2010 J.Coordination Chemistry Reviews, **256** 2601-2627.

[5] Mongwaketsi N , Klumperman B, Sparrow S , Maaza M, Synthesis and characterization of porphyrin nanotubes/rods for solar radiation harvesting and solar cells,2012 J. Physica B, **407** 1-5.

[6] Panagiotis T, Angaridis A , Athanassios C. Functionalized porphyrin derivatives for solar energy conversion, 2014 Polyhedron, **82**.

[7] Narayana Rao D , Excited state dynamics in porphyrins in relevance to third-order nonlinearity and optical limiting, 2003 J.Optical Materials, **21** 45-49.

[8] FUKUZUMI S , Electron transfer chemistry of metalloporphyrins and related metal complexes, 2000 Journal of Porphyrins and Phthalocyanines, **04** 398-400.

[9] Akiharu Satake M, and Yoshiaki K, Porphyrin-Carbon Nanotube Composites Formed by Noncovalent Polymer Wrapping,2005 J.Chem. Mater.**17** (1-9.

[10] Su Young Ryu, Jin Ho C, Sung Ho H, Akihiro Frube, Tsuyoshi A, and Hiroshi M, Optical Properties of Oxotitanium (IV) Meso-tetrakis(4-sulfonatophenyl)porphyrin Intercalated into the Layered Double Hydroxides (LDH) Studied by Laser Spectroscopy, 2003 Bull. Korean Chem. Soc., **24**.

[11] Khenfouch M ,Wéry J, Baïtoul M, Maaza M , Photoluminescence and dynamics of excitation relaxation in graphene oxide-porphyrin nanorods composite, 2014 Journal of Luminescence, **145** (33-37.

[12] S. Stankovich, D.A. Dikin, R.D. Piner, K.A. Kohlhaas, A. Kleinhammes, Y. Jia, Y. Wu, S.T. Nguyen, R.S. Ruoff, Synthesis of graphene-based nanosheets via chemical reduction of exfoliated graphite oxide, 2007 Carbon, **45** 1558-1565.

[13] Zhang L , Liang J ,Huang Y , Ma Y ,Wang Y, Size-controlled synthesis of graphene oxide sheets on a large scale using chemical exfoliation,2009J.Carbon, **47** 3365-3368.

[14] Remyamol T, John H, Gopinath P, Synthesis and nonlinear optical properties of reduced graphene oxide covalently functionalized with polyaniline,2013 J.Carbon, **59** 308-314.

[15] Sterlin Leo Hudson M , Seema A, Sadhasivam T, Ashish B, Satoru S, Sankar S, Hydrogen uptake of reduced graphene oxide and graphene sheets decorated with Fe nanoclusters, 2014 international j ournal of hydrogen energy, **39**.

[16] Thuy Hoa L , Van Hoang L, Jin Suk C, Seung Hyun H, Fabrication of a novel 2D-graphene/2D-NiO nanosheet-based hybrid nanostructure and its use in highly sensitive NO2 sensors, 2013 Sensors and Actuators B, **185**.

[17]Aydin M, DFT and Raman spectroscopy of porphyrin derivatives: Tetraphenylporphine (TPP), 2013 Vibrational Spectroscopy, **68**141-152.

[18] Song W, He C , Zhang W, Chen Y, Li X, Dong Y, Synthesis and nonlinear optical properties of reduced graphene oxide hybrid material covalently functionalized with zinc phthalocyanine,2014 Carbon, **77** 1020-1030.

[19] Aydin M , DFT and Raman spectroscopy of porphyrin derivatives: Tetraphenylporphine (TPP), 2013 Vibrational Spectroscopy, **68** 141-152.

[20] Alben J , Choi S, Adler A, Caughey W, INFRARED SPECTROSCOPY OF PORPHYRINS*, 1973 Annals of the New York Academy of Sciences, **206** 278-295.

[21] Bao M, Bian Y , Rintoul L , Wang R, Arnold D, Ma C , Jiang J, 2004Vibrational spectroscopy of phthalocyanine and naphthalocyanine in sandwich-type (na)phthalocyaninato and porphyrinato rare earth complexes: (Part 10) The infrared and Raman characteristics of phthalocyanine in heteroleptic bis(phthalocyaninato) rare earth complexes with decreased molecular symmetry, Vibrational Spectroscopy, **34**283-291.

[22] Gómez-Navarro C, Weitz R , Bittner A, Scolari M, Mews M, Kern K, Electronic Transport Properties of Individual Chemically Reduced Graphene Oxide Sheets, 2007 J.Nano Letters, **7** 3499-3503.

[23] Wang G, Shen X, Wang B , Yao J , Park J , Synthesis and characterisation of hydrophilic and organophilic graphene nanosheets, 2009 Carbon, **47** 1359-1364.

[24] Lebold T, Yeow E, Steer E, Fluorescence quenching of the S1 and S2 states of zinc meso-tetrakis(4-sulfonatophenyl)porphyrin by halide ions, 2004 Photochemical & Photobiological Sciences, **3** 160-166.

[25] Gonçalves P, Corrêa D, Franzen P, De Boni L, Almeida L, Mendonça C, Effect of interaction with micelles on the excited-state optical properties of zinc porphyrins and J-aggregates formation, Spectrochimica Acta Part A: 2013 Molecular and Biomolecular Spectroscopy, **112** 309-317.

[26] Mongwaketsi N, Alexander N, Malik M, and Raymond S, Ionic self-assembly of porphyrin nanostructures on the surface of charge-altered track-etched membranes, 2010 Journal of Porphyrins and Phthalocyanines, **14**1-7.

IOP Conf. Series: Materials Science and Engineering **186** (2017) 012004 doi:10.1088/1757-899X/186/1/012004

Application of the high temperature series expansions method Green's function theory to the spinel systems: $A_xA'_{1-x}B_2X_4$

A El Grini[1], S Salmi[1], A Marzouk[2], N Benzakour[1], K Bouslykhane[1] and A Hourmatallah[2]

[1] Laboratoire de Physique du Solide-USMBA-FSDM-Fès

[2] Equipe de Physique du Solide-USMBA-ENS

Email: elgrini1@yahoo.fr

Abstract. The magnetic properties of Mn-Cu ferrites $Mn_xCu_{1-x}Fe_2O_4$, have been studied using the many-body Green's function theory (GFT) and high temperature series expansion theory (HTSE). The thermal magnetization and the magnetic susceptibility are given for different values of magnetic field and dilution x. The transition temperature T_C is calculated as a function of Mn concentration. The obtained results are in good agreement with experimental results. The critical exponents associated with the magnetic susceptibility (γ) and the correlation lengths (ν) have been deduced. The obtained values are comparable to those of 3D Heisenberg model.

1. Introduction

Spinel ferrites have been studied extensively due to easy to synthesis and abundant uses in technological and industrial applications [1-3]. The useful properties of the spinel ferrites mostly depend upon the chemical composition, nature of the additives and their distribution i.e. tendency to occupy tetrahedral (A) or octahedral (B) site. In the diluted spinel systems $Mn_xCu_{1-x}Fe_2O_4$, the cation distribution for each concentration x indicates that both $Mn^{2+}$and Cu^{2+} ions have the preference at octahedral site since both Mn^{2+}, Cu^{2+} ferrites belong to the family of inverse spinels [4].

The objective of this paper is to study the magnetic behaviour of diluted spinel systems $Mn_xCu_{1-x}Fe_2O_4$.Using the Green's function theory (GFT)[5-14],with the random-phase approximation (RPA) [12,13]. The thermal magnetization $<S^z>$ and the magnetic susceptibility χ_{zz} are given for different values of external magnetic field and dilution x.The nearest neighbor and the next-neighbor super-exchange J_1(Fe – Fe) and J_2(Mn(Cu) – O – Fe), respectively, in the range $0 \leq x \leq 1$ for $Mn_xCu_{1-x}Fe_2O_4$ have been obtained by using the mean field theory. The values of the intra-plane and inter-plane interactions $J_{aa}(x)$, $J_{ab}(x)$and $J_{ac}(x)$ respectively, are deduced. The high temperature series expansions combined with padé approximants is given to determine the critical T_C versus x. We have used this technique to determine the critical exponents γ and ν associated with the magnetic susceptibility χ and the correlation length ξ, respectively.

2. Theoretical methodologies

2.1. Green function method

The Hamiltonian of the spin-S Heisenberg exchange interaction J between nearest neighbourlattice sites and under an external magnetic field $\mathbf{B}(Bx, By, Bz)$ is

$$H = -\sum_{\langle kl \rangle} J_{kl}(S_k^x S_l^x + S_k^y S_l^y) - \sum_k (B^x S_k^x + B^y S_k^y + B^z S_k^z) \quad (1)$$

Where k and l denote the lattice sites and <kl> indicates summation over the nearest neighbours only. The z-axis is perpendicular to the (x-y) plane. The external field component in the z-direction enhances the effect of the transverse field components $B^{\pm}$.
The transformations $S_i^{\pm} = S_i^x \pm iS_i^y$ and $B^{\pm} = B^x \pm iB^y$ are introduced.
As in reference [10], we introduce the set of thermal Green's functions in the spectral representation.

$$G_{ij(\eta)}^{\alpha mn}(\omega) = \langle\langle S_i^{\alpha}; (S_j^z)^m (S_j^-)^n \rangle\rangle_{\omega} \quad (2)$$

Where ω denotes the energy and i, j the lattice sites and

$$\begin{cases} \alpha = +, -, z \\ \eta = \pm 1 \\ m \geq 0 \\ n \geq 1 \end{cases}$$

In order to treat the problem, one needs the following Green's functions

$$G_{ij(\eta)}^{\alpha mn}(\omega) = \langle\langle S_i^{\alpha}; (S_j^z)^m (S_j^-)^n \rangle\rangle_{\omega} \quad ; \alpha = +, -, z \quad (3)$$

Where $\eta = \pm 1$ refer to the commutator ($\eta = -1$) or anti-commutator ($\eta = 1$) Green's functions, respectively, n≥1 and m≥ 0 are positive integers, *i* and *j* denote lattice sites.
The applied Green function method in this work was described in detail in ref [13, 14]. After using the spectral theorem, we finally obtain the component of magnetisation for general spin S.

$$\langle S^z \rangle = \frac{(S - \phi_k)(1 + \phi_k)(1 + \phi_k)^{2S+1} + (1 + S + \phi_k)\phi_k^{2S+1}}{(1 + \phi_k)^{2S+1} - \phi_k^{2S+1}} \quad (4)$$

Where
$$\phi_k = \frac{1}{e^{\beta E_k} - 1} = \frac{1}{2} coth\left(\frac{\beta E_k}{2} - 1\right) \quad (5)$$

The z component of magnetic susceptibility is defined by:

$$\chi_{zz} = \frac{\langle S^z(B^z) \rangle - \langle S^z(0) \rangle}{B^z} \quad (6)$$

2.2 High temperature series expansion

In addition, The mean field theory, leads to simple relations between the paramagnetic Curie-temperature θ_p and the critical temperature T_C (or the Néel temperature T_N), respectively. Following, the method of Holland and Brown [17], and using the experimental values of J_1(x=0)=10.75K and J_1(x=1)=11.5K [18], we have deduced the values of the exchange integrals J_2(x=0), and J_2(x=1).To determine J_1(x) and J_2(x) in the whole range of concentration we have used the probability law [19].

We give also the values of the intra-plane and inter-plane interactions $J_{aa}(x) = 2J_1, J_{ab}(x) = 4J_1 + 8J_2$ and $J_{ac}(x) = 4J_2$, respectively and the ratio of inter to intra-planar interactions $\frac{(J_{ab} + J_{ac})}{J_{aa}}$.

In recent work [19, 20], a relation between the correlation length $\xi(T)$ and the correlation functions is given in the case of the case of diluted magnetic B-spinel lattice with a particular ordering vector $\vec{q}(0,0,q)$.

$$\left(\frac{\xi}{a}\right)^2 = \frac{1}{8S(\vec{q}_0)}\left(-\gamma_{ac} + \frac{1}{16}\frac{\gamma_{ab}^2}{\gamma_{ac}}\right) \quad (7)$$

With $S(\vec{q}_0)$ is the structure factor at $\vec{q} = \vec{q}_0$. In the ferrimagnetic case we get $\vec{q}_0 = \vec{0}$.
The simplest assumption that one can make concerning the nature of the singularity of the magnetic susceptibility χ is that the neighbourhood of the critical point the above two functions exhibit an asymptotic behaviour:

$$\chi(T) \propto (T - T_C)^{-\gamma} \quad (8)$$

$$\xi^2(T) \propto (T_c - T)^{-2\nu} \quad (9)$$

Estimates of T_C or T_N and the critical exponents γ and ν associated with the magnetic susceptibility magnetic χ(T) and correlation length ξ (T), respectively for $Mn_xCu_{1-x}Fe_2O_4$in the concentration range $0 \leq x \leq 1$, have been obtained using the Padé approximant method (P. A). The [M,N] P. A to the series $F(\beta)$ is a relational fraction P_M/Q_N with P_M and Q_N, polynomials, of degree M and N in β, satisfying: $F(\beta) = P_M/Q_N + \theta(\beta^{M+N+1})$. The sequence of [M,N] P. A to both the Log(χ(T)) and Log (ξ^2(T)), was found to be convergent. The simple pole correspond to the residues to the critical exponents γ and ν for different dilutions are given for $Mn_xCu_{1-x}Fe_2O_4$.The value of critical exponents γ and ν associated to the magnetic susceptibility χ(T) and the correlation length ξ(T), respectively, have been estimated in the range of the composition $0 \leq x \leq 1$. The obtained central values of γ and ν are $\gamma = 1.327 \pm 0.02$ and $\nu = 0.729 \pm 0.01$.These values are insensitive to the dilution in ordering phase.

Table 1. The Néel temperature T_N(K) obtained respectively; in ref [12], and by Green's function theory (GFT), and by (HTSE) theory ,and the values of the first, second, intra-plane, and inter-plane exchange integrals obtained by mean field theory of the $Mn_xCu_{1-x}Fe_2O_4$ system in this work.

x	$T_N(K)$ [18]	$T_N(K)$ (G.F.T	$T_N(K)$ (H.T.S.E)	$\frac{J_1}{k_B}(K)$	$\frac{J_2}{k_B}(K)$	$\frac{J_{aa}}{k_B}(K)$	$\frac{J_{ab}}{k_B}(K)$	$\frac{J_{ac}}{k_B}(K)$	$\frac{(J_{ab}+J_{ac})}{J_{aa}}$
0	290	294	299.14	-10.75	7.05	-21.50	13.4	28.20	-1.93
0,1	---	324	329.11	-10.82	7.79	-21.64	19.04	31.16	-2.30
0,2	---	348	353.05	-10.89	8.59	-21.78	25.16	34.36	-2.73
0,25	385	365	370.12	-10.93	9.01	-21.86	28.36	36.04	-2.95
0,3	---	378	383.10	-10.97	9.45	-21.94	31.72	37.80	-3.16
0,4	---	400	405.16	-11.04	10.39	-22.08	38.96	41.56	-3.64
0,5	415	419	424.11	-11.12	11.41	-22.24	46.80	45.64	-4.15
0,6	---	444	449.12	-11.19	12.51	-22.38	55.32	50.04	-4.71
0,7	---	465	470.08	-11.27	13.69	-22.54	64.44	54.76	-5.29
0,75	475	479	484.06	-11.31	14.32	-22.62	69.32	57.28	-5.59
0,8	---	493	498.15	-11.35	14.97	-22.70	74.36	59.88	-5.91
0,9	---	524	529.17	-11.42	16.34	-22.84	85.04	65.36	-6.58
1	550	554	559.08	-11.50	17.82	-23.00	96.56	71.28	-7.29

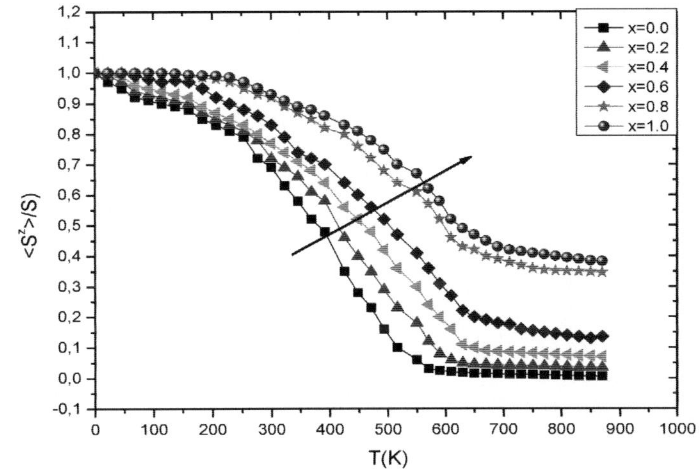

Figure 1. Thermal normalized magnetization in $\langle\frac{S^z}{S}\rangle$ spinel system $Mn_xCu_{1-x}Fe_2O_4$ for B^x= 0, and with the values of dilution x= 0.0 to 1.0.

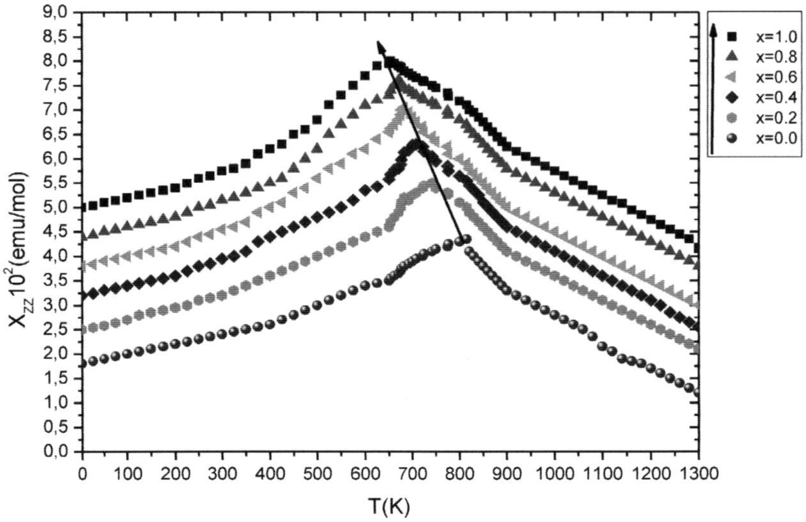

Figure 2. Temperature dependencies of the normalized magnetic susceptibilities χ_{ZZ} in spinel system $Mn_xCu_{1-x}Fe_2O_4$ for B^x= 0, and with the values of dilution x= 0.0 to 0.1

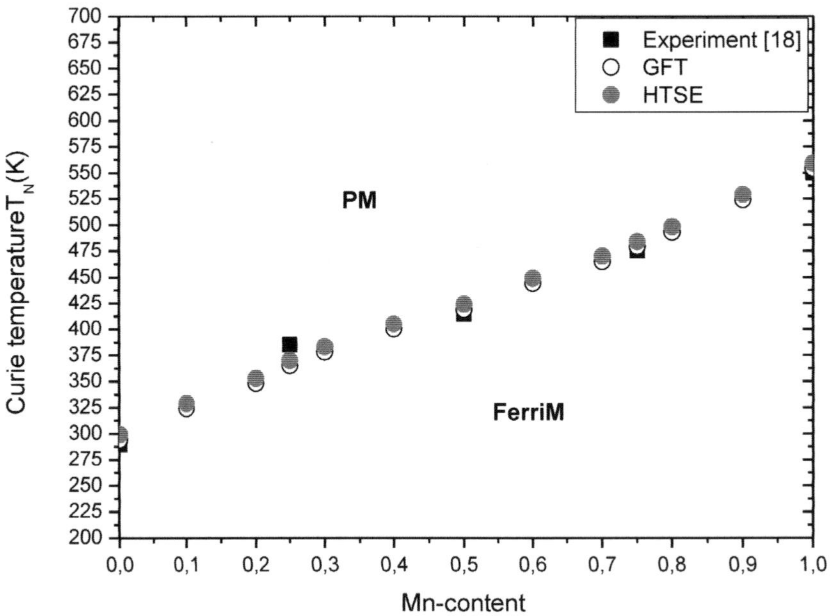

Figure 3. Magnetic phase diagram of $Mn_xCu_{1-x}Fe_2O_4$. The solid circles are the results of (H.T.S.E) method. The open circles are the results of (G.F.T) theory. The solid squares represent the experimental points deduced by magnetic measurements [16].

3. Discussion and Conclusion

We investigate the magnetic reorientation within the Heisenberg model for the ferrimagnetic system in the random-phase approximation (RPA). The applied Green's function method in this work was described in detail elsewhere [12-14].The thermal magnetization $<S^z>$ and the magnetic susceptibility χ_{zz} are deduced for different values of external magnetic field and dilution x and are presented in **figure 1** and **figure 2**respectively.

The high temperature series expansions extrapolated combined with Padé approximants method [17, 19] is shown to be a convenient method to provide valid estimations of the critical temperatures for a real system. By applying this method to the magnetic susceptibility $\chi(x)$, we have estimated the critical temperature T_C for each dilution x in the $Mn_xCu_{1-x}Fe_2O_4$ system. The values of T_C obtained by experiment measurements, (GFT) and (HTSE) theories are represented in **figure 3**.

References

[1] El-Sayed H M, *Journal of Alloys and Compounds*, 474 (2009) 561-564.
[2] Igaraski H, Okazi K, *J. Am. Ceram. Soc.* 60 (1977) 51.
[3] Satyamurthy N S, Natera M G, Youssef S I, Begum R, Sirivatava C M, *Phys. Rev.* 181 (2)(1969)
[4] Bland J A C, Heinrich, B,*Ultrathin Magnetic Structures.Springer*, Berlin (1994).
[5] Lugert G, Robl W, Pfau L, Brockmann M, Bayreuther G,*J. Magn. Magn. Mater.*121 (1993)498.
[6] Schulte O, Klose F, Felsch W,*Phys. Rev.* B **52**, (1995) 6480.
[7] Farle M, Mirwald-Schulz B, Anisimov A N, Platow W,Baberschke, K,*Phys. Rev.* B **55**(1997) 3708.
[8] Bland J A C, Heinrich B : *Ultrathin Magnetic Structures.Springer*, Berlin (1994).
[9] Lugert G, Robl W, Pfau L., Brockmann M, Bayreuther G,*J.Magn. Mater.* **121** (1993) 498.
[10] Schute O, Klose F, Felch, W, *Phys. Rev.* B **52**(1995) 6480.
[11] Farle M, Mirwald Schulz B, AnisimovA N, Platow W, Baberschke K, *Phys. Rev.* B **55** (1997) 3708.
[12] Frobrich P et al, *Eur. Phys. J.* B **13**, 477 (2000).
[13] Frobrich P et al, *Eur. Phys. J.* B **18**, (2000) 579.
[14] El Grini A. Hamedoun,M., Benyoussef,A., Masrour,R., Mounkachi,O., Hourmatallah,A., N. Benzakour,N and Bouslykhane K, *Phase Transitions*,**86** (2013) 1186-1203
[15] Tyablikov S V, *Ukr. Mat. Zh.* **11**(1959)289.
[16] Tyablikov SV,*Methods in the Quantum Theory of Magnetism*.Plenum, New York (1967).
[17] Hamedoun M, Houssa, Benzakour N,Hourmatallah A, *J. Phys: Condens. Matter***10** (1998) 3611.
[18] Mazhar U. Rana Misbah-ul-Islam, Tahir Abbas, *Solid State Communications***126** (2003) 129–133.
[19] Hamedoun M, Bakrim H,. Bouslykhane K, Hourmatallah A, Benzakour N, Filali A, Masrour R, *J. Phys Condens Matter***20** (2008) 125216.
[20] Hamedoun M, Houssa M, Cherriet N Y, Bakkali FZ, *Phys. Stat. Sol.*(b) **214** (1999) 403

NIEL calculations for estimating the displacement damage introduced in GaAs irradiated with charged particles

E El Allam[1], C Inguimbert[2], S Addarkaoui[1], A Meulenberg[3], A Jorio[1] and I Zorkani[1]

[1] Solid State Physics Laboratory, Faculty of Sciences Dhar el Mahraz, University Sidi Mohamed Ben Abdellah, P.B 1796 Atlas-Fez, Morocco
[2] ONERA-DESP, 2 avenue E. Belin, 31055 Toulouse, France
[3] Science for Humanity Trust, Inc. Tucker, GA, USA

E-mail: elmehdi.elallam@usmba.ac.ma

Abstract. The application of Non-Ionizing Energy Loss (NIEL) in estimating the impact of electron, proton, and heavy ion irradiations on Gallium Arsenide is presented in this paper. The NIEL for deuteron, alpha particle, lithium ion and oxygen ion is computed using the SR-NIEL and NEMO codes. The NIEL calculations are compared with the introduction rate of displacement damage measured in n-type GaAs. Very good agreement is found between the NIEL and experimental results for protons (< 20 MeV), electrons, and a variety of ions. However, a discrepancy can be observed for high-energy protons.

1. Introduction

The space radiation environment includes high energy particles such as electrons, protons, and heavy ions. These charged particles may degrade the electrical performance of many devices used in space applications. Estimation of degradation introduced by the radiation in these devices is not always reliable and needs to be improved.

A variety of experimental techniques has been used to study the effect of radiation in gallium arsenide (GaAs). The type and density of the produced defects, can be investigated by means of photoluminescence spectroscopy (PL) and deep level transient spectroscopy (DLTS) [1, 2]. It has been shown that various defects can be produced in silicon doped GaAs [1]. Some simple defects such as gallium vacancies (V_{Ga}), silicon impurity at the arsenic site (Si_{As}), and more complex defects can be introduced when charged particles pass through n-type GaAs doped with silicon. These microscopic defects are the so called displacement damage responsible for the degradation of electrical and optical properties of semiconductor materials that leads to a degradation of semiconductor device performance (e.g., LED's, laser diode, solar cells etc.). It has been reported that the introduction rate of these displacement damages in GaAs is inversely proportional to the energy of heavy charged particles, such as protons and heavy ions, and proportional to the energy of incident electrons [3]. That is to say that the number of defects is proportional to the amount of atomic displacements produced by the incident radiation field and thus to the Non-Ionizing Energy Loss (NIEL). The amount of atomic displacements produced can be estimated from the Non-Ionizing Energy Loss which gives the amount of energy deposited by an incident particle that can lead to displacement damage [4]. NIEL has been calculated for many incident particles on different materials by different authors [e.g., 4, 5, 6, 7]. In many cases, NIEL is demonstrated to correlate with experimental degradation measurements. But in some cases, the NIEL fails to give good degradation predictions, for example, in GaAs irradiated with protons having energies greater than 10 MeV [8, 9, 10].

This paper proposes a comparison between experimental data of displacement damage effects due to electron, proton and heavy ion irradiations in n-type GaAs and NIEL calculations for these incident particles in order to investigate the limit of the NIEL scaling approach for GaAs semiconductor.

Calculation of NIEL

Non-Ionizing Energy Loss (NIEL) is defined as the part of the incident particle energy consumed through Coulombic, nuclear elastic, and nuclear inelastic interactions that can occur between the incident particle and the lattice atoms, to produce vacancy-interstitial pairs (i.e., displacement damage). NIEL Can be calculated as [6]

$$NIEL(E) = \frac{N_A}{A} \int_{T_d}^{T_{max}} Q(T)T\left(\frac{d\sigma}{dT}\right)_E dT \qquad (1)$$

Where N_A is Avogadro's number, A is the atomic mass of the lattice atom, E is the incident particle kinetic energy, T is the kinetic energy transferred to the target atom, $Q(T)$ is Lindhard partition function, $(d\sigma/dT)$ is the differential cross section (Coulombic and nuclear elastic/inelastic) for atomic displacements. The integral is calculated from the threshold energy for displacement T_d, i.e., the energy needed to displace the target atom from its lattice position, to the maximum energy T_{max} which can be transferred by the incident particle to target atoms. The unit of NIEL is MeV.cm^2/g.

In this paper, two tools are used to calculate the NIEL for heavy ions in GaAs. The first is the NEMO code developed by the French Aerospace Lab (ONERA) [4] and the second is the SR-NIEL code developed by National Institute for Nuclear Physics (INFN) [7, 11].

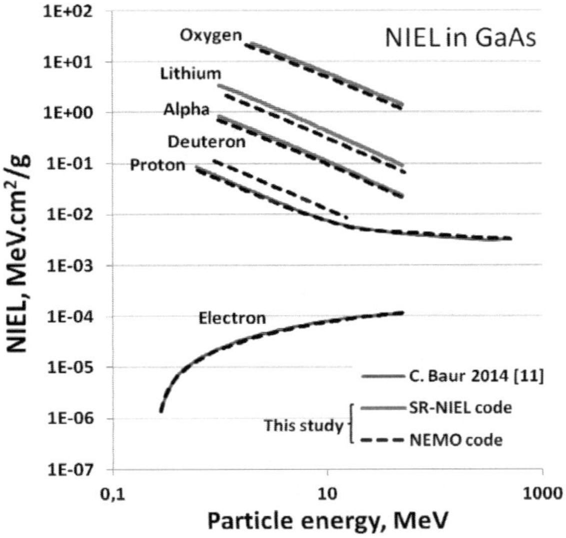

Figure 1. NIEL for charged particles in GaAs. For, protons and electrons, the data were taken from C.Baur et al [11]. For deuterons, alpha particles, lithium ions, and oxygen ions, the NIELs were calculated in this study by SR-NIEL [7] and NEMO code [4].

The two methods of calculation (and the results) are similar. Both methods use the $Q(T)$ Lindhard-Robinson partition function [12], with a threshold displacement energy T_d = 10 eV for both Ga and As. In the case of electrons, the SR-NIEL code [7], employs the Mott differential cross section [13]. In the case of protons, the NIEL is the sum of the contribution from the Coulomb scattering and the nuclear interactions. NIEL (Coulomb) is calculated [14] by using the elastic Coulomb cross section discussed by Boschini, and NIEL (nuclear) [15] was evaluated by Jun. Concerning the NEMO code; electron NIEL is computed [4] using McKinley and Fesbash scattering cross section. For protons, the calculation is based on cross-section calculations, which combine both Coulombic and nuclear elastic scattering, and by solving the Schroedinger equation for a given optical potential [4]. The nuclear inelastic contribution is estimated [4] by using the GEANT4 toolkit. For heavy ions, in both methods only the Coulombic contribution is taken into account. It is estimated according to the ZBL screened Coulomb potential discussed by Messenger [16]. The inelastic contribution remains negligible in comparison to the Coulombic process in the energy domain of interest (< 30 MeV) of our study. The results of NIEL for GaAs as a function of particle energy are shown in Figure 1. It can be seen that the NIEL curves obtained from SR-NIEL and NEMO code are quite similar.

2. Experimental data and comparison

A lot of work has been done on the effect of irradiation with charged particles in semiconductors such as GaAs. The authors of [1, 3], by using the technique of photoluminescence (PL), concluded that radiation introduces two type of defects in n-type GaAs doped with silicon. The first is the gallium vacancy defect (V_{Ga}) and the second is silicon at the arsenic site (Si_{As}). The latter is an acceptor impurity (i.e., the silicon dopant atom is transferred from the gallium site (Si_{Ga}), where it is a donor, to the arsenic site (Si_{As}) under the effects of radiation).

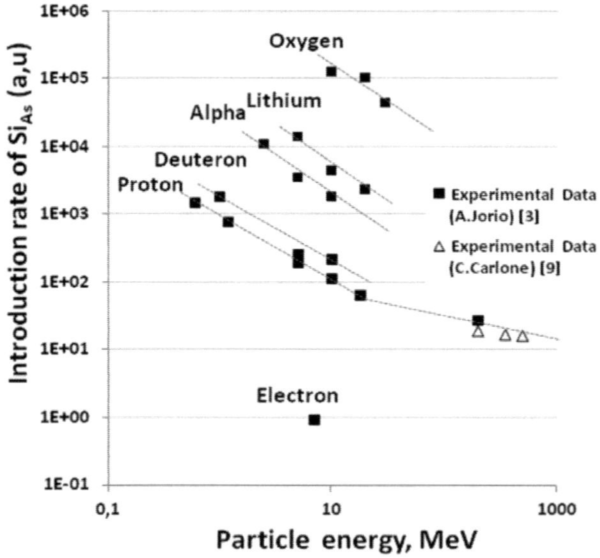

Figure 2. Introduction rate of displacement damage (Si_{As}) measured in GaAs as a function of energy for different particles.

In irradiated n-type GaAs, the degradation of Hall mobility is demonstrated to be correlated to the introduction of two types of microscopic defects V_{Ga} and Si_{As} [17]. This degradation depends on the type and the energy of incident particles and is related with the defect introduction rate associated with each type of particle and therefore to their respective NIELs. Figure 2 shows the defect introduction rate (e.g., Si_{As}) in n-type GaAs irradiated with 7 MeV electrons, protons with energies ranging from 0.6 MeV up to 200 MeV, alpha particles with energies in the range [2.5 MeV, 10 MeV], lithium ions with energies going from 5 MeV up to 20 MeV, and oxygen ions from 10 MeV up to 30 MeV. These measurements, performed by Jorio et al [3], are in good agreement with the defect introduction rate measured by Carlone et al [9] for protons in the range [200 MeV to 500 MeV]. These experimental data can be compared to the NIEL values (Figure 3).

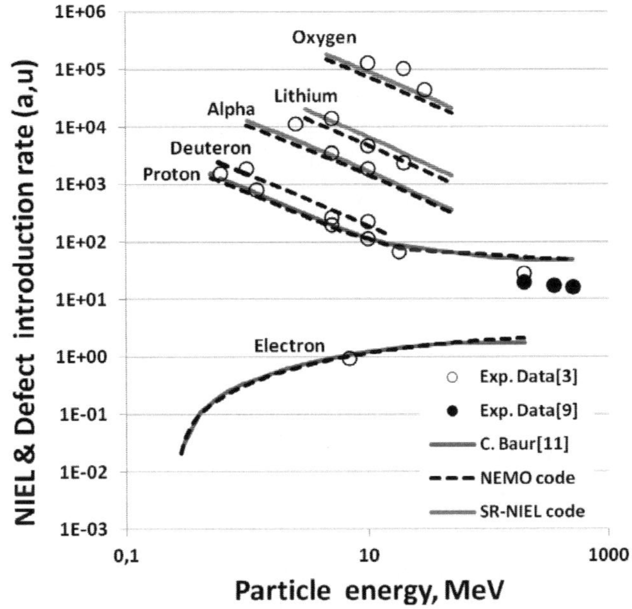

Figure 3. Comparison between NIEL calculations and the measured values of introduction rate of displacement damage (Si_{As}) as a function of energy for n-type GaAs irradiated with various incident particles. The NIEL values are normalized as a function of the experimental value for 10 MeV proton irradiation.

The NIEL values shown in figure 1 have been compared to the measured defects introduction rate (Si_{As}) for the various particles presented in Figure 2. To make this comparison, the NIEL values have been normalized to the experimental defect introduction rate of 10 MeV protons. Good agreement is found between the NIEL and the defect introduction rate in most cases except for protons of energies greater than 20 MeV. The relative damage rates of 7 MeV electrons and heavy ions are quite well estimated. But, for proton energies above 20 MeV, the NIEL seems to overestimate the damage level introduced by the incident radiation.

3. Discussion

Most studies of radiation effects on semiconductor materials and devices, typically compare NIEL with some change in device characteristics (e,g, changes in carrier lifetime of GaAs-LEDs [8], degradation of the maximum power of GaAs solar cells [11], etc.). As shown in the previous section, the microscopic defects are responsible for these changes. The overall good agreement between experimental data and NIEL values presented in the literature and in this work validates the NIEL scaling approach except for high energy protons for which a systematic discrepancy is observed (GaAs).

By looking at Figure 3, one observes that for 7 MeV electrons, the measured values of the introduction rate (7 MeV) agree well with the corresponding NIEL calculation. Similarly, a good correlation can be observed for protons of energies up to 20 MeV. The displacement damage produced by 10 MeV protons is approximately 2 orders of magnitude larger than for 7 MeV electrons. However, for protons, at high-energy the NIEL seems to overestimate the real damage produced. This discrepancy could be attributed to the production of clusters of defects by high energy PKAs (Primary Knock on Atoms) in which annealing processes can reduce the amount of stable defects [18].

Table 1. Introduction rate of displacement damage (Si_{As}). Comparison with the predicted values of NIEL for both SR-NIEL code [7] and NEMO code [4]. The NIEL values are normalized to 1 for proton NIEL.

	Defect introduction rate		NEMO values		SR-NIEL values	
Particles	5 MeV	10 MeV	5 MeV	10 MeV	5 MeV	10 MeV
Proton	1	1	1	1	1	1
Deuteron	1,4	1,9	2	1.7	***	***
Alpha	17	16	15,5	13	16,4	15
Lithium	70	40	50	42	65	58
Oxygen	***	1132	764	654	854	806

It is assumed that the particle energies are those corresponding to the energy of the particle in the sensitive region of the test devices. The strong dependence of damage with particle path length in a test sample, makes miscalculation of the actual energy values in the sensitive region a possible source of systematic error in the model. The good representation of most defect introduction rates by the NIEL calculations indicates the validity of this portion of the general radiation-damage approach. The non-random variations can lead to a search for specific deviations from the model. The greatest source of error would now appear to be that of accurately determining the energy profile(s) of the radiation environment within the sensitive regions of both the test devices and the devices being studied. However, the possibility of radiation annealing or other moderating effects in the high damage regions of PKAs would need to be explored. As a practical matter, few environments and spacecraft component locations would provide a significant component of radiation damage from this portion of the spectrum and the NIEL calculation appears to err on the safe side. For heavy ions, even if there is a limited set of data, the results presented here indicate that there is a proportionality between NIEL and experimental results at the energies considered in this study. In Table 1, a comparison of the relative damage rate of protons and heavy ions is presented for two energies (5MeV, and 10 MeV). The case of protons is taken as reference, and both the experimental defect introduction rate and theoretical NIEL values are arbitrarily equated to 1. As can be seen in this table, in the case of 10 MeV incident energy, the relative experimental damage rate for the oxygen ions, lithium ions, alpha particles and deuterons are respectively 1132, 40, 16 and 1,9 times greater than the one measured for the protons. These values are close to those predicted by NIEL. This validates the NIEL scaling approach for these particles in this domain of energy.

4. Conclusions

This paper presents a comparison of the NIEL with some measured defect introduction rate in GaAs. Two different toolkits have been tested (NEMO [4] and SR-NIEL [7]). They provide NIEL values very close to each other. Good agreement is found between experimental data and theoretical NIEL calculations for various types of particles and energies. Protons with energies in the range [0.6 MeV, 200 MeV], 7 MeV electrons and a variety of ions with energies going from some MeV up to 30 MeV have been compared with the corresponding NIEL values. The introduction rate of displacement damage for high-energy protons (>20 MeV) is lower than the predicted values of NIEL. Even if there are some discrepancies for high-energy protons, the NIEL factor is demonstrated to be a relevant parameter for predicting degradation of GaAs-semiconductor devices in a complex space radiation environment.

Acknowledgements
One of the authors, El Mehdi El Allam, would like to thank the Abdus Salam International Centre for Theoretical Physics (Federation Scheme, Trieste, Italy) for its support and kind hospitality.

References

[1] Khanna S M, Jorio A, Carlone C, Parenteau M, Houdayer A and Gerdes Jr J W 1995 *IEEE Trans. Nucl. Sci.* **42** 2095-2103
[2] Warner J H, Cress C D, Messenger S R, Walters R J, Ringel S A and Park J 2010 *IEEE Trans. Nucl. Sci.* **57** 1940–45
[3] Jorio A 1995 Effets de L'irradiation sur L'arsenure de Gallium Dope au Slicium de Type n : Observation Optique de la Lacune de Gallium. PHD thesis (Sherbrooke: Sherbrooke University Canada)
[4] Inguimbert C and Gigante R 2006 *IEEE Trans. Nucl. Sci.* **53** 1967–72
[5] Summers G P, Burke E A, Shapiro P, Messenger S R and Walters R J 1993 *IEEE Trans. Nucl. Sci.* **40** 1372–79
[6] Akkerman A, Barak J, Chadwick M, Levinson J, Murat M and Lifshitz Y 2001 *Radiat. Phys. Chem.* **62** 301–10
[7] SR-NIEL website. [Online]. Available: http://www.sr-niel.org/
[8] Barry A L, Houdayer A J, Hinrichsen P F, Letourneau W G and Vincent J 1995 *IEEE Trans. Nucl Sci.*, **42** 2104–07
[9] Carlone C, Parenteau M, Houdayer A, Hinrichsen P and Vincent J 1997 *IEEE Trans. Nucl. Sci.* **44** 1856–61
[10] Srour J R 2003 *IEEE Trans. Nucl. Sci.* **50** 653–70
[11] Baur C, Gervasi M, Nieminen P, Pensotti S, Rancoita P G and Tacconi M 2014 *Proc. 13th ICATPP Conf. on Astroparticle, Particle, Space Physics and Detectors for Physics Applications (Como, Italy)* vol 8 (Singapore: World Scientific) p 698–713
[12] Norgett M, Robinson M and Torrens I 1975 *Nucl. Eng. Des.* **33** 50–54
[13] Moth N F 1929 *Proc. Roy Soc. Lond.* Series A. **124** 425–42
[14] Boschini M J, Consolandi C, Gervasi M, Giani S, Grandi D, Ivanchenko V, Nieminem P, PenBOtti S, Rancoita P G, Tacconi M, ESA E and Noordwijk A G 2011 *Proc. 12th ICATPP Conf. on Cosmic Rays for Particle and Astroparticle Physics, (Como, Italy)* vol 6 (Singapore: World Scientific) p 9–23
[15] Jun I, Xapsos M, Messenger S R, Burke E A and Walters R 2003 *IEEE Trans. Nucl. Sci.* **50** 1924–32
[16] Messenger S R, Burke E A, Xapsos M A, Summers G P, Walters R J, Jun I and Jordan T 2003 *IEEE Trans. Nucl. Sci.* **50** 1919–23
[17] Jorio A, Parenteau M, Aubin M, Carlone C, Khanna S M and Gerdes Jr J W 1994 *IEEE Trans. Nucl. Sci.* **41** 1937–1944
[18] Warner J H, Inguimbert C, Twigg M E, Messenger S R, Walters R J, Romero M J and Summers G P, 2008 *IEEE Trans. Nucl. Sci.* **55** 3016–24

Microscopic study of the structure of the Steel Ni-based Alloy: Hastelloy G35 Alloy

F Sabir[1, 4], O Ben Lenda[1],S Saissi[1], K Marbouh[1], B Tyouke[1], L Zerrouk[1], A Ibnlfassi[2], K Ouzaouit [3]and S Elmadani[1]

[1] Laboratoire Physico-Chimie des Procédés et des Matériaux, Faculté des Sciences et Techniques, Université Hassan 1er, Route de Casablanca,BP577, 26000, Settat, Maroc
[2] Sciences de l'Environnement Et Développement, Faculté des Sciences et Techniques, Université Hassan 1er, Route de Casablanca,BP577, 26000, Settat, Maroc
[3]Research Center REMINEX, Managem Group, Marrakech, Morocco

Email: sabir.fouzia890@gmail.com

Abstract. The study of the influence of heat treatment on changes of mechanical and structural properties of Steel Ni-based Alloy is a highly interdisciplinary topic at the interface of the physical chemistry of metallic materials, which also helps in environmental and economic protection.After heat treatment, the structural and micro-structural studies for the different transformation temperature led to identify phases formed and the morphology. This work has been carried out using different techniques: X-ray diffraction, optical microscopy and scanning electron microscopy.
Keywords. Steel Ni-based Alloy, Heat treatment, Optical Microscope, Scanning electron microscopy, X-ray diffraction.

1. Introduction

The hastelloy G35 superalloy is a nickel alloy with a high proportion of chromium, austenitic matrix and a face-centered cubic structure [1]. Steel nickel-based has improved mechanical properties and durability in its employment conditions, especially environmental ones [2][3]. The goal of this work is to treat this alloy at well determined temperatures to have different precipitates / phases, and to study the kinetics of the transformations that characterize the hardening structural mechanism of the nickel-based alloy [4].

2. Hastelloy G35 Alloy in the initial state

The superalloyHastelloy G35 Alloy is a nickel-based alloy, resistant to a variety of corrosive environments by forming a passivation film; it is one of the commercial alloys developed by Haynes International, Inc for use in the chemical process industry.
The measurements of the G35 steel hardness at the receiving state before heat treatment give anaverage of 50.75 HRF.After the various steps for manufacturing steel, there is the occurrence of phase transformation.

Figure 1.Optical microscopy of the steel G35 before heat treatment

3. Heat treatment influence on hardening steel G35

3.1. X-ray diffraction of the steel G35 alloy to different temperatures studied

We observe that as treatment temperature increases, the intensity of the phases during overagingbecomes increasingly higher[4]; this is due to rearrangements of atomic positions (Figure 2). Miller indices (HKL) calculated are similar to those obtained in the HASTELLOY C-22HSAlloy [5].

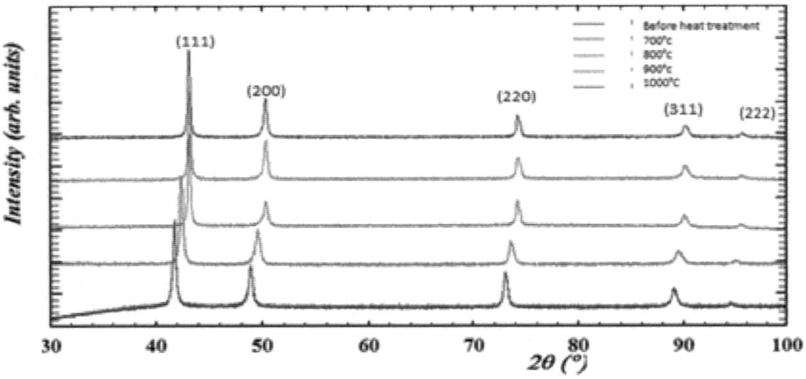

Figure 2. X-ray Diffraction of steel G35 before heat treatment and treated at700°C, 800°C, 900°C, 1000°C during overaging

3.2. Metallographic examination

3.2.1. Optical microscope. The observation of the structure of the alloys is corresponding to the continuous /discontinuous transformations is made by using the optical microscope, the samples must undergo mechanical polishing by a series of abrasive from the abrasive 120 to 1200, chemical polishing using aqua regia; a mixture of hydrochloric acid (HCl) and concentrated nitric acid (HNO_3) in a proportion of 4 volumes of hydrochloric acid to 1 nitric acid.

The structural hardening mechanisms of the SteelsG35 treated at 700°C, 800°C,900°C and 1000°C are characterized by a discontinuous transformation which characterizes the aging and a coarse discontinuous precipitation characterized by the presence of polyhedral (figure 3-4-5-6). Indeed, the solutebecomes poorer in the matrix to feed the precipitates that contribute to the formation of gross precipitation, because of softening.

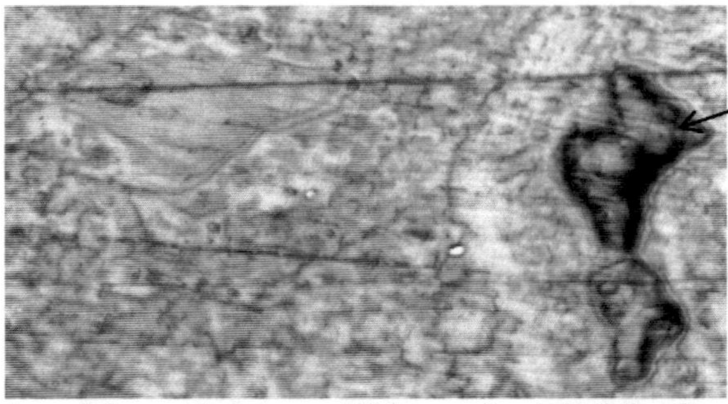

Figure 3. Visualization of the steel structure G35 treated at 700 ° C, after 1 month 10 days

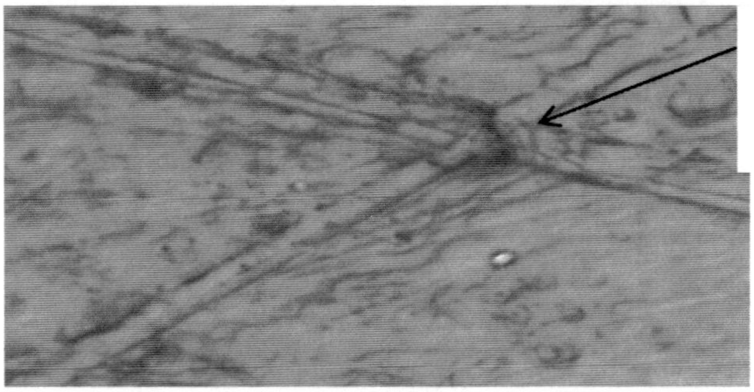

Figure 4. Visualization of the steel structure G35 treated at 800 ° C, after 1 month 15 days

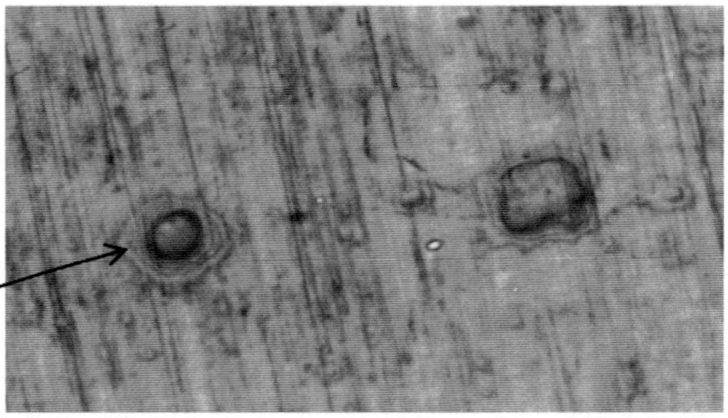

Figure 5. Visualization of the steel structure G35 treated at 900 ° C, after 1 month 8 days

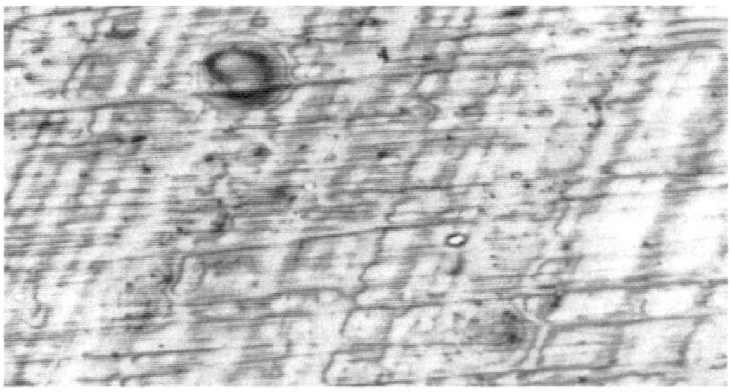

Figure 6. Visualization of the structure of the steel G35 treated at 1000 °C, after 1 month 15 days

3.2.2. Scanning electron microscopy. The microscope used is mark Philips XL30 and has a maximum resolution of 100 nm. The microscope is coupled to a type detector EDXS (Energy Dispersive X-rays Spectroscopy) which allows local quantitative analysis of the composition.

Observations by scanning electron microscope of steel based Nickel, treated at 700 °C show the matrix is the type austenite. By EDXS analyzes performed on the steel shows that the precipitates are very rich in carbon (table 1), against analyzes performed on the matrix confirm the simultaneous presence of Ni, Fe and Cr (Figure 7, 8).

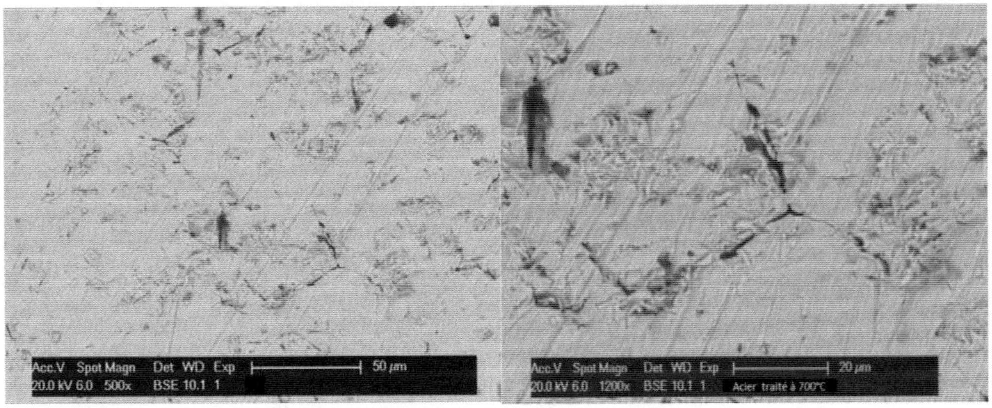

Figure 7.Visualization by scanning electron microscope of the steel based on Nickel, treated at 700°C, after 5 month of aging

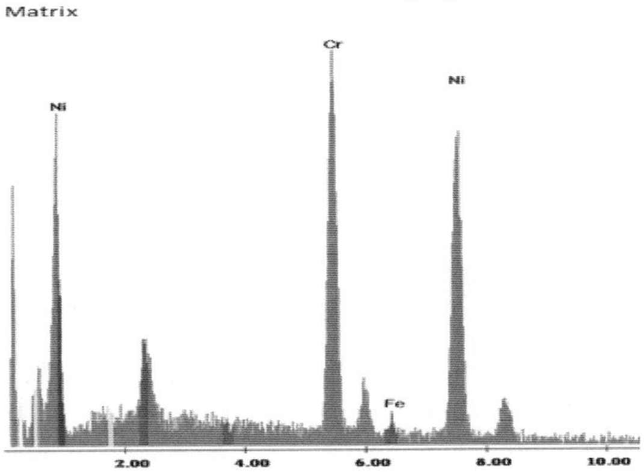

Figure 8.Chemical compositions of the alloy matrix for nickel-based Steel treated at 700 ° C, after 5 month of aging

Table 1. Chemical Composition of the target points

Zone/ Element	Fe		Ni		Cr		C	
	Wt%	At%	Wt%	At%	Wt%	At%	Wt%	At%
Matrix	1.97	1.97	61.90	59.09	36.13	38.94	0	0
Carbure	1.16	0.53	42.23	18.35	23.93	11.74	32.67	69.38

Heat treatment at 900 °C to 1000 °C shows that the matrix is formed of Fe, Mo, Ni and Cr.Observations with a scanning electron microscope show that the matrix is austenite type. Besides, analysis by EDXS performed on the precipitates show that these are formed mainly of Ni, Cr and C as revealed by the spectra of Figures 10 and the results of Table 2.

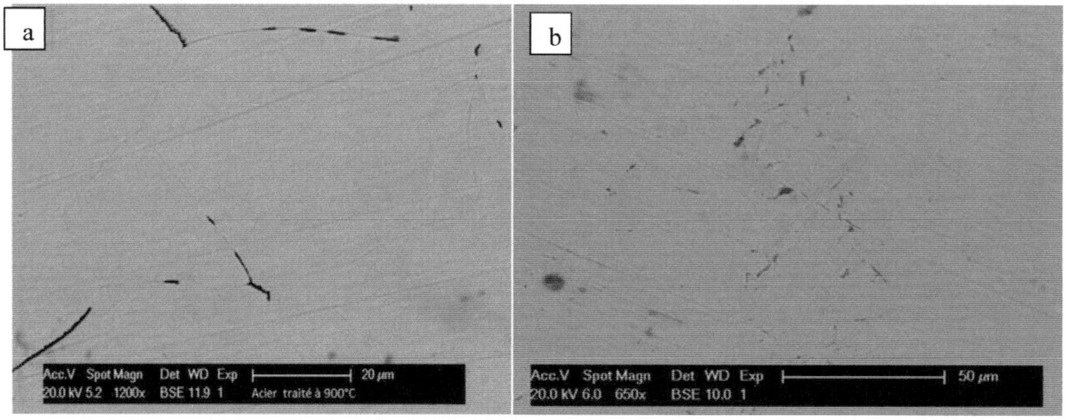

Figure 9. Visualization by scanning electron microscope of the steel based on Nickel,
(a) treated at 900°C, after 5 month of aging
(b) treated at 1000°C, after 5 month of aging

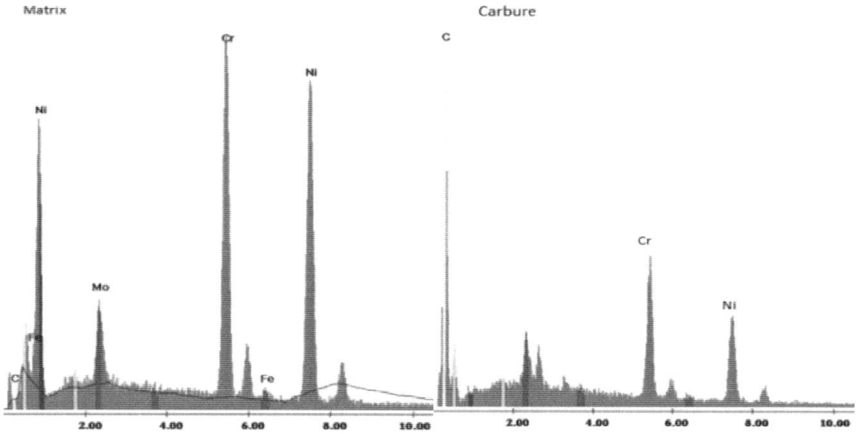

Figure 10. Identification of phases with different chemical compositions by EDXS and SEM for nickel-based steel treated at 900 ° C, after 5 month of aging

Table 2. Chemical Composition of the target points, nickel-based steel treated at 900 ° C, after 5 month of aging

Zone/ Element	Ni		Cr		Mo		Fe		C	
	Wt%	At%	Wt%	At%	Wt%	At%	Wt%	At%	Wt%	At%
Matrix	56.47	55.75	33.76	37.63	8.10	4.90	1.67	1.73	0	0
Carbure	41.77	22.04	25.13	14.97	9.23	2.98	0.76	0.42	23.10	59.58

4. Conclusions

The structural and micro-structural studies for the heat treatment of the different temperatures: 700°C,800°C, 900°C and 1000°C; have led to identify the formed phases and morphology. After heat treatment, we have followed theSteel nickel-based quenching structure, indeed, after the successive chemical attacks we observe; on aging, the movement of fronts which characterize the discontinuous transformation; it is followed by an overaging which consists on a coarse discontinuous processing.

The scanning electron microscope is coupled to EDXS which allows local quantitative analysis of the composition. The EDXS analyzes carried out on the steel show that for lower temperatures (700°C and 800°C), the precipitates are very rich in carbon and the matrix consists of Mo, Ni and Cr. For Heat treatment at 900 °C to 1000 °C shows that the matrix is formed of Mo, Ni and Cr.Besides, analysis by EDXS performed on the precipitates show that these are formed mainly of Ni, Cr and less C.The X-ray diffraction spectra obtained after the treatment revealed that the structure remains austenitic.

References

[1] G.W. Meetham, The Development of Gas Turbine Materials 1st ed. Applied Science, London, 1981.

[2] AMY Zahia, Caractérisation de l'acier inoxydable X5CrNi18-10 et étude de son oxidation à température, République Algérienne Démocratique et Populaire, Ministère de l'enseignement supérieur et de la recherche scientifique, 2013.

[3] ASM speciality handbook Nickel, Cobalt and their alloys, ASM international, United stated of America, 2000.

[4] Sabir.F et all, Journal of Science and Arts, 3(32), pp. 269-284, 2015.

[5] Huang, E.W., Clausen, B., Wang, Y., Choo, H., Liaw, P.K., Benson, M.L., Pike, L.M., Klarstrom, D.L., International Journal of Fatigue, 29(9–11), 1812, 2007.

IOP Conf. Series: Materials Science and Engineering **186** (2017) 012007 doi:10.1088/1757-899X/186/1/012007

Structural and optical properties of hydrothermally synthesized vanadium oxides nanobelts

I Derkaoui[1,3*], M Khenfouch[2,3], I Elmokri[1], B. M Mothudi [2], M. S Dhlamini[2], S. J Moloi[2], I Zorkani[1], A Jorio[1], M Maaza[4,5]

[1]University Sidi Mohammed Ben Abdellah, Faculty of Sciences Dhar el Mahraz, Laboratory of Solid state Physics, Group of Nanomaterials and Renewable Energies, PO Box 1796 Atlas Fez 30 000, Morocco
[2]UNISA university of South Africa, Department of Physics, College of Science, Engineering and Technology, Science Campus, Cnr Christiaan de Wet & Pioneer Avenue Florida 1709, Johannesburg, South Africa
[3]Africa Graphene Center, 2 Boekenhout street, Florida, Johannesburg, 1709 South Africa.
[4]Nanosciences African Network (NANOAFNET), iThemba LABS-National Research Foundation, 1 Old Faure road, Somerset West 7129, PO Box 722, Somerset West, Western Cape-South Africa
[5]UNESCO-UNISA Africa Chair in Nanosciences Nanotechnology, College of Graduate Studies, University of South Africa, Muckleneuk ridge, PO Box 392, Pretoria-South Africa

E-mail: derkaouiissam@gmail.com

Abstract. Nanostructured metal oxides have attracted a lot of attention recently owning to their unique structural advantages, and demonstrated promising chemical and physical properties for various applications. In this study, we report the structural and optical properties of vanadium oxide nanoparticles (VO-NPs) prepared via a hydrothermal method. Our results are revealing that the components of VO-NPs films are having a belt-like shape with a uniform size distribution. The nanobelts with very smooth and flat surfaces have a typical length of up to 1μm and a width of about 9-14 nm. Moreover, The UV-Visible spectroscopy revealed an excellent optical propertiesshowing clearly the good stoichiometry and crystallinity of our VO-NPs films. This synthesis method provides a new simple route to fabricate one-dimensional nanostructured metal oxides which is suitable for a large field of applications.

1. Introduction

Among the transition metal oxides, vanadium oxides have attracted special interest because of their coordination polyhedral and a wide range of oxidation states (from +2 to +5) in the vanadium-oxygen system [1]. Nanostructured vanadium oxides materials, such as nanorods, nanowires, nanotubes and nanobelts, have attracted considerable attentions because to both their distinctive geometries, specific physical and chemical properties that are different from the bulk materials due to their limited size and high density of corner or edge surface sites [2].

In the light of previous reports, vanadium oxides nanoparticles possessing unique electrical and optical properties, it was found to show high electronic conductivity arising from a mixed-valence V

$3^{+}/5^{+}$ and structural stability arising from increased edge sharing and the consequent resistance to lattice shearing during cycling [3]. Moreover, it can undergo a first-order transition from a high-temperature metallic phase to a low temperature semiconductor phase at around 68 °C that is the result of an atomic structural rearrangement, this temperature is increasing with decreasing the size of particles [4]. Owing to their important properties they exhibit potential application in fabricating nanoscale electronic, optical, electrochemical, optoelectronic, and electromechanical devices [5,6]. Because of these attractive properties and applications, many different synthesis techniques have been developed for the synthesis of vanadium oxide nanoparticles. For instance, sol–gel [7], chemical vapor deposition [8] and pulsed laser ablation [9]. Hydrothermal synthesis is found to be a suitable method to obtain vanadium oxide nanoparticles. This technique is ideal for the processing of very fine powders and films having excellent reproducibility, high purity, narrow particle size distribution and controlled morphology [10].

In this work, we are presenting the structural and optical properties of nanostructured metal oxide nanoparticles which include growth mechanism description and interactions discussion. In this sense, transmission electron microscopy (TEM), X-ray diffraction (XRD), Fourier transform infrared spectroscopy (ATR-FTIR)and ultraviolet-visible spectrum (absorbance/reflectance) were investigated.

2. Samples preparation

2.1 Synthesis of VO-NPs

All of the chemical reagents were analytical grade. They were purchased from Acros Organics and used without further purification. The detailed for the synthesis was as follows.

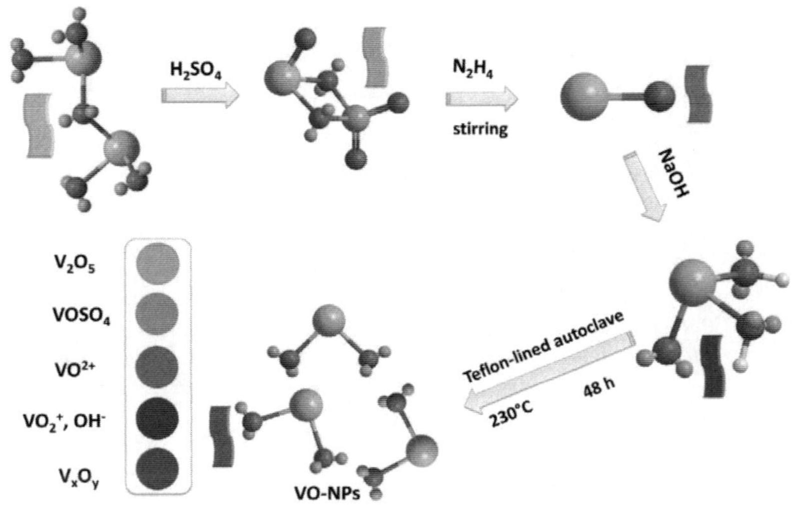

Figure 1. Growth mechanism of vanadium oxides nanoparticles.

VO-NPs were synthesized using a hydrothermal procedure. In a typical synthesis addition of 0.75 ml of sulphuric acid, H_2SO_4, into an aqueous suspension of 0.45g of V_2O_5 was followed by the addition of hydrazine hydrate(N_2H_4-$2H_2O$). After stirring thoroughly, the solution's color changed from yellow (V^{5+} valence state) to blue, indicating the reduction of V^{5+} to yield V^{4+} ions in solution. The pH of the resultant strongly acidic blue VO^{2+} solution was then adjusted to pH≈5 by adding NaOH solution. Heat could also be applied to the solution while adding NaOHsolution. A gray to brown precipitate was formed during the addition of NaOH solution. Afterward, the suspension was transferred into a 25 mL Teflon-lined stainless autoclave. The autoclave was maintained at 230°C for 48 h and then air cooled to room temperature. The resulting dark blue precipitates were collected and washed with distilled water and ethanol several times and then dried at 60°C under vacuum for few hours(Figure 1).

2.1 Characterization

In this study, transmission electron microscopy (TEM) is employed to observe the morphology of VO-NPs, XRD pattern was obtained from a Rigaku Smart Lab system using Cu Kα (λ=1.54178 Å), ATR-FTIR spectrum was obtained using a Broker Vertex 70 with ATR system, Raman spectrum was obtained using a Bruker Multi Ram with and excitation wavelength of 1064 nm. Finally, Uv-Vis (absorbance/ reflectance) spectrum was carried out using Perkin Elmer lambda 1050 UV/Vis/NIR spectrometer.

3. Results and discussion

3.1TEM measurements

To investigate the morphology of vanadium oxides nanoparticles, TEM images have been taken and shown in Figure 2a. Figure 2a shows a typical panoramic TEM image of the VO-NPs film, where the components are nanobelts with a uniform size distribution. The nanobelts with very smooth and flat surfaces have a typical length of up to 1µm and a width of about 9-14 nm.

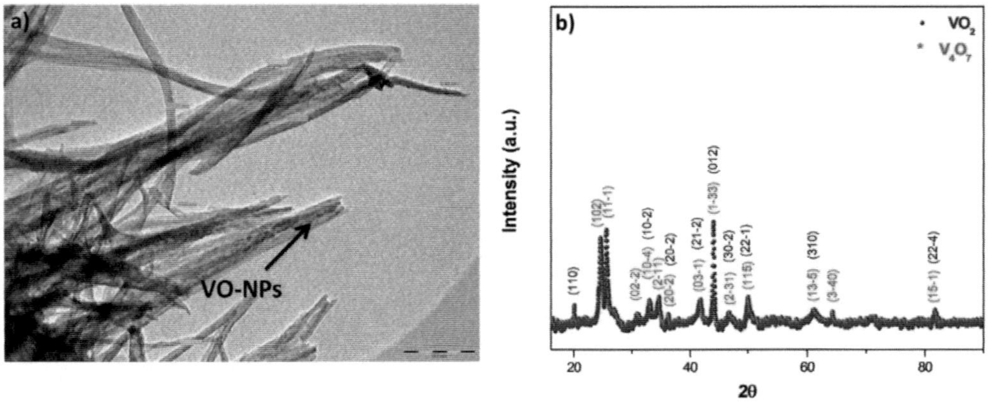

Figure 2. (a) TEM images of VO-NPs nanobelts, (b) X-rays diffraction pattern of VO-NPs.

3.2X-ray diffraction

Figure 2 b shows the XRD pattern of the as-prepared vanadium oxide nanoparticles. This pattern revealed the existence of the two phases tetravanadium septoxide (V_4O_7) and vanadium dioxide (VO_2). Strong diffraction peaks were observed at Bragg angles 24.61°, 25.53°, 30.98°, 33.03°, 34.63°, 36.35°, 41.65°, 43.97°, 46.59°, 49.79°, 61.29° and 81.79°, which they could be assigned to the (102) $(11\bar{1})$ $(02\bar{2})$ $(10\bar{4})$ $(2\bar{1}1)$ $(20\bar{2})$ $(03\bar{1})$ $(1\bar{3}3)$ $(2\bar{3}1)$ (115) $(13\bar{5})$ and $(15\bar{1})$ planes, which can be readily indexed to the Triclinic crystalline phase (Space group 2 : A^{-1}) of V_4O_7 with the lattice constants a=5.456640 A°, b=6.941393 A° and c= 12.139513 A° (α =95.099998 β =122.599998

γ =109.250000) (ICDD card N° 1008024).Furthermore, the diffraction peaks at Bragg angles 20.16°, 33.03°, 36.35°, 41.65°, 43.97°, 46.59°, 49.79°, 61.29° and 81.79° are assigned to the (110) (10$\bar{2}$) (20$\bar{2}$) (21$\bar{2}$) (012) (30$\bar{2}$) (22$\bar{1}$) (310) and (22$\bar{4}$) planes, this series can be perfectly indexed to the Monoclinic(b) crystalline phase (Space group 14 : P121/c[1]) of VO_2 with lattice constants a=5.849727 A°, b=4.600944 A°, c= 5.474888 A° and β=122.599998(ICDD card N° 9009089), revealing that the V^{5+} ions in V_2O_5 have been reduced to V^{4+} ions by the hydrazine in the reaction. No peaks of any other phases or impurities were detected in the spectra, which mean that the products are mainly composed of vanadium oxides nanoparticles. In addition, the plot in Figure 3 present an estimation of these V_4O_7 andVO_2 nanoparticles size calculated using Hall method where the obtained size average was found to be 96 (56) and 94 (93) respectively.

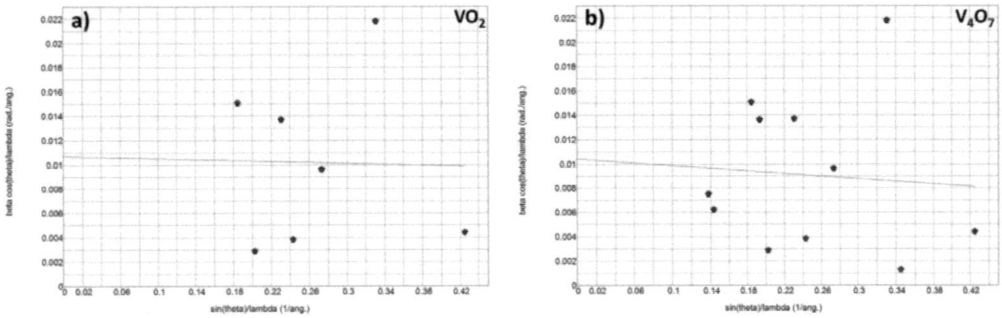

Figure 3. Size estimation in vanadium oxide nanostructures (a) and (b) of VO-NPs.

3.3 Fourier transform infrared spectroscopy

ATR-FTIR spectroscopy is one of the powerful tools for detecting the species on the nanostructures surface. As we can see VO-NPs (Figure 4) infrared spectrum has multiple bands, the two ones at 617 and 670 cm^{-1} are associated to the characteristic V-O-V octahedral bending modes [11]. Moreover, the peak located at 893 cm^{-1} is attributed to the coupled vibration V=O and V-O-V [12]. The characteristic peak at 910 cm^{-1} corresponds to V^{4+} terminal oxygen bonds (V^{4+} =O) and the peak at 1005 cm^{-1} which corresponds to V^{5+} terminal oxygen bonds (V^{5+} =O) becomes stronger [11,13]. The stretching vibrations of the V=O (vanadyl) bond is observed at 990 cm^{-1} while the band located at 1070 cm^{-1} is assigned to the oxidation state from V^{5+} to V^{4+}[12]. Therefore, the IR results are consistent with the formation of VO-NPs nanobelts which is in good agreement with the TEM observations.

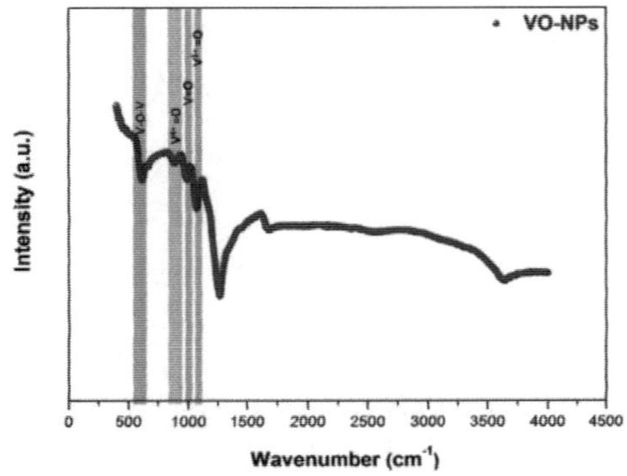

Figure 4. The ATR-FTIR spectrum of vanadium oxides nanoparticles.

3.4 UV-Vis spectroscopy

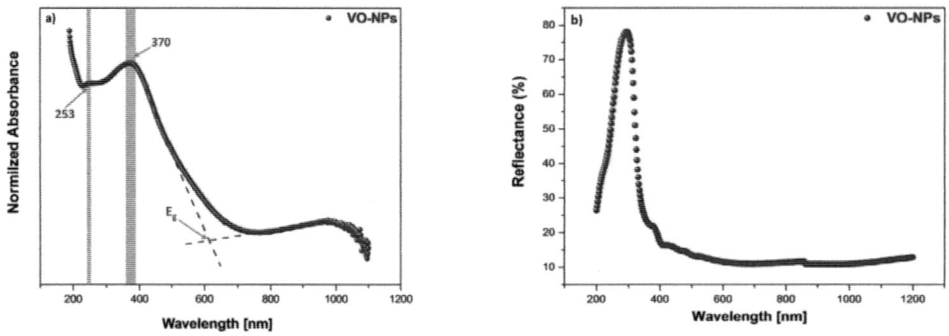

Figure 5. Uv-Vis optical spectra of the (a) VO-NPs absorbance and (b) VO-NPs reflectance.

Figure 5 shows the ultraviolet-visible spectra of VO-NPs films on glass substrates measured in the wavelength range 200-1000 nm. The absorbance spectrum of VO-NPs (Figure 5a) has two bands located at around 253 nm and 370 nm, the first should be assigned to charge transfer transition, involving oxygen and vanadium (IV) in tetrahedral coordination, present in isolated species [14,15]. The second broad band at 370 nm is attributed to octahedral V^{4+} species [14].The gap energy from Uv-visible spectrum absorbance can be determined at the tangent to the absorption front, and the width of the band gap is strongly dependent on the conjugation length. The optical band gap of VO-NPs nanobelts was found to be 3.33 eV which is higher than the band gap of vanadium oxide nanosheets reported by Wang (2.1 eV) [16]; This increment in the optical band may be attributed to the dimensional confinement effects of the VO-NPs [17,18].Figure 5b shows the reflectance spectrum of VO-NPs films. Reflectivity change was observed distinctly in the wavelength region 200-400 nm and slightly above 400 nm. In the wavelength region 200-400 nm, the VO-NPs films has a high reflectance (78%) showing clearly the good stoichiometry and crystallinity of the nanobelts which have an influence to enhanced the optical properties of vanadium oxides nanoparticles films.

4. Conclusion

To summarize, nanostructured vanadium oxides were successfully synthesized on glass substrates via a simple hydrothermal synthesis method. The structural investigations revealed the existence of two phases of VO-NPs while the vibrational properties are confirming the formation of these nanoparticles which assisted to the understanding of the growth mechanism. Moreover, our results showed that the geometrical shape of these vanadium oxides nanostructures is a belt with a uniform size distribution. The length and diameter of these nanobelts with very smooth and flat surfaces are 1μm and about 9-14 nm, respectively. Furthermore, the growth mechanism was found to have a great impact on the optical properties and strong absorption of Uv-Vis light of our VO-NPs nanobelts films. This study is focused on its offering of information which may assist to control the preparation of this nanostructured vanadium oxides and modulating their properties to enhance their efficiency then improve the performance of their applications including storage devices.

Acknowledgements

Special thanks to the Innovation city of USMBA (Morocco), Abdus Salam International Centre for Theoretical Physics (Trieste-Italy), Nanosciences African Network, University of South Africa Department of Physics, iThemba Labs (South Africa), African Laser Center and Africa Graphene Center.

References

[1] Livage J 2010*Materials*, **3** 4175-4195
[2] Fernandez-Garcia M, Martinez-Arias A, Hanson J C and Rodriguez J A 2004 *Chemical Reviews* **104** 4063-4104
[3] Lampe-Önnerud C, Thomas J O, Hardgrave M and Yde-Andersen S 1995 *Journal of Electrochemical Society* **142** 3648-3651
[4] Soltane L, SediriF and Gharbi N 2012 *Materials Research Bulletin* **47** 1615-1620
[5] Ji S D, Zhao Y G, Zhang F and P. Jin 2010 *Journal of* Crystal Growth **312** 282-286
[6] Ganganagappa N and Siddaramanna A 2012 *Materials Characterization* **68** 58-62
[7] Yin D C, Xu N K, Zhang J Y and Zheng X L 1996 *Journal of Physics. D: Applied Physics* **29** 1051-1057
[8] Manning T D, Parkin I P, Pemble M E, Sheel D and Vernardou D 2004 *Chemistry of Materials* **16** 744-749
[9] Kim D H and Kwok H S 1994 *Applied Physics Letters* **65** 3188-3190
[10] Byrappa K and Adschiri T 2007 *Progress in Crystal Growth and Characterization of Materials* **53** 117-166
[11] Sediri F and Gharbi N 2007 *Materials* Science *and* Engineering B **139** 114-117
[12] Pavasupree S, Suzuki Y, Kitiyanan A, Pivsa-Art S and S. Yoshikawa 2005 *Journal of Solid State Chemistry* **178** 2152-2158
[13] Lavayen V, O'Dwyer C, Cardenas G, Gonzàlez G and Sotomayor Torres C M 2007 *Mater. Res. Bull* **42** 674-685
[14] Murgia V, Torres E M F, Gottifredi J C and Sham E L 2006 *Applied Catalysis A* **312** 134–143
[15] Liu J and Xue D 2010 *Nanoscale Research Letters* **5** 1619-1626
[16] Wang Y T and Chen C H 2013 *Inorganic Chemistry* **52** 2550-2555
[17] Wang Y, Zhang Z, Zhu Y, Li Z, Vajtai R, Ci L and Ajayan P M 2008 *ACS Nano* **2** 1492-1496
[18] Liu J and Xue D 2008 *Advanced. Materials* **20** 2622-2627

IOP Conf. Series: Materials Science and Engineering **186** (2017) 012008 doi:10.1088/1757-899X/186/1/012008

Influence of dipolar interactions on the superparamagnetic relaxation time of γ-Fe_2O_3

A Labzour[1], A Housni[2,3], K Limame[2,4], A Essahlaoui[1] and S Sayouri[2]

[1]LSI, Faculté polydisciplinaire de Taza, USMBA-Fès, Maroc.
[2]LPTA, Département de Physique, Faculté des Sciences–DM, BP. 1796, Fès-Atlas-Fès, Maroc.
[3]Institut Supérieur des Professions Infirmières et Techniques de Santé-Fès, Maroc.
[4]CRMEF, BP 49, Fès, Maroc.

Email : alabzoura@yahoo.fr

Abstract. Influence of dipolar interactions on the Néel superparamagnetic relaxation time, τ, of an assembly of ultrafine ferromagnetic particles (γ-Fe_2O_3) with uniaxial anisotropy and of different sizes has been widely studied using Mössbauer technique. These studies, based on different analytical approaches, have shown that τ decreases with increasing interactions between particles. To interpret these results, we propose a model where interaction effects are considered as being due to a constant and external randomly oriented magnetic field $\mathbf{B}(\psi,\phi)$. The model is based on the resolution of the Fokker-Planck equation (FPE), generalizes previous calculations and gives satisfactory interpretation of the relaxation phenomenon in such systems.

1. Introduction

Due to their interest for technological applications, a great number of both experimental and theoretical studies have been devoted to the study of dynamical properties of nanoparticles [1-28].
Different experimental techniques are used to study superparamagnetic relaxation of fine particles: alternative susceptibility, [3-8], Mössbauer spectroscopy [6, 8, 10, 23], neutrons diffraction [9] and ferromagnetic resonance [11].
In the superparamagnetic domain, the relaxation time of isolated particles is generally expressed by Néel-Brown law [1, 2]:

$$\tau = \tau_0 \exp\left(\frac{\Delta E}{K_B T}\right) \qquad (1)$$

where τ_0 is of the order of 10^{-10}-10^{-13} s^{-1}, and weakly dependent on temperature, ΔE is the energy barrier separating two easy magnetization directions, K_B is the Boltzmann constant and T the temperature.
Eq. (1) constitutes a good approximation if $\Delta E>2KT$. For a particle of uniaxial symmetry, the magnetic anisotropy energy is given by $E(\theta) = -KV\cos^2\theta$, where K is the anisotropy constant, V is the volume of the particle and θ is the angle between the magnetization vector and the easy magnetization axis; in this case, the energy barrier is equal to KV.

2. Previous results

Magnetic interactions between nearest neighbours nanoparticles have a great influence on the superparamagnetic relaxation phenomenon. The latter has been the subject of numerous experimental and theoretical studies [3-28].

In the presence of a magnetic field, B, parallel to the easy magnetization direction, the energy of the particle becomes

$$E = KV\left(\sin^2\theta - 2h\cos\theta\right) \tag{2}$$

where h=μB/2KV. Two situations may be considered depending on the value of h. For h≥1, a stable state exists in which E has a minimum if θ=0 and a maximum if θ=π. For h<1, E has two minima at θ=0 and at π, separated by a maximum at θ=arcos(-h). The two barriers $\Delta E^{\pm}=E_B(1\pm h)^2$, corresponding to the crossing from θ=0 to π, and vice versa, give rise to two relaxation time constants:

$$(\tau_{\pm})^{-1} = \frac{1}{2\tau_0}(1\pm h)\left(1-h^2\right)^{\frac{1}{2}} \exp\left[-\alpha(1\pm h)^2\right] \tag{3}$$

where $h = \dfrac{\mu B}{2KV} \prec 1$ and $\alpha = \dfrac{KV}{K_B T} \succ\succ 1$.

The global relaxation time is given by

$$\frac{1}{\tau} = \frac{1}{\tau_+} + \frac{1}{\tau_-} \tag{4}$$

Shtrikmann and Wohlfarth [12] have considered two cases corresponding to weak and strong couplings, and have represented the action of weak interactions as that of a statistically averaged magnetic field H_I, which leads to

$$E_B = KV + H_I MV\, th\left(H_I MV/K_B T\right) \tag{5}$$

and so gives the Vogel-Fulcher law

$$\tau = \tau_0 \exp\left[\frac{KV}{K_B}\left(\frac{1}{T-T_0}\right)\right] \tag{6}$$

valid for $T>>T_0$ with $T_0=H_I^2M^2V/KK_B$

In the case of a strong coupling, they introduced an effective volume with an order temperature

$$V_{eff} = V_0\left(T - T/T_0\right)^{-1} \tag{7}$$

Chantrell and Wohlfarth [13] have proposed a modified version of the model of Shtrikmann and Wohlfarth [12], and introduced interaction effects; short range interactions and long range ones are represented using an effective volume $VT(T-T_0)^{-1}$, and an effective field H_i respectively. The energy barrier is then given, in the presence of a magnetic field H, by

$$E_B = \frac{KVT}{(T-T_0)}\left(1 \pm h \pm \frac{\mu H_i}{2KV}\right)^2 \tag{8}$$

where h=μH/2KV

Dormann et al. have also proposed a model which predicts that the energy barrier increased with increasing interactions [5]

$$E_B = E_{Ba} + E_{B\,\mathrm{int}} \tag{9}$$

and $E_{B\,\mathrm{int}} \approx n_I M^2 V b_I\, L\left(M^2 V b_I/K_B T\right)$

n_I is the number of nearest neighbors and $b=\langle V\rangle/d^3$ (regular arrangement).

This model has been successfully applied to interpret magnetic measurements of nanoparticles dispersed in an alumina matrix.

However, Mössbauer measurements on weak interacting Fe_2O_3 nanoparticles [13-15] revealed a decrease of the relaxation time with increasing interactions between particles.

The increase of the relaxation time with interactions is in accordance with magnetic measurements but disagrees with Mossbauer ones performed on γ-Fe_2O_3 nanoparticles. Dormann et al. have explained this discrepancy as due to the decrease of the damping parameter η_r with increasing interactions between particles, appearing in the expression of τ_0 [16-18, 21-25]

$$\tau_0 = \frac{1}{\eta_r}\left[\frac{|m(0)|}{E_{Ba}\gamma_0}\sqrt{\frac{\pi}{4}}\alpha^{-\frac{1}{2}}\left(1+\frac{1}{\alpha}\right)\right] \tag{10}$$

Variations of the parameter [18] η_r may be a consequence of two contributions. The first one comes from the spin disorder at the surface of the particles which decreases significantly with increasing interactions and then reduces η_r. The second one is generated by the spatial variation of the dipolar field acting on each particle due to its neighbours.

An other approach has been proposed to interpret Mossbauer experimental observations [15], in which the authors considered that the factor τ_0 as independent of interactions and an effective energy barrier which decreases with increasing interactions.

$$\tau^* \approx \tau_0 \exp\left[\alpha - \frac{\xi^2}{3}\left(1 - \frac{3}{4}\alpha^{-1}\right)\right] \tag{11}$$

This approach has given results that are in contradiction with experience [23]. Indeed, if the preexponential does not vary, T_B could increase with interactions [14, 22, 23]. However, Mössbauer measurements on weak interacting Fe_2O_3 nanoparticles [13-15, 22, 23] indicate that relaxation is more fast and T_B more low for interacting particles. Attribution of this systematically observed effect to a decrease of E_B [15] seems to be incorrect. Besides, this model does not take into account of the aleatory character of the field B_I.

3. A new approach

We have considered interactions influence on relaxation time, and have developed calculations which give satisfactory interpretation of experimental results.

Consider a monodomain particle of uniaxial symmetry, with magnetic moment μ_i. The other particles of the system act on a particle (i) via a dipolar field B_I at an angle ψ with e_z considered as the easy magnetization axis (Fig.1).

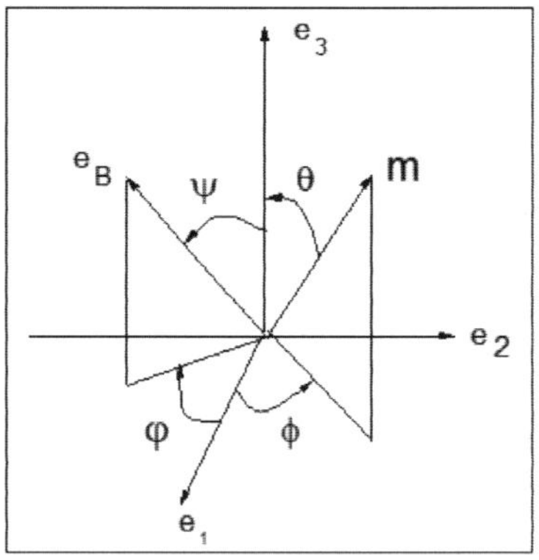

Figure 1. Dipolar field B_I and magnetic moment in spherical coordinates

The energy anisotropy can be expressed as:

$$E = KV\sin^2\theta - \vec{\mu}_I\vec{B}_I \tag{12}$$

With the following variables changes, $\beta=1/K_BT$, $\alpha=KV/K_BT$ and $\zeta=\mu_iB_I/K_BT$

this energy may be written in the form:

$$\beta E = \alpha\sin^2\theta - \varsigma[\cos\theta\cos\psi + \sin\theta\cos\phi\sin\psi\cos\varphi - \sin\theta\sin\phi\sin\psi\sin\varphi] \tag{13}$$

We can express the global minimum energy and the local maximum energy in the limit of weak interactions in which $\zeta \ll 1$. In this limit, minimum and maximum values are obtained for $\theta \approx 0$ and 90 respectively, as shown in fig.2. giving the energy anisotropy (βE) versus θ, φ, ϕ and h.

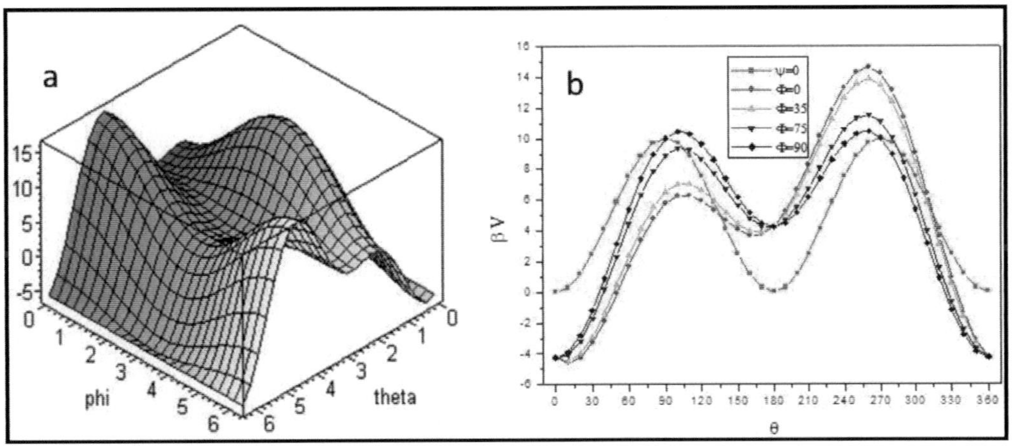

Figure 2. Energy anisotropy βE as a function of the parameters θ and φ:
a- $\psi=45°$, $\phi=0$ and h=0.3; **b**- $\varphi = 0$, $\psi = 0$ and h=0;
$\varphi = 0$, $\psi = 45$, h=0.3, and $\phi = 0$, $\phi = 35$, $\phi = 75$, $\phi = 90$.

When θ is close to zero, $\sin\theta \approx \mathrm{tg}\theta$

Expressing the derivative of Eq. 13 with respect to θ, $(\partial\beta E/\partial\theta)=0$, gives:

$\sin\theta \cong h\,[(\sin\psi\cos\phi)/(1+h\cos\psi)]$, where $h=\xi/2\alpha$. $h \ll 1$, so $\sin\theta \approx h\sin\psi\cos\phi$,

Consequently the minimum value of energy when θ is close to zero is:

$$\beta E_{\min} = -\alpha\left[h^2 \sin^2\psi \cos^2\phi + 2h\cos\psi\right] \tag{14}$$

When θ is close to $90°$, a similar approximation to that used above leads to $\cos\theta \approx -h\cos\psi$, and then:

$$\beta E_{\max} = \alpha\left[1 + h^2\cos^2\psi - 2h\cos\varphi\sin\psi\cos\phi + 2h\sin\varphi\sin\psi\sin\phi\right] \tag{15}$$

The energy barrier, $\Delta E=E_{max}-E_{min}$, can be written as:

$$\beta\Delta E = \alpha\left[1 + h^2\left(1-\sin^2\psi\sin^2\varphi\right) + 2h\cos\psi - 2h\cos\phi\sin\psi\cos\varphi + 2h\sin\phi\sin\psi\sin\varphi\right] \tag{16}$$

The transition probability to overcome the energy barrier is given by

$$f(\psi,\varphi,\phi) = (2\pi\tau_0)^{-1}\exp(-\beta\Delta E) \tag{17}$$

Integration of $f(\psi,\varphi,\phi)$ with respect to ϕ gives:

$$f(\psi,\varphi) = \tau_0^{-1}\exp-\alpha\left[1+h^2\left(1-\sin^2\psi\sin^2\varphi\right)\right]\times\left\{1-\varsigma\cos\psi+\left(\frac{\varsigma}{2}\right)^2\left(1+\cos^2\psi\right)\right\} \tag{18}$$

And the relaxation time is given by:

$$\tau^{-1} = \frac{\iint f(\psi,\varphi)\sin\psi\, d\psi\, d\varphi}{\iint \sin\psi\, d\psi\, d\varphi} \tag{19}$$

$$\tau^{-1} = \frac{\tau_0^{-1}}{4\pi}\int_0^{2\pi}\exp-\alpha\left(1+h^2\cos^2\varphi\right)\times\left[2+\left(\frac{\langle\varsigma\rangle}{2}\right)^2\left(\frac{8}{3}-\frac{2}{3}\alpha h^2\sin^2\varphi\right)\right]d\varphi \tag{20}$$

Hence

$$\tau = \frac{\tau_0}{I_0\left(\alpha h^2/2\right)} \exp \alpha \left[1 - \frac{4}{3}\langle h^2\rangle\left(\alpha - \frac{3}{8}\right)\right] \tag{21}$$

where I_0 is the Bessel's function

Finally τ can be expressed as

$$\tau = \frac{\tau_0}{1 + \left(\frac{\alpha\langle h^2\rangle}{4}\right)^2} \exp \left[\alpha - \langle \xi^2\rangle\left(1 - \frac{5}{24}\alpha^{-1}\right)\right] \tag{22}$$

Expression (22) may be written as a function of (11).

$$\tau = \tau^* \tau' \tag{23}$$

where

$$\tau' = \exp\left[-\frac{4\alpha^2 h^2}{3}\left(2 + \frac{1}{8}\alpha^{-1}\right) - \left(\frac{\alpha h^2}{4}\right)^2\right] \tag{24}$$

We can write with a good approximation

$$\tau = \frac{\tau^*}{1 + \frac{\alpha h^2}{6} + \frac{8\alpha^2 h^2}{3} + \left(\frac{\alpha h^2}{4}\right)^2} \tag{25}$$

$$\tau = \frac{\tau_0}{1 + \frac{\alpha h^2}{6} + \frac{8}{3}\alpha^2 h^2 + \left(\frac{\alpha h^2}{4}\right)^2} \exp \alpha \left[1 - h^2\left(\frac{4}{3}\alpha - 1\right)\right] \tag{26}$$

where $\alpha \leq 20$, $0 < h \leq 1$, In this expression, contributions of the denominator are not negligible.

This relation clearly shows the effect of the relaxation time variation as a consequence of the variation of interactions between particles. τ decreases (as a consequence of the increase of the damping factor) with increasing interactions as revealed by Mössbauer experiment and dynamic measurements [23], and hence is in contradiction with the supposition that the parameter τ increases as a function of interactions.

The blocking temperature T_B can be deduced from our approach from equation (21), by writing that the relaxation time value is equal to that of the measuring time, which is given by:

$$T_B \approx \frac{KV}{C K_B}\left[1 - h^2\left(4C - \frac{5}{6}\right)\right] \tag{27}$$

with $C = Ln(\tau_m/\tau_0)$, $\tau_0 = 10^{-10}$s, $\tau_m = 10^{-8}$s

For a given value of h, T_B increases linearly, and is a decreasing function of h for a fixed value of the parameter KV (Fig. 3).

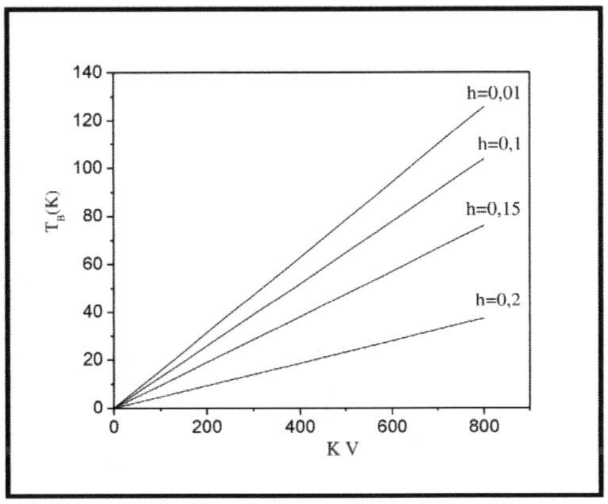

Figure 3. Blocking temperature, as a function of the average parameter (KV)

4. Conclusion

Superparamagntic relaxation in the presence of dipolar interactions has been studied. The results are in good agreement with experimetal ones, in particular Mössbauer measurements, and show a decrease of the relaxation time with increasing interactions, contrary to previous results indicating an increase of this parameter. The decrease of τ may be imputed to the decrease of the pre exponental factor.

References

[1] Néel L 1949 Ann. Geophys.**5** 99.
[2] Brown W F 1963 Jr Phys Rev. **130** (5) 1677.
[3] El-Hilo M, O'Grady K and Chantrell R W 1992 Journal of Magnetism and Magnetic Materials, **117** 1-2, 21-28.
[4] Luo W et al. 1991 Phys Rev Lett. **67** 2721.
[5] Dormann J L, Bessais L and Fiorani D 1988 J Phys C. **21** 2015.
[6] Johansson C, Hanson M, Hendriksen P V and Mørup S 1993 Journal of Magnetism and Magnetic Materials, **122** 1–3 125-128.
[7] Prené P et al. 1993 IEEE Trans, Magn. **29** 2658.
[8] Dickson D P E, Reid N M K, Hunt C, Williams H D, El-Hilo M and O'Grady K 1993 Journal of Magnetism and Magnetic Materials. **125** 3 345-350.
[9] Mirebeau I, Bellouard C, Hennion M, Dormann J L, Djega-Mariadassou C and Tessier M 1992 Journal of Magnetism and Magnetic Materials. 104–107 1560-1562.
[10] Morup S 1990 Hyperfine Interact. **60** 959.
[11] Ziethen H M, Trautwein A X, Winkler H and Schmidt F 1989 Studies in Surface Science and Catalysis. **49** 1043-1052.
[12] Shrikmann S, and Wohlfarth EP 1981 Phys. Lett. **85** A, 467.
[13] Chantrel RW and Wohlfarth EP 1983 Magnetism and Magnetic Materials. 40 1.
[14] Prené P, Tronc E, Jolivet J P, Livage J, Charkaoui R, Nogués M, Dormann J L and Fiorani D 1993 IEEE Trans. Magn. 29 2658.
[15] Morup S and Traoc E 1994 Phys. Rev. Lett. 72 3278.
[16] Jiang J.Z, Morup S, Jonsson T and Svedlidh P i I 1996 Ortalli(Ed), Proc, ICAME 95, SIF, Bologna, 529.
[17] Prené P, Tronc E, Jolivet J P and Dormann J L 1995 in:I. Ortalli, (Ed), Proc, ICAM'95, SIF, Bologna, 485.
[18] Tronc E 1996 Nuovo Cimento. 18 D 163.
[19] Dormann J L, D'Orazio F, Lucari F, Tronc Prené P, Jolivet J P, Fiorani D, Charkaoui R and Nogués M 1996 Phys. Rev. B **53** 14291.

IOP Conf. Series: Materials Science and Engineering **186** (2017) 012008 doi:10.1088/1757-899X/186/1/012008

[20] Tronc E and Jolivet J P 1997 Mater. Sci. Forum. **659** 235-238.
[21] Dormann J L, Fiorani D and Tronc E 1997 Adv Chem. Phys. **98** 283.
[22] Dormann J L, Cherkaoui R, Spinu L, Noguès M, Lucari F, DÕOrazio F, Fiorani D, Garcia A, Tronc E and Jolivet J P 1998 Journal of Magnetism and Magnetic Materials. **187** L139-L144.
[23] Charkaoui R. 1999 Ph-D thesis, Rabat Univ. (Morocco).
[24] Tronc E, Fiorani D, Nogués M, Testa A M, Lucari F, D'Orazio F, Grenèche J M, Wernsdorfer W, Galvez N, Chanéac C, Mailly D and Jolivet J P 2003 Magnetism and Magnetic Materials. **262** 6-14.
[25] Dormann J L, Spinu L, Tronc E, Jolivet J P, Lucari F, D'Orazio F And Fiorani D 1998 Magnetism and Magnetic Materials. **183** L255-260.
[26] Garcia-Otero J, Markus Porto, Rivas J and Armin Bunde 2000 Phys Rev Lett. **84** 167-170.
[27] Zysler R D, Fiorani D, Testa A M 2001 Magnetism and Magnetic Materials. **224** 5-11.
[28] Poddar P, Telem-Shafir, T, Fried T, Markovich G 2002 Phys Rev. B **66** 060403 (R) 1-3.

Preparation and characterization of Miocene clay powders in the region of Taza (Morocco) after doping with metal oxides Al_2O_3

L Mesrar[2], M Lakrim[1], M Akdim[1], A. Benmar[2], N. ES –Sbai[3] and R Jabrane[1]

[1] Department of Earth Sciences, Laboratory of Georesources and Environment, Faculty of Science and Technology of Fez, BP.2202 Road Immouzzer, Fez, Morocco.
[2] Department of Genie Civil, Laboratory LOMC UMR CNRS, University Havre 75 rue Bellot, 76600 Le Havre, France.
[3] Department Electrical Engineering, Signals and Systems Laboratory Component (USL), Faculty of Science and Technology of Fez, BP.2202 Road Immouzzer, Fez, Morocco.

Email: mesrarlaila@gmail.com

Abstract. The marl's minerals are abundant untreated material in several areas worldwide. They are often under-valued for human use. However, due to demands of the society in terms of sustainability and energy saving, the valuation of these resources to develop new materials, most environmentally friendly has become a concern both scientific and industrial aims. Ceramics are the growing research to obtain materials with good chemical stability and good hot properties [1]. The balance between these properties and industrial requirements allowed clay materials uses at craft departure (pottery, tile), to progress towards high-tech applications such as electrical and thermal insulation, electric candle, sound insulation [2]. The behavior of the doping (Al_2O_3), which has more scientific research interest, has been a renewed interest since 1980 [3] with the emergence of alumina very high purity. Miocene marl is one of the widespread geological substrates in Fez-Taza vicinity (Central Morocco). In this study we proceed by a physicochemical characterization of the marl after doping with metal oxides, by various analytical techniques, namely the X-ray fluorescence, the mineralogical analysis and geotechnical test. The doping of these marl was conducted by solid oxides of Al_2O_3 at different percentages (5%, 10% and 15%). The results of chemical analysis showed the Al_2O_3 increase during doping. So, the mineralogical analysis of doped clays shows peaks' increases for kaolin. The marl doped acquired the property of their good plasticity and good mechanical resistance compared to crass marl.
Index Terms. Clay, Pozzolan, Doping, Chemical, Geochemical, Mechanical characteristic.

1. Introduction

In Morocco, a well-developed mining activity since the beginning of this century has allowed Moroccan citizens to acquire know-how in the areas of operations and mineral exploration. This was accompanied by the necessary infrastructure development in the implementation and development of this activity [1].

In particular, The Miocene clay substrates are abundant in the region of Taza, yet they remain insufficiently valued. However, considering the new demands of the society in terms of sustainable development and energy conservation, recovery of the clay and its use in building m aterials may be more environmentally sustained and become a scientific and industrial concern.

Ceramics are the growing research to obtain materials with good chemical stability and good hot properties [3]. The balance between these properties and industrial requirements allowed clay materials uses at craft departure (pottery, tile), to progress towards high-tech applications such as electrical and thermal insulation, electric candle, sound insulation, etc [4]. The behavior of the doping (Al_2O_3), which has more scientific research interest, has been a renewed interest since 1980 [5] with the emergence of alumina very high purity. The present study focuses on the influence of doping on the microstructure of marl, based on the results obtained by X-ray fluorescence and X-Ray.

In this perspective, the doping of the marl clays with alumina oxide could increase the added value of the clay used in the technology and building industry [6].

One of the major problems encountered by users of these materials is the fragility and high shrinkage induced by cookery causing defects in the briquette [7], and can go until it bursts. To improve performance, efficiency and the quality in the manufacture of this marl, we opted doping dry voice of these materials by the alumina oxide it has considering percentages of 5, 10 and 15%.

The findings allow us to conceive and suggest new improvements of the physicochemical and mechanical parameters for marl, to rise the value and use of this low cost material.

2. Materials and Methods

2.1.Materials

The basic raw material is a clay marl Miocene the region of Taza (Figure 1) which outcrop in the eastern area of Taza (Figure 2) .

The importance of clay-based materials in the ceramic industry is due to their plastic properties [8] allowing final ceramic products to retain their shapes after firing. The process is complex because usually the shaped product depends on the mineralogical composition of the raw material and the nature and quantity of impurities in clay.

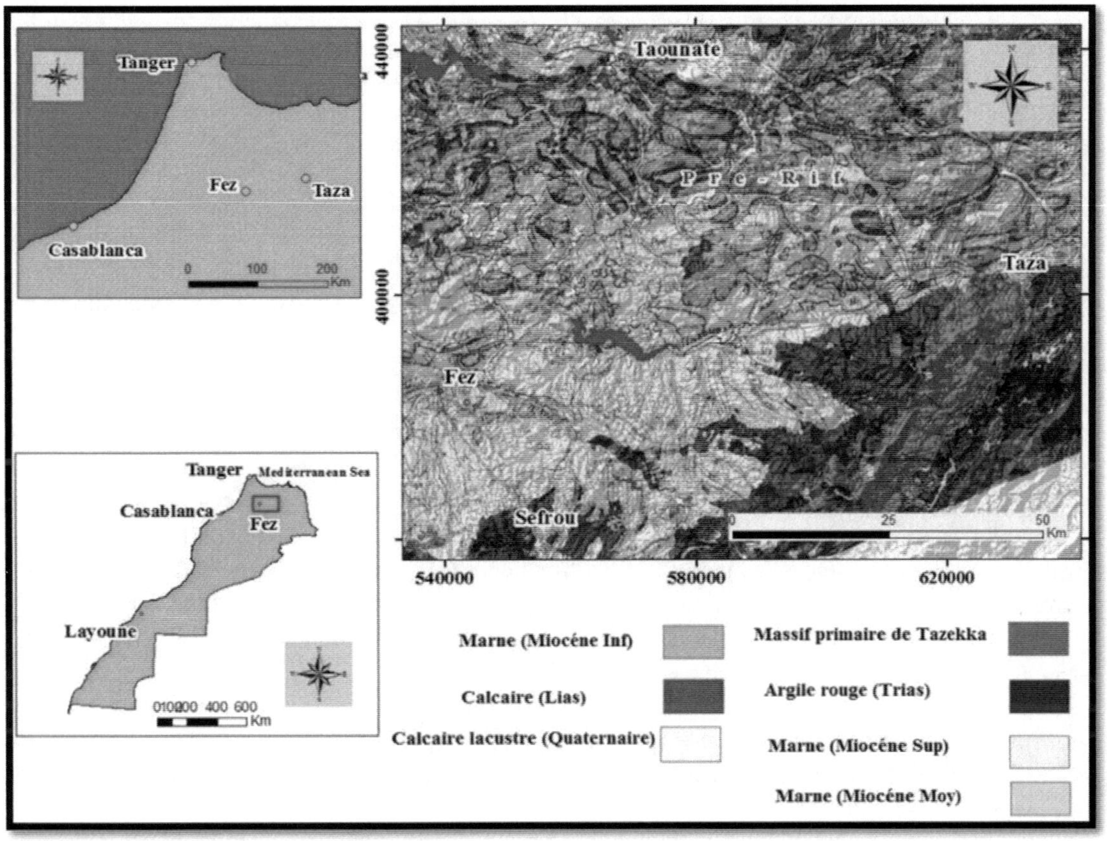

Figure 1. The location and geological map of Fez-Taza area (Morocco).

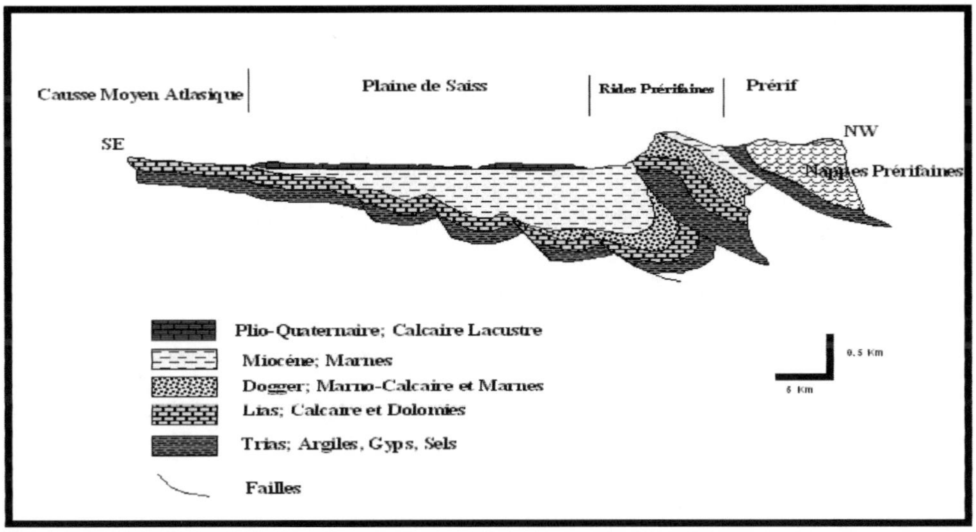

Figure 2. Geological cross section of South Rif (DRH et ONI, 1971-1977).

2.1.Preparation of composite clay / Alumina oxide

The Al_2O_3/clay system was achieved by solid route at room temperature. The doping ranged from 5%, 10%, to 15%. The implementation of composite clay / Alumina oxide, consisted of, first, a 20-minute grinding our base material, then a direct mixture in the solid state of the clay with the oxide of alumina, different percentages of 5, 10 and 15%, and grinding in a ball mill for 20 minutes agate. It allows obtaining homogeneous powders relatively fine-grained and narrow particle size distribution and that at room temperature.

2.3. Methods

In the aim to study the homogeneity and the concentration of Al_2O_3 incorporated in the clay sheets, several analysis techniques were used:

Chemical analysis: the quantitative chemical analysis of clays was performed by X-ray fluorescence spectrometry (WD-XRF) spectrometer type Axios sequential brand Panalytical. This analysis was made after grinding the material and taking 10 g of each sample. Then we took 10g of each powder sample and pressed it in a hydraulic press, in order to make the pellets.
X-ray diffraction DRX study was done with the Philips diffractometer XPERT-PRO" PW 3064, with the Kα1,2 radiation of copper. X-ray diffraction (XRD) data were obtained using oriented samples prepared on glass slides in the 2θ range of 2–40° at a scanning speed of 2° min−1. The geotechnical property was performed in the laboratory of geosciences and environment at the FST (Sidi Mohamed Ben Abdellah University). The evaluation of plasticity was performed by the Casagrande method. The Atterberg limits are conventional physical constants indicating the clay transition from plastic to liquid state (WL) and from the plastic to solid state (WP). The plasticity index, which represents the difference between the liquid limit and plastic limit, defines the extent of the plastic field of clay.
Bricks production is carried out using a metal mold, provided for the operation. It has a length of 50 mm and a width of 10mm. Bricks are previously dried in the open air until the stabilization of the withdrawal. They are fired in electric oven programmable to the temperatures of 900°C, with a bake cycle of 30 minutes and a bearing of one hour and a half to the maximum firing temperature. The rate temperature of firing rise 10°C/min.
The parameters resulting from the bricks firing are observed as the materials withdrawal along drying and firing, the ignition loss, and the mechanical resistance to flexural strength.

3. Results and Discussion

3.1. Chemical analysis

We note that the predominant components in all samples are silica and alumina, respectively, for the raw clay and doped (Table.1).
The easiest way to determine the influence of a doping is to make measurements of essay material doped before and after doping. Thus, analysis of the chemical composition by XRF clays doped with Al_2O_3, reveals a significant enrichment in alumina. Its concentration increases as the doping level is increasing.

Table 1. Percentage major elements of our crude marl (T1) and doping with Al_2O_3.

Sample	SiO_2	Al_2O_3	CaO	Fe_2O_3	MgO	K_2O	Na_2O	TiO_2
T3 Brut	47	13.3	7.63	6.13	2.71	0.92	0.92	0.31
T3 doped with 5%	45.3	16.5	7.06	5.93	2.13	0.86	0.92	0.31

T3 doped with 10%	44.4	19.4	6.65	5.65	2.53	0.86	0.77	0.29
T3doped with 15%	42.1	25	6.06	5.21	2.22	0.77	0.74	0.25

However, doping causes a slight decrease in some major element remains. It is varying quantitatively. There is also an increase in the content of alumina oxide in combination with increasing doping content (5.10 and 15%). Its exceeds twenty percent (20%).

2.2. Mineralogical analysis

The mineralogical analysis of crude and doped samples by Alumina at different percentages (5, 10 and 15%), shows that kaolin, illite, quartz, dolomite and calcite are the main crystalline phases (Fig. 3). The increase of the peak intensity of the kaolin doped marl can be explained by the incorporation of alumina in the silica tetrahedral site (Fig. 4). So, the intensity of the highest peak is manifested by a 15% Al_2O_3 doped marl. Furthermore, there is thus a continuous and positive correlation between increasing the Al_2O_3 content of the doping and the elevation of the peak intensity (Fig. 4). This result is conform with the results of chemical analysis.

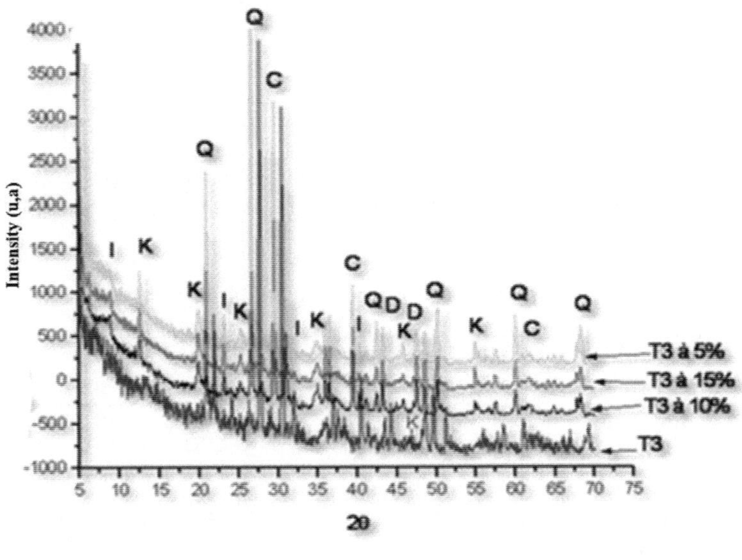

Figure 3. X-Ray of crude marl T3 before and after doping with oxide alumina at different percentages 5, 10 and 15%. With: K = kaolin, Q = quartz, I =Illite, C = calcite, D = dolomite.

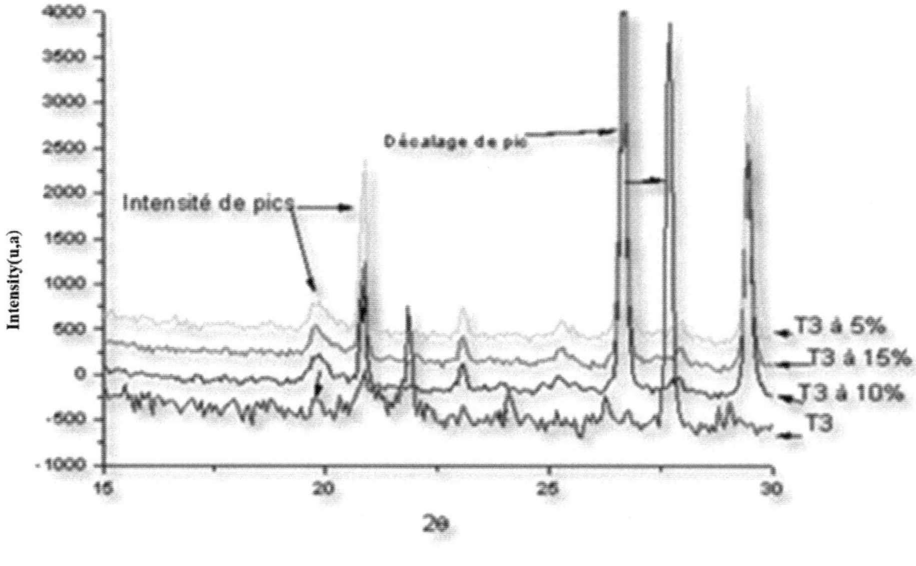

Figure 4. X -Ray comparison of peaks intensity of Kaolin between crude and differently doped marls.

3.2. *Technological testing (*Atterberg limits)

Based on the results of Table 2, we note the raw material T3 has an average plasticity later increase with the level of doping. However the greater plasticity is observed in the doped marl with 15% alumina.

Table 2. The Results of Plasticity Analysis and Water Content in the Samples.

Sample	Liquid Limit (Wl)	Plastic Limit (Wp)
T3 Brut	45.65	18.59
T3 doped with 5%	50.07	23.90
T3 doped with 10%	46.74	22.30
T3 doped with 15%	46.86	27.94

3.3.Effect of Firing

During the rise of the firing temperature, the physicochemical transformations of the phase's marls affect the material (Fig.5). Each mineral in the mixture behaves as if it was alone .The decomposition products or new structural varieties, which take place under the influence of heat, react together to form new compounds [9].

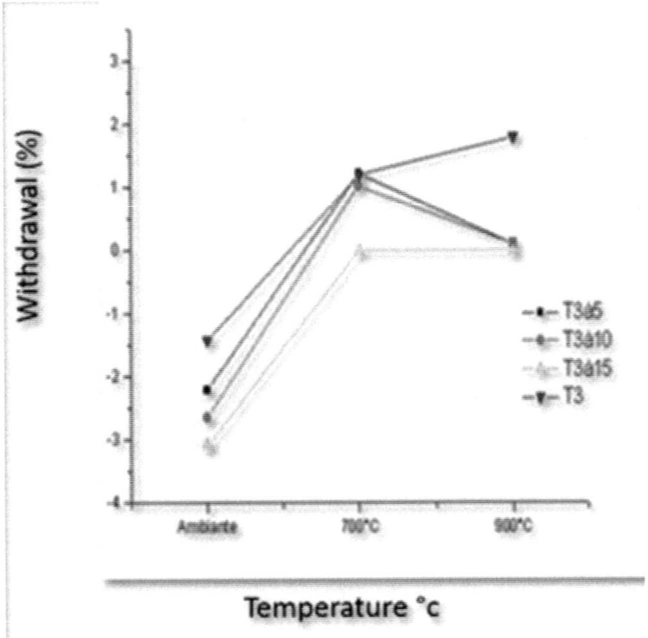

Figure 5. Evolution the withdrawal of raw samples and doped according to the firing temperature.

However, upon baking, the initial crystalline minerals disappeared and make space in varying proportions to an amorphous phase and new minerals [10]. Shrinkage accompanies this solid reaction, and is associated with changes in porosity and particle size. The proportions of the clay phase and / or associated minerals play an important role in these changes as well as the particle size removing charged in the raw material is very important. This is due to the expansion case by new neoformed minerals.

In this study all doped materials has attended a withdrawal reduction until its cancellation. Therefore be concluded the doping plays a role in limiting internal withdrawal.

3.4.Mechanical properties

The marl doped 5% of Al_2O_3 have a relatively high resistance up to 25 KN compared with bricks doped 10% and 15% Al_2O_3 (Fig.6). We find that the effect of doping alumina is obvious. Indeed, the higher content of doping increasing the mechanical strength of bricks decreases.

This resistance is in fact related to the increase of the porosity induced by the increase of the content of the doping. This causal relationship is in full agreement with the results obtained in the study of the porosity of the bricks.

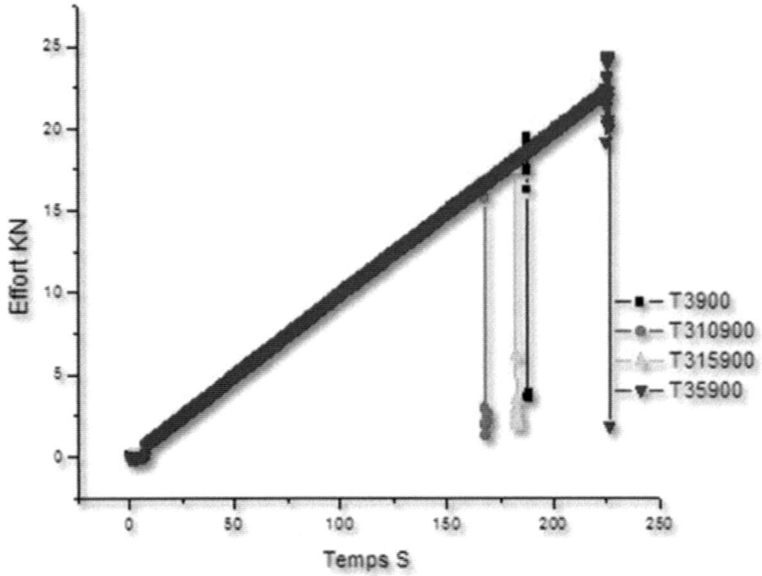

Figure 6. The Firing Impacts on the samples characteristics: loss on ignition and compression strength for the temperature 900°C.

4.Conclusion

The results of chemical analysis showed the higher rate of Al_2O_3 doping increases, its content increases in the marl.

Mineralogical analysis indicated the presence of illite and kaolin clay minerals such as essential clay.The intensity of kaolin peaks increase with increasing doping level. This can be attributed to the combination of alumina with free silica of kaolin.

Technological tests showed suitable industrial properties of these marls, their doping with aluminum oxide induced an increase in their plasticity. The alumina acts as a degreaser and decrease shrinkage. However, the withdrawal response is correlated with the content of Al_2O_3 doping. Indeed, the degreaser plays the role of internal framework and limits the withdrawal.

The mechanical resistance of the doped bricks marl is independent of quantity Al_2O_3 doping. The marl doped with 5% Al_2O_3 have a relatively high resistance compared with bricks doped with 10% and 15% Al_2O_3. So, additional content of doping decreases mechanical strength of bricks.

This resistance is in fact related to the increase of the porosity induced by the increase of the content of the doping. This causal relationship is in full agreement with the results obtained in the study of the porosity of the bricks. The bricks of marls fired at 900°C and doped with 5% gives better mechanical strength of up to 25 kN. So, suggesting the use of this composite in ceramics, in construction and in pottery.

Acknowledgement

The study benefits from several individuals and institutions support In the CNRST. X-ray analyzes and Chemical analyzes were carried out at CNRST Rabat.

References

[1] Mesrar, L. El Aroussi O., Lakrim M., Lahrach A., Chaouni A., et Jabrane R., Characterization of Miocene formations in the region of Taza for valuations and industrial operations, MATEC WEB OF CONFERENCES, n° 01014, 20-30, 2012.

[2] Mesrar .L, El Aroussi O., Lakrim M., Lahrach A., Chaouni A., et Jabrane R., Technological valorisation of the miocene clay in the region of Fez (Marocco) : caracterisation and

exploitation possibilities. Present Environment and Sustainable Development (PESD), VOL. 7, no.1, 311-317p, 2013.

[3] Fiori, C., Fabbri, B., Donati, G., Venturi, I., 1989. Mineralogical composition of the clay bodies used in the Italian tile industry. Appl. Clay Sci 4, pp:461 –473, 1989.

[4] Mesrar. L., Chahboun .A,. Lamcharfi .T,. Elbasset. A,. and Jabrane. R., Characterization of Miocene Marl of Fez Regions of Marocco after doping with MnO_2 . Clay Recearch, Vol, 31, No. 2, 79-83 p, 2012.

[5] Mesrar .L, El Aroussi O., Lakrim M., Akdim.M., Akhrif . I., and Jabrane. R., The physic-mecanical characteristics of the clays in and after doping with metal oxide Al_2O_3 in the region of fez (Morocco). Present Environment and Sustainable Development (PESD), VOL. 9, no. 1, 2015.

[6] Mesrar. L., R.Jabrane,. 2015. Characterization physic chemical and mechanical of clays in the sillon south Rif (Morocco). Clay Research, Vol 33, No.2, pp 65-73.

[7] Baccour, A. Zghal, A. Medhioub, M et Mhiri T., Caractérisation physicochimique et mécanique de matériaux céramiques obtenus à partir des argiles tunisiennes, Verres, Céramiques/ Composites Vol.1. N°2, pp 25-33, 2011.

[8] Jouenne. C.A., Traité de céramiques et de matériaux minéraux. Ed. Septima, Paris, 620p, 1984.

[9] Trimdade, M.J., Transformations of calcareous rich clays with firing: a comparative study between calcite and dolomite, Applied Clay Science 42 (2009) 345– 355, 2009.

[10] Ferrari S., (2006). The use of illitic clays in the production of stoneware tile ceramics Applied Clay Science 32 (2006) 73–81.

IOP Conf. Series: Materials Science and Engineering **186** (2017) 012010 doi:10.1088/1757-899X/186/1/012010

Determination of the structural properties of the aqueous electrolyte $LiCl6H_2O$ at the supercooled state using the Reverse Monte Carlo (RMC) simulation

M. ZIANE[1,3], M. HABCHI[2,3], A. DEROUICHE[1], S.M. MESLI[2,3], F. BENZOUINE[1] and M. KOTBI[4]

[1] LPPPC, Physics Department, Sciences Faculty Ben Msik P.O. 7955, Casablanca, Morocco
[2] École Préparatoire en Sciences et Thecniques, BP 165RP, Bel Horizon, 13000 Tlemcen, Algeria
[3] ISP group, NMSE division, URMER, Abou Bekr Belkaid University of Tlemcen, Algeria
[4] Department of Physics, LPT, Abou Bekr Belkaid University, BP 119 Tlemcen, Algeria

E-mail: zianesmp@gmail.com

Abstract. A structural study of an aqueous electrolyte whose experimental results are available. It is a solution of $LiCl6H_2O$ type at supercooled state (162K) contrasted with pure water at room temperature by means of Partial Distribution Functions (PDF) issue from neutron scattering technique. The aqueous electrolyte solution of the chloride lithium LiCl presents interesting properties which is studied by different methods at different concentration and thermodynamical states: This system possesses the property to become a glass through a metastable supercooled state when the temperature decreases. Based on these partial functions, the Reverse Monte Carlo method (RMC) computes radial correlation functions which allow exploring a number of structural features of the system. The purpose of the RMC is to produce a consistent configuration with the experimental data. They are usually the most important in the limit of systematic errors (of unknown distribution).

1. Introduction

Aqueous electrolyte solutions are of great interest in physicochemistry of materials, they possess the property of forming a glass through a metastable supercooled liquid state when the temperature decreases [1, 2, 3, 4, 5, 6]. Many properties of aqueous electrolyte solutions are due to the hydrogen bridge between molecules of water, resulting from orientations and interactions of water molecules and from the other interactions between water-ion and ion-ion, under influence of the salt concentration [2, 3, 7, 8, 9, 10, 11, 12, 13, 14].

Understanding of thermodynamic liquid and glass states is, in our view, advanced enough compared to the inevitable metastable supercooled state which ensures the passage between the two states. Several persuasive theoretical and experimental studies have been proposed over the last decades to explain and spell out the behaviour of this state. The purpose of this article is to add a piece in the puzzle of understanding anomalies of water and aqueous electrolyte solutions by studying the solution of LiCl in water (R=6) at the supercooled state (162K). The aim is to complement previous works by displaying all of pair distributions functions and computing some of physical properties such as numbers and the average positions of coordination.

We use the Reverse Monte Carlo (RMC) simulation in conjunction with neutron diffraction (ND) data [15]. This technique has the advantage to be applied without any specified inter-atomic and/or intermolecular interactions, it can be also, hybridized by introducing interactions potentials as additional constraint [9, 11, 16, 17]. RMC simulation allows the construction of a 3-dimensional model on the atomic level based on both experimental data and some geometric constraints. This simulation method completes the experiment by computing the pair correlation functions between each two components of the studied system.
Implementation method and details of the simulation performed here are described in Section 2. In Section 3 we give obtained results and their discussion. Finally in Section 4, conclusion is drawn.

2. Simulation details

The aim of the RMC simulation [18, 19, 20] is to produce three dimensional structural models of ordered or disordered systems consistent with the available experimental data (in particularly those of diffraction) within fixed standard deviation. This technique is based on the same principle of the metropolis Monte Carlo (MMC) method [21], but instead of minimizing the potential term, the difference between the computed and the experimental total or partial distribution functions $G(r_i)$ is the quantity to be minimized χ^2 , which is given by

$$\chi^2 = \sum_i (G^{RMC}(r_i) - G^{Exp}(r_i))^2/\sigma(r_i)^2 \tag{1}$$

Where $G^{RMC}(r_i)$ and $G^{Exp}(r_i)$ are the total or partial distribution functions obtained from the RMC configurations , $\sigma(r_i)$ is an estimate of the experimental error.
The initial configuration depends on the type of the simulated materials; when studying ordered samples, atoms are positioned in the average crystallographic positions, and several unit cells are used to produce the simulation box. If studying disordered materials, a random distribution of atoms without unreasonably short inter-atomic distances is generated by using an initial code. A new configuration is obtain by random choices and moves of atoms, after each move, the new $G(r_i)$ is calculated as well as the χ^2. If χ^2_{new} is less then χ^2_{old}; the agreement between experimental and the current configuration is improved by the move, thus the move is accepted and another move is made, if χ^2_{new} is increased, it is not rejected outright but accepted with the conditional probability

$$\exp(-(\chi^2_{new} - \chi^2_{old})/2) \tag{2}$$

The process is then repeated until χ^2 fluctuates around an equilibrium value. When satisfactory agreement between experimental and calculated data sets is obtained, the resulting configuration should be a three-dimensional structure compatible with the experimental data and the detailed structural information such as the number of coordination, the nearest neighbors positions and bond angle distribution functions can be calculated from atomic networks.
Four available experimental partial distribution functions $G^{EXP}_{XX}(r), G^{EXP}_{XH}(r), G^{EXP}_{HH}(r), G^{EXP}_{Cl_\alpha}(r)$ obtained by the neutrons scattering technique from the isotopic substitution are used [2, 4, 5, 6, 8]. Where the subscript X defines all atom species except the hydrogen one, while $Cl\alpha$ represents the correlation between Cl and all the other species.

$$\begin{gathered} G_{XX}(r) = 0.685g_{OO}(r) + 0.37g_{OCl}(r) - 0.085g_{OLi}(r) + 0.05g_{ClCl}(r) \\ -0,023g_{ClLi}(r) + 0.026g_{LiLi}(r) \end{gathered} \tag{3}$$

$$G_{XH}(r) = 0.828g_{OH}(r) + 0.224g_{HCl}(r) - 0.051g_{HLi}(r) \tag{4}$$

$$G_{HH}(r) = g_{HH}(r) \tag{5}$$

$$G_{Cl\alpha}(r) = 0.28g_{OCl}(r) + 0.624g_{HCl}(r) + 0.104g_{ClCl}(r) - 0.025g_{ClLi}(r) \tag{6}$$

The initial configuration is a cubic box of 2880 atoms with periodic boundary conditions, where the oxygen atoms forms initially a face-centered cubic lattice and the chlorine and lithium atoms are placed in the interstices. The new configurations are obtained by random choices and moves of atoms under geometric constraints and they are accepted or rejected according the same conditions of MMC algorithm. Simulation parameters are summarized in Tables 1 and 2.

Table 1. Closest approach distances $S_{ij}(\AA)$ between atom pairs in Å . $S^0_{ij}(\AA)$ Are used for initially setting up a disordered hard sphere fluid before introducing the experimental curves in the RMC simulation.

	OO	OH	OCl	OLi	HH	HCl	HLi	ClCl	ClLi	LiLi
$S_{ij}(A)$	2	0.92	2.1	1.75	0.9	1.85	2.1	3.75	2.9	3.2
$S^0_{ij}(A)$	3	0.94	2.8	2.0	1.55	2.25	2.7	4.9	4.8	3.2

Table 2. Number of atoms of each species, the total number of atoms N , the total atomic density ρ , the simulation box length L and the temperature of the $LiCl6H_2O$ system for three thermodynamic states; liquid, supercooled liquid and glass

	N_O	N_H	N_{Cl}	N_{Li}	N	ρ	L	T
Liquid						0.09394	31.297	300
Supercooled	864	1728	144	144	2880	0.09575	31.0993	162
Glassy						0.09599	31.0734	120

3. RESULTS AND DISCUSSION

3.1. Partial correlation functions H(r)

The Partial Correlation Functions (PCF), $H(r)$ equivalent to the Partial Distribution Functions $G(r)$ given by $H(r) = G(r) - 1$, are shown in figures 1. The experimental PCFs and also those computed by RMC for the supercooled liquid state, the difference ΔH between them is also drawn to permit an easy comparison. A good agreement and a clear concordance are highlighted between all of the computed and the experimental results. Consequently, the generated 3-dimensional configurations are compatible with the experimental data and thus allow displaying all of the pair distribution functions.

3.2. Pair distribution functions

The effect of LiCl on the structure of a solution will be discussed as usually in terms of the water-water correlations.

3.2.1. Water-water

Main intermolecular correlations of the H_2O molecule are observed in the Fig.2. In the pair distribution function $g_{OO}(r)$ (Fig.2.a) a describing water-water correlations. The pair correlation functions between two atoms of the H_2O molecule three thermodynamic states are compared to those of the pure water. The principal peak of $g_{OO}(r)$ is not affected by ions and has kept the

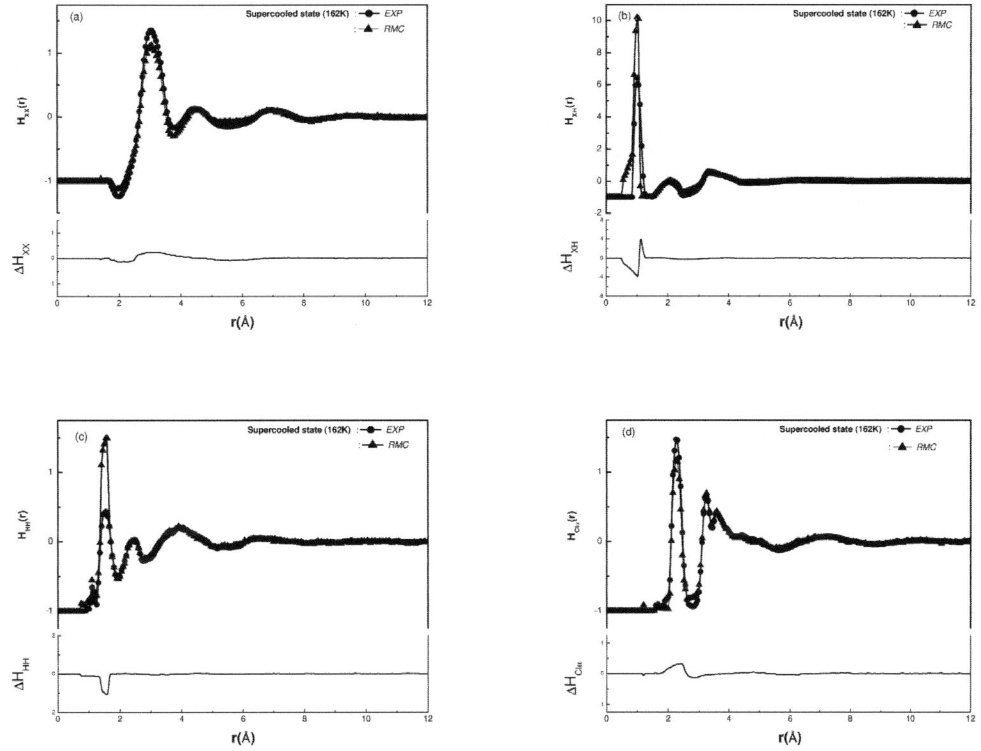

Figure 1. Experimental and computed Partial Correlation Functions $H_{ij}(r)$ for the supercooled liquid state, ΔH_{ij} is the difference between them.

same position. The peak of the second coordination (Fig.2.a) shows an interesting behavior. In the glassy state this pick oscillates in phase and practically with the same intensity of the pure water showing a significant order, while no structure is visible in the liquid. We can also estimate that the coordination numbers of water and the glassy solution are practically the same.
The supercooled state possesses an intermediate structure between liquid and the glassy state. It suggests that the lattice hydrogen bond beginning to reorganize at the supercooled state when the temperature decrease, the solution passes from a state where no meaningful order is observed, to another state more ordered via the metastable supercooled state. This suggests that this structure is broken when the temperature increases.
For the other functions, the first intramolecular peak of both $g_{OH}(r)$ (Fig.2.b) and that of $g_{HH}(r)$ (Fig.2.c) are identical in OH the three thermodynamic states, suggesting that the internal structure of the water molecule didn't change in any case.
Hence, neither the state changes nor the presence of ions affect the well know structure of the water molecule. However, a small shift has been observed for the first and the second coordination of $g_{OH}(r)$ (situated at $r = 2$Å and $r = 3.36$Å respectively) in the solution with respect to the pure water, probably due to the presence of ions. For the same reason the peak of the first and the second coordination in $g_{HH}(r)$ (located at 1.56Å and 2.49Å respectively) are raised in the pure water case. Otherwise, for the third coordination (situated at 3.9Å), the peak of the

solution is more intense. This is synonym of the greater role of the ions in the raising of the long-range order.

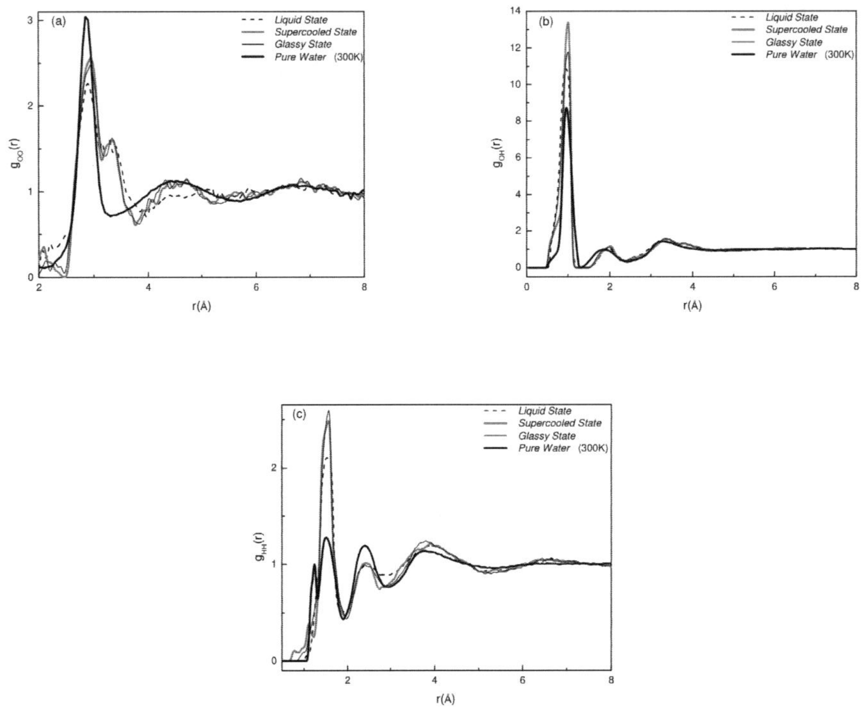

Figure 2. Pair distribution functions water-water: $g_{\mathrm{OO}}(r)$ (a), $g_{\mathrm{OH}}(r)$ (b) and $g_{\mathrm{HH}}(r)$ (c) at Glassy, Liquid and Supercooled Liquid state contrasted to the pure water at room temperature.

4. CONCLUSIONS

In this paper, we have investigated the structure of the chloride lithium aqueous solution $LiCl6H_2O$ at the supercooled liquid state by the Reverse Monte Carlo simulation on the basis of available neutron diffraction data. The main peaks of some PCFs for the supercooled state are more intense instead to be between the liquid and the glassy states. We suggest through the inspection of the linear combination of these functions, that the causes of this behavior are the water-water discussed in the framework of this work and the chlorine-water correlations in the restructuring of the hydrogen bonds and the hydration shells from the liquid to the glassy state.

References

[1] Winkel K, Seidl M, Loerting T, Bove L E, Imberti S, Molinero V, Bruni F, Mancinelli R and Ricci M.A 2011 *J. Chem. Phys.* B **134** 024515
[2] Prvel B, Jal J F, Dupuy-Philon J and Soper A K 1995 *JJ. Chem. Phys.* B **103** 1886
[3] Kotbi M and Xu H 1998 *J. Mol. Phys.* **94** 373
[4] Jal J F, Soper K, Carmona P and Dupuy J 1991 *J. Phys: Condens. Matter.* **3** 551
[5] Dupuy-Philon J, Jal J F and Prvel B 1995 *J. Mol. Liq.* **64** 13

[6] Aouizerat-Elarby A, Dez H, Prevel B, Jal J F, Bert J and Dupuy-Philon J 2000 *J. Molec. Liq.* **84** 289
[7] Fei L, Junsheng Y, Dongchan L, Shenyu L and Zhen H 2015 *J. Molecular Structure.* **1081** 38-43
[8] Chandra A 2003 *J. Phys. Chem. B.* **107** 3899-3906.
[9] Habchi M, Mesli S M, Kotbi M and Xu H 2012 *Eur. Phys. J. B.* **85** 255.
[10] Xu J J, Yi H B, Li H J and Chen Y 2013 *J. Molecular Physics.* 860244.
[11] Kotbi M, Xu H, Habchi M and Dembahri Z 2003 *J. Phys. Lett. A.* **315** 463.
[12] Harsnyi I and Pusztai L 2012 *J. Chem. Phys.* **137** 204503.
[13] Bouazizi S and Nasr S 2014 *J. Mol. Liq.* **197** 77-83.
[14] Pluharova E, Fischerb H E, Masona P E, and Jung- wirth P 2014 *J. Molecular Physics.* **112** 1230-1240.
[15] Mc Greevy R L, Howe M A and Wiks J D 1993 *A General Purpose Reverse Monte Carlo.* **3**
[16] Mesli S M, Habchi M, Kotbi M and Xu H 2013 *Condens. Matter. Phys.* **16** 1-8.
[17] Opletal G, Petersen T C, O'Malley B, Snook I K, Mc Culloch D G and Yarovsky I 2008 *Comput. Phys. Commun.* **178** 777.
[18] Mc Greevy R L and Pusztai L 1988 *J. Molec. Simul.* **1** 359.
[19] Mc Greevy R L 2001 *J. Cond. Matter.* **13** R877.
[20] Jing-Xiang Z, Hui L, Jie Z, Xi-Gui S and Xiu-Fang B 2009 *J. Chin. Phys. Soc.* **18** 1674-1056.
[21] Metropolis N, Rosenbluth A W, Rosenbluth M N, Teller A H and Teller E 1953 *J. Chem. Phys. Chem.* **21** 1087.

IOP Conf. Series: Materials Science and Engineering **186** (2017) 012011 doi:10.1088/1757-899X/186/1/012011

Key micromechanics issues in integrated material design

M Bennoura and A Aboutajeddine

Mechanical engineering laboratory, Faculty of Science and Technology, University of Sidi Mohamed Ben Abdellah, Fes, Morocco

E-mail: mounia.bennoura@usmba.ac.ma

Abstract. Nowadays the acceleration of material discovery is essential more than ever to hold the fast evolving requirements of innovative products. This acceleration depends on our ability to set up a material design process for tailoring materials from targeted engineering performances. One of the important building block passages, in the material design journey, is the bridging of micro-scale to meso-scale through micromechanical models. Unfortunately, these models include a lot of uncertainties resulting from their inbuilt ad-hoc assumptions, which inevitably impacts the material design process performance. In the present paper, robust design methods are reviewed and subsequently applied to quantify uncertainty in micromechanical models and mitigate its impact on material design performances. This includes examining principles for evaluating the level degree of uncertainty on material design process, and their use in micromechanical models. Also, developing robust design approaches to alleviate uncertainty effects and improve the quality of the design performance. Ultimately, the limitations of these approaches are discussed and the research opportunities, to overcome the shortness of actual approaches in respect to micromechanical models, are clarified.

1. Introduction

Engineering materials and materials science have witnessed several evolution phases. Primarily, Materials have been discovered by chance then new materials have been developed with trial-and-error techniques, these techniques allowed materials to be treated as a black box subjected to repeated experiments [1]. Experimental results then were gathered in materials databases. Thus, engineering materials has been taught using the paradigm of selecting materials on the basis of tabulated databases of properties [1]. Consequently, the performance of many engineering products and systems is limited fundamentally by the properties of available constituent materials. Evidently, one can no longer limit the performance of systems to available materials. Thus, the focus shifted towards the design of materials with superior properties for particular applications. The goal is to tailor material microstructure for desired applications that satisfy performance requirements on the system level. Often, however, these performances are in conflict when choosing the microstructure. It is important to note, that these desired material properties and performance characteristics usually depend on phenomena that operate at different length and time scales. Material design is then built on a multi-scale journey from the smallest scale to the system level. One of the important multi-scale passages is the bridging of micro-scale to meso-scale through micromechanical models. The primary objective of this work is to discuss overall challenges in incorporating micro-mechanical models into materials design process.

2. Leading challenges in integrating micromechanical models in material design process

Micromechanical models play a crucial role in the achievement of materials design initiative. These models aim to predict accurately the response of a material under uniform load conditions from the properties of their constituents. Unfortunately, virtually all micromechanical models are embryonic from material design point view. It is important to emphasize that micro-mechanical models inevitably incorporate assumptions and approximations that impact the precision and accuracy of predictions. When added to the material design process, they bring their drawbacks and the uncertainty may be magnified when a model is utilized near the limits of its intended domain of applicability or when information propagates through a series of models [2]. Furthermore, microstructure itself has a random character, such as sizes and distributions of grains, phases, and so forth. These sources of uncertainty give rise to the need for considering sensitivity of properties and responses of interest to variation of microstructure at various scales, as well as propagation of model uncertainty through multiscale model chains. So one of the most important and challenging problems is the uncertainty in these models associated to various sources, including [3]: (i) insufficient knowledge about microstructure. (ii) natural variability and randomness in materials (phases introduced by processing conditions) that are not taken into account. (iii) Ad-hoc parameters used by these models to presumably describe the microstructure

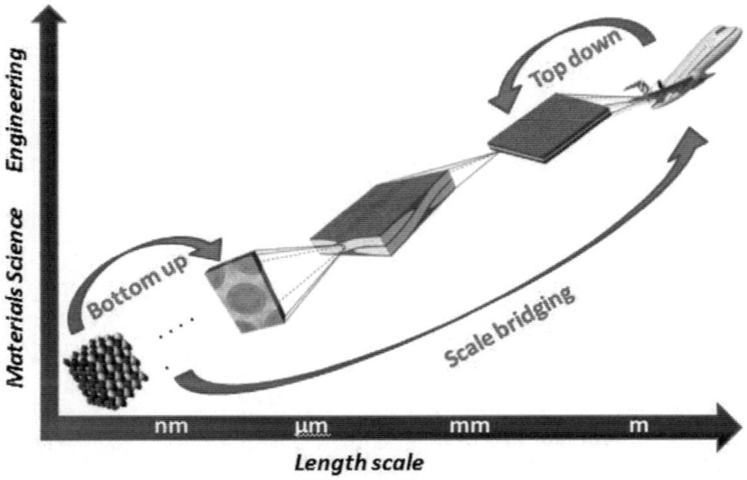

Figure 1. Hierarchy of levels from atomic scale to system level in concurrent materials design.

Uncertainty is prevalent in most facets of micro-mechanical models, the uncertainty of microstructure and model parameters impact significantly material design process. Consequently, a combined strategy of bottom–up and top–down modeling (figure 1) to target performance requirements will be very difficult [2,3]. In this respect, managing uncertainty and its propagation through a model is of paramount importance. The only way to gain this goal is by having a good understanding of microstructure (approximations and simplifications), in order to let the models be more realistic and transmit information without loss, by combining the approach with simulation tools, and finally by easing the decision making. To account of uncertainty, a practical approach includes many aspects [2,4], mathematical techniques for evaluating the level degree of uncertainties on material design journey, and design methods for performing and generating robust design, by seeking the robust solutions that are relatively insensitive to variation of microstructure and various other sources of uncertainty. In later sections of this paper all these methods will be investigated.

3. Uncertainty quantification

To properly quantifying uncertainty, it is necessary to categorize the types of uncertainty in a model.

The uncertainty quantification is *the process of identifying different sources of uncertainty and developing corresponding mathematical representations, by estimating output uncertainty (probability distribution of system response) with variability in input parameters* [2]. One must mention that quantification of aleatory uncertainty (inherent in the physical system and can only be quantified in a statistical sense) is relatively easier than quantification of epistemic uncertainty (arises from incomplete knowledge of a system) if measuring a system behavior is feasible [5, 6, 7, 8].

As presented in table 1, depending on the types of system model deterministic (there are no random errors in a system response) or non-deterministic and the type of input parameters are uncertain (without probability density functions due to lack of information) or may be presented as Probability distribution, uncertainty analysis methods have been classified as non-probabilistic and probabilistic methods. The probabilistic methods are further devised on statistical and non-statistical approaches. We can note from the table that the Monte Carlo simulation is the only method for quantifying non-deterministic system response variation with the probability distributions of input parameters. If the input variables are non-parametric, it is impossible to quantify uncertainty.

As aforementioned, the aleatory system uncertainty is quantifiable; however, epistemic uncertainty is virtually impossible to quantify [2]. Therefore, reducing uncertainty (instead of being quantified) is feasible when designer has large amounts of data and by increasing the knowledge or getting more information about system. Instead, researchers [9, 10, 11, 12] have focused on procedures to reduce uncertainty and validate a model's accuracy. However, the restriction of these methods is that designers must have some true results. In some cases, we cannot obtain any better data than current data due to severe uncertainty in a system or restrictions of computational or experimental expenses, which is the case of multiscale materials design specially micromechanical model. Uncertainty analysis for nondeterministic simulation is indispensable for multiscale materials design since the simulation models for heterogeneous material behavior tend to be stochastic [4].

Table 1. Uncertainty analysis methods versus characteristics of a system [2].

System model	Deterministic		Non-deterministic	
Input	Uncertain	Probability distribution	Uncertain	Probability distribution
Available uncertainty analysis methods	Non-probabilistic methods : Interval analysis Fuzzy logic	Probabilistic methods: 1.Statistical methods -Monte carlo simulation -Latin Hypercube sampling 2.Non- statistical method -First and second order moment methods -Polynomial chaos expansion -Stochastic reponse surface methods	N/A	Probabilistic methods -statistical methods: Monte Carlo

4. Uncertainty management

Uncertainty management process allows designing a system to be unresponsive to uncertainty without removing the underlying sources, by using robust design approaches. In other words, these approaches aim to alleviate uncertainty effects and improve the quality of a system performance [13], by making the system response insensitive to uncontrollable system input variations. This is also called parameter design. In robust design literature, design parameters are split into three categories: (i) control factors, parameters that a designer adjusts, (ii) noise factors, exogenous parameters that are not under a designer's control, (iii) responses, are performance measures for the product or process. It is essential for designers to identify where the uncertainty sources are located in a system model in order to employ a relevant uncertainty management approach.

Robust design approaches are categorizing as: Type I robust design, originally proposed by Taguchi [14], focuses on designing a systems that fulfill the performance despite variations in noise factors which are parameters that designers cannot control in a system. Type II robust design [15] is used to design a system that fulfills the performance despite variations associated with parameters that a designer can control in a system. A method for Types I and II robust design has been proposed, namely the Robust Concept Exploration Method [16]. Briefly, it based on design of experiments for generating set points to achieve trade-off between performance measures and uncertainty components in the compromise decision support problem, for more detail of this method, see [17].These types of robust design have been extended to include Type III [2], which considers sensitivity to uncertainty set in a model which is due to a combination of limited knowledge and data of nonparametric system noise or un-configured system noise. Type III robust design considers not only the objective function but also the two uncertainty bounds due to the non-parametric variability, un-configured variability, and model parameter uncertainty.

The final type of robust design is for multidisciplinary system which takes into account uncertainty and managing the uncertainty generated in the design process chain, this type of uncertainty arises from accumulated and propagated errors in decisions made by designers in subsequent series of uncertain subsystem models. To deal with this kind of uncertainty a method called The Inductive Design Exploration Method (IDEM) had been proposed by Choi et al [2,18], to identify adjustable ranges of design variable values under different types of uncertainty which is rise from natural variability, model uncertainty and uncertainty propagation in a design process chain. IDEM is schematically illustrated in figure 2. IDEM has two major objectives: (1) to explore top–down, requirements driven design, guiding bottom–up modeling and simulation, and (2) to manage uncertainty in model chains. It is based on three steps:

Step 1: It is necessary to define the design space exploration (x space in figure 2), the interdependent space (y space), and the performance space (z spaces). Discrete points are generated in each of these spaces (Discretizing).

Step 2: The discrete points which are generated are evaluated based on the mapping models (models f and g in figure 2) and the evaluated data sets which are composed of a discrete input point and output range are stored in a database.

These two steps are called parallel discrete function evaluation in IDEM, process as follows:

Discretizing: all possible combinations of discrete input of associated input variables are created

Grouping: The discretized points created by the discretization process are grouped as input sets for mapping models.

Mapping: The points in each group are evaluated in parallel by each mapping model and the evaluation results are stored with the input points.

Merging: Sets of input points and corresponding outputs obtained in the mapping step are merged to form a set of evaluated original discrete points.

Step 3: Feasible regions in y and x spaces are successively identified using a metric called Hyper-Dimensional Error Margin Index (HD-EMI), for determining if a discrete point from an input space maps to a feasible design solution in the output space. It is the Inductive Discrete Constraints Evaluation (IDCE).

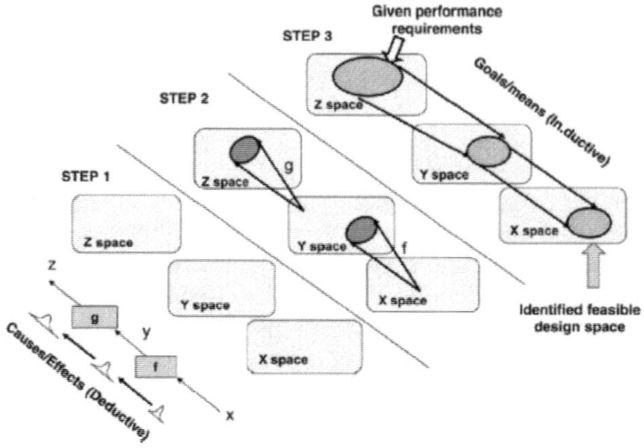

Figure 2. A schematic procedure for IDEM [19].

IDEM has been extended to account for model parameter uncertainty [19]. Additional work needs to be performed to account for other representations of epistemic uncertainty.

5. Critical evaluation

After presenting an overview of the actual methods and approaches of the design under uncertainty process, which comprise uncertainty quantification in engineering systems and uncertainty management methods, we critically evaluate the capabilities of those methods and approaches from material design perspective. Then we identify the research opportunities for facilitating the integration of micromechanical models in material design process.

As discussed, several methods have been performed to quantify and estimate the degree of aleatory uncertainty in a system, but the uncertainty analysis methods for non-deterministic system models have not been widely investigated in literature except for Monte Carlo method [2]. Monte Carlo method is the only method for quantifying a non-deterministic response variation with input parameters probability distributions; however, it is computationally too intensive to use into material design. In the case where the input variables are non-parametric, it is impossible to quantify uncertainty. Therefore, an uncertainty analysis method for non-deterministic simulation is indispensable. The epistemic uncertainty is reduced rather than quantified, but in the case of multiscale materials design especially micromechanical model, it is impossible to employ the actual approaches because of the limited knowledge about the material microstructure [2].

For managing uncertainty many robust design methods for designing a system to be insensitive to uncertainty were reviewed in the previous section. Type I robust design is not valid if noise factors cannot be quantified [19]. However the uncertainty in material design tends to be unparameterizable. Neither the types I and II robust design can design a system to be insensitive to uncertainty in system model such as the morphology variability in material, even if Type III robust design approach have been proposed, but It is nearly impossible to use this approach in materials design until any computationally inexpensive uncertainty analysis method have been performed [2].

The inductive design exploration method has been presented as potential approach for the multidisciplinary robust design methods; however, it's still short, first because it cannot overcome the uncertainty in material models due to idealization associated with a lack of complete knowledge of a given phenomenon and its depiction. For example, there is uncertainty in the constitutive models which are used to represent behavior of individual phases. Second IDEM may be computationally intensive if the number of design variables is large or the simulation model is expensive [2], also if the design process is not sufficiently constrained, it may give large sets of feasible processing spaces in

the material design hierarchy and it becomes difficult to choose a single solution from these sets of solution [19].
The problem of uncertainty in material design process has been widely studied, but The main conclusion to be drawn is that none of the outlined approaches for mitigation uncertainty had satisfactory results to deal with this endemic uncertainty in constitutive models, which are used to represent behavior especially in micromechanical modeling, that why the improvement of the accuracy of this models and facilitate their incorporating in material design process is our primary proposes.

6. Conclusion
In this work, a number of issues related to integrating micromechanical models in material design process were identified. These issues were mainly associated to uncertainty quantification and uncertainty management in micromechanical models. This uncertainty is due mainly to idealization resulting from the lack of complete knowledge of microstructure and its description. The key methods and approaches for dealing with this uncertainty were reviewed and critically evaluated; first to identify their drawbacks and second to draw the research opportunities in this field. The main conclusion is that further work is needed to deal with uncertainty in micromechanical models for the ultimate goal to incorporate them in the material design process. The best way to do this, in our point of view, is by building a rational material design exigencies framework for developing efficient micromechanical models for the material design journey.

References
[1] McDowell D L and Story T L 1998 New directions in materials design science and engineering *Report of a Workshop Sponsored by the US National Science Foundation* Georgia Institute of Technology and Morehouse College Atlanta GA.
[2] Jitesh H Panchal, Surya R Kalidindi and David L. McDowell 2013 Key computational modeling issues in Integrated Computational Materials Engineering *Computer-Aided Design* **45** 4–25
[3] David L. McDowell and Olson G B 2008 Concurrent design of hierarchical materials and structures *Springer* 10.1007/s10820-008-9100-6
[4] Choi H J 2005 *A robust design method for model and propagated uncertainty* Ph.D. Dissertation, The GW Woodruff School of Mechanical Engineering, Georgia Institute of Technology, Atlanta, GA.
[5] Isukapalli S S, Roy A and Georgopoulos P G 1998 Stochastic Response Surface Methods (SRSMs) for Uncertainty Propagation: Application to Environmental and Biological Systems *Risk Analysis* **18** 351-63.
[6] Der Kiureghian A 1989 Measures of Structural Safety under Imperfect State of Knowledge J. *Structural.Eng* **115** 1119-39.
[7] Haukass T 2003 Types of Uncertainties, Elementary Data Analysis, Set Theory, Reliability and Structural Safety *Lecture Notes*, University of British Columbia.
[8] Nikolaidis E 2005 Types of Uncertainty in Design Decision Making *Engineering Design Reliability Handbook* CRC Press New York, pp. 8-1.
[9] Jin R, X Du and Chen W 2003 The Use of Metamodeling Techniques for Optimization under Uncertainty *Journal of Structural & Multidisciplinary Optimization* **25** 99-116.
[10] Sargent R G 2003 Verification and Validation of Simulation Models *Winter Simulation Conferences* New Orleans LA 37- 48.
[11] Kennedy M C and O'Hagan A 2000 Predicting the output from a complex computer code when fast approximations are available *Biometrika* **87** 1-13.
[12] Qian Z , Seepersad C C , Joseph V R, Wu C F J and Allen J K 2004 Building Surrogate Models based on Detailed and Approximate Simulations *ASME DETC04* Salt Lake City UT DETC2004/DAC57486.
[13] Allen JK, Seepersad CC, Choi H J and Mistree F 2006 Robust design for multidisciplinary and

multiscale applications *ASME Journal of Mechanical Design* **128** 832–43.
[14] Taguchi G 1993 Taguchi on Robust Technology Development Bringing Quality Engineering Upstream *ASME Press* New York.
[15] Chen W, Simpson T W, Allen J K and Mistree F 1996 Using Design Capability Indices to Satisfy a Ranged Set of Design Requirements *Advances in Design Automation* Irvine CA 96-DETC/DAC-1090.
[16] Chen W 1995 *A robust concept exploration method for configuring complex systems* Ph.D Dissertation, GW Woodruff School of Mechanical Engineering, Georgia Institute of Technology Atlanta Georgia.
[17] Chen W and all 1997 A robust concept exploration method for enhancing productivity in concurrent systems design *Concurrent Engineering-Research and Applications* **5***203-17*
[18] Panchal J H, Choi H J, Shepherd J, Allen J K, McDowell D L and Mistree F 2005 A strategy for simulation-based multiscale,multifunctional design of products and design processes *ASME Design Automation Conference* Long Beach CA Paper Number DETC2005-85316.
[19] Sinha A, Panchal JH, Allen JK and Mistree F 2011 Managing uncertainty in multiscale systems via simulation model refinement *ASME 2011 international design engineering technical conferences and computers and information in engineering conference American Society of Mechanical Engineers* 471-83

Electromagnetic Properties of Composites CoFeNiBSiMo, CoFeBSiCr and CoMnSiB in wide frequency bande

My C Elboubakraoui[1], S Bri [2]and J Foshi[1]

[1]Electronique, Instrumentation et de Mesure Physique: EIMPFSTE – Moulay Ismail University, Meknès, Morocco.
[2] Matériaux et Instrumentations: MIN, ESTM, Moulay Ismail University, Meknès, Morocco.

E-mail : charif2060@gmail.com

Abstract. In this paper, our study is based on composites with ferromagnetic wire with a frequency region in the microwave regime with scattering spectra strongly dependent on an external magnetic field and the microwire periodicity. Three types of composites made of grids of continuous and short-cut wires CoFeNiBSiMo, CoFeBSiCr and CoMnSiB are considered to employ different types of spectra of the permittivity and permeability in the frequency band 1 GHz-12 GHz. The complex permittivity increases remarkably with the increase of microwire periodicity, with negative real permittivity is observed over frequency range for wires. With this increase in the microwire periodicity, first we note decreased minimum reflection loss and maximum absorption.

1. Introduction

The knowledge of intrinsic material properties (permittivity and permeability) is required in most digital applications including monolithic microwave circuits using wide frequency range. There are various techniques for extracting these properties; some are single frequency while others offer broadband characterization. Composite structural materials containing periodic or random microwires of conducting scattering elements have received much attention because they make it possible to engineer a selective spectral response to the electromagnetic radiation. It has been demonstrated that composites with conducting wires possess a number of unique properties at microwave and optical frequencies, such as anomalous large photonic band gaps, and resonance like anomalous dispersion of the effective dielectric permittivity also associated with negative values. A recent trend is to achieve the adjustability of these materials [1], [2], [3].

The composite wire with metal incorporated can show a high dispersion of the effective permittivity in the microwave range. By using ferromagnetic wire, it is possible to adjust the sensitivity of this dispersion by changing the size, the magnetic structure of the wire with mechanical or thermal external magnetic stimuli.

2. Mathematical formulation of the proposed method

2.1. Modelling effective permittivity for composite with wire

The microwave tunable properties of a microwire composite are the response of effective permittivity to the electromagnetic wave through the surface impedance. Therefore, to explain this phenomenon, one needs to understand the basics of wave interactions with the materials and the effective medium approaches to characterize heterogeneous composites. The effective permittivity can be treated in two different types of composite [1]:

- Drude model: is applicable to composite materials containing long wire.
- Lorentz model: for composite containers short wire, this model is effective and applicable to all insulation materials.

The effective permittivity is defined [3], [4] :

$$\varepsilon_{eff}=\varepsilon_d+4\pi p\langle\alpha\rangle \tag{1}$$

α is the polarisability of an individual inclusion and εd is the dielectric permittivity matrix, and $p=\frac{\pi a^2}{b^2}$ is the surface concentration of wire embedded in matrix.

The interaction between the short wire inclusions is neglected; the average polarization of a wire defined in equation (1) is given by [5]:

$$\langle\alpha\rangle=\frac{1}{2\pi\ln\left(\frac{b}{a}\right)(Ka)^2}\left(\frac{2}{Kb}\tan\left(\frac{Kb}{2}\right)-1\right) \tag{2}$$

With

$$K=\frac{\omega\sqrt{\varepsilon_d}}{c}\left(1+\frac{ic\varsigma_{xx}}{\omega a\ln\left(\frac{b}{a}\right)}\right)^{1/2} \tag{3}$$

For continuous composite wire, the effective permittivity also depends on the wire surface impedance. In this case, there are no charge distribution along the wire and associated dipole resonances. As a result, the effective permittivity for waves with the electric field polarization along the wires has a characteristic plasma dispersion behavior [5]. The effective permittivity:

$$\varepsilon_{eff}=\varepsilon_d-\left(\frac{\omega_p}{\omega}\right)^2 \tag{4}$$

With ω_p the plasma frequency:

$$\omega_p^2=\frac{2\pi c^2}{b^2\ln\left(\frac{b}{a}\right)} \tag{5}$$

In the case of a strong skin effect and due to the Giant magneto-impedance effect GMI, It is possible to write the solution of the electric field in the system which depends on the wire conducting and magnetic properties through the surface impedance. The effective permittivity of a plasma dispersion behavior is given by:

$$\varepsilon_{eff}=\varepsilon_d-\frac{p\omega_p^2}{\omega^2\left(1+i\frac{c\varsigma_{xx}}{\omega a\ln\left(\frac{a}{b}\right)}\right)} \tag{6}$$

In the case of a strong skin effect in the inclusions and a scalar permeability μ, the surface impedance is determined by the Leontovich condition [6].

$$\varsigma_{xx}=(1-i)\sqrt{\frac{\omega\mu}{8\pi\sigma}} \tag{7}$$

Following the low-field GMI phenomena utilized for field sensing, this section is focused on the high field absorption that can be potentially used for stress sensing. Here we treated theoretically the GMI phenomenon in the microwave frequency range by analyzing the ferromagnetic resonance (FMR), which is considered to be related to the high-frequency GMI. Using the classic Landau– Lifschitz–Gilbert equation [7], in the case of ferromagnetic wires or composite containing arrays of wires, one can obtain equation (8) to calculate the variation of the complex component of permeability with varying applied magnetic fields. [8].

$$\mu=1+\frac{\mu_0\gamma M_s[\mu_0\gamma(H_{dc}+H_k)-i\omega\alpha]}{-\omega^2+\omega_{fmr}^2-i\omega\alpha\mu_0\gamma[2(H_{dc}+H_k)+M_s]} \tag{8}$$

Where μ_0 is the vacuum permeability, α is the damping parameter, γ is the gyromagnetic parameter, M_s is the saturation magnetization, H_{dc} denotes the applied magnetic field, H_k is the anisotropy field and ω_{fmr} denotes the angular resonance frequency which is given by:

$$\omega_{fmr}=\mu_0\gamma\sqrt{(H_{dc}+H_k+M_s)(H_{dc}+H_k)} \tag{9}$$

It is reasonable to infer that, in addition to the magnetic field, stresses also produce a similar effect by modifying the magnetoelastic anisotropy of the microwires as expressed in equation:

$$H_k=\frac{3|\lambda_s|}{M_s}(\sigma_{zz}-\sigma_{\varphi\varphi}+\sigma_{app}) \tag{10}$$

Where λs is the magnetostriction constant, σ_{zz}, $\sigma_{\varphi\varphi}$ and σ_{app} are the longitudinal internal stress, radial internal stress, and applied stress, respectively [1], [8].

2.2. Microwave absorption

Absorption is a phenomenon that occurs when microwaves interact with the materials. The absorption or attenuation originates from the dielectric loss and the magnetic loss. For a single layer composite, reflection loss RL [9] is given by:

$$RL=20\log\left|\frac{\sqrt{\frac{\mu}{\varepsilon}}\tanh\left(i\frac{\omega d}{c}\sqrt{\mu\varepsilon}\right)-1}{\sqrt{\frac{\mu}{\varepsilon}}\tanh\left(i\frac{\omega d}{c}\sqrt{\mu\varepsilon}\right)+1}\right| \tag{11}$$

With d is thickness of the absorber. The attenuation constant α is the real part of the propagation constant, given by:

$$\alpha=Re\left(\frac{i\omega\sqrt{\mu\varepsilon}}{c}\right) \tag{12}$$

ε and μ are the relative permittivity and permeability of the absorbing material, respectively. For a magnetic absorber, the matching frequency can be regarded as the ferromagnetic resonance frequency, which can be given as:

$$\omega=\gamma H_k \tag{13}$$

3. Results and discussions

The microwave properties of wire composites were investigated for different types of samples wire glass coated amorphous CoFeNiBSiMo, CoFeBSiCr and CoMnSiB with radius of a=5 μm were glued in paper to form wire-lattices with separation b. Calculations are performed in the frequency range of 1GHz-12GHz. The flowchart of calculating the complex effective permittivity and effective permeability are given by proposed algorithm:

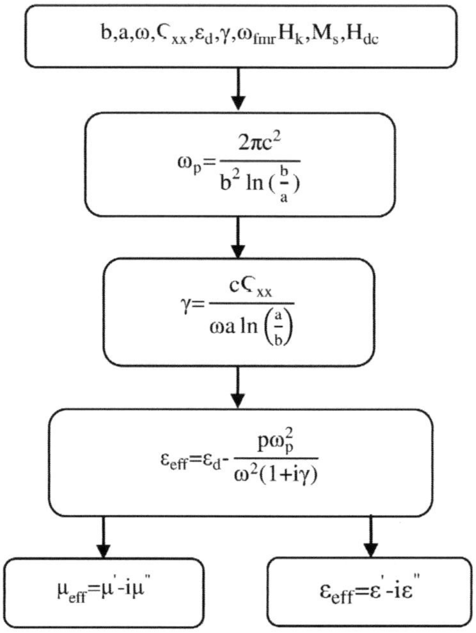

Table1. Material parameters obtained from tests magnetization

Composite wire	Wire radius a (μm)	Anisotropy field H_k (A/m)	Gyromagnetic ratio γ ($mA^{-1}.s^{-1}$)	Saturation magnetization $4\pi M_s$ (A/m)	Damping parameter α	Conductivity σ (s/m)
CoFeNiBSiMo	5	880	$2,21x\ 10^5$	10^6	0.02	10^{16}
CoFeBSiCr	5	500	$3,76x10^4$	5.10^5	0.01	$7,7x\ 10^5$
CoMnSiB	5	880	$2,51x10^5$	5.10^5	0.02	$7,6x10^{15}$

3.1. The effective permittivity

Figures (1), (2) and (3) illustrate the variation of the effective complex permittivity of the frequency as a function of the periodicity thickness for composite wire in a magnetic field. The material parameters obtained from tests magnetization are given in table 1.

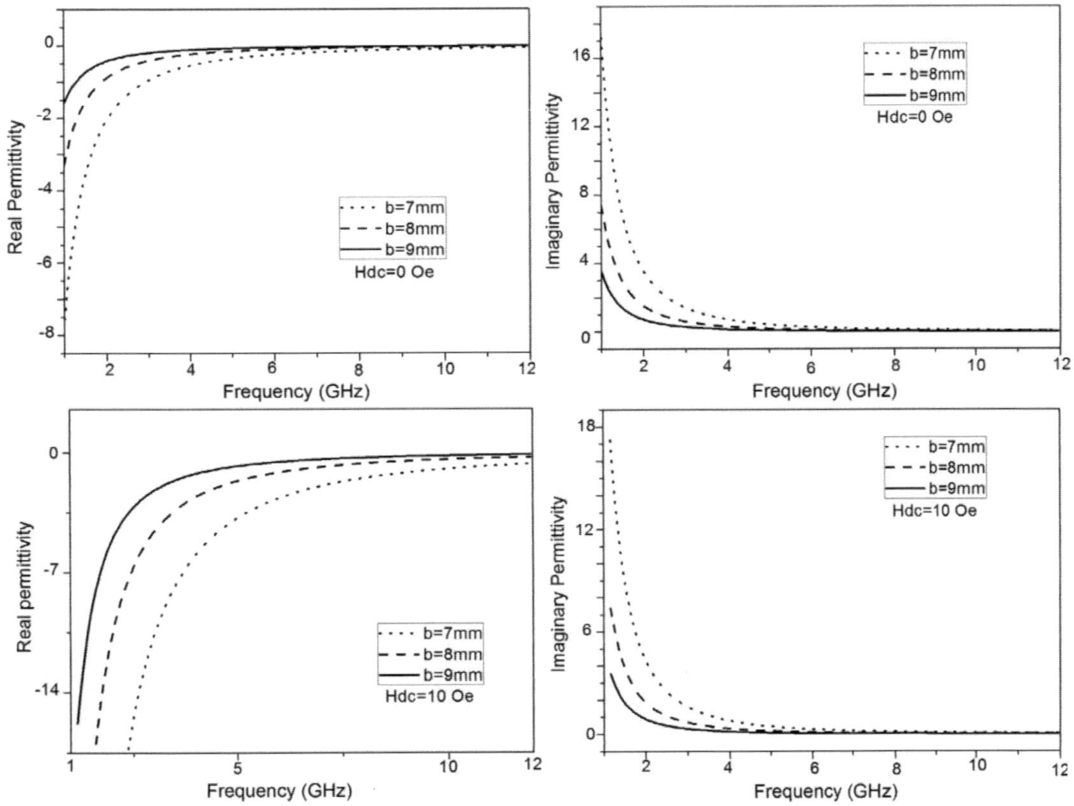

Figure1. Variations of the real part and the imaginary part of the effective dielectric permittivity depending on the frequency for CoFeNiBSiMo

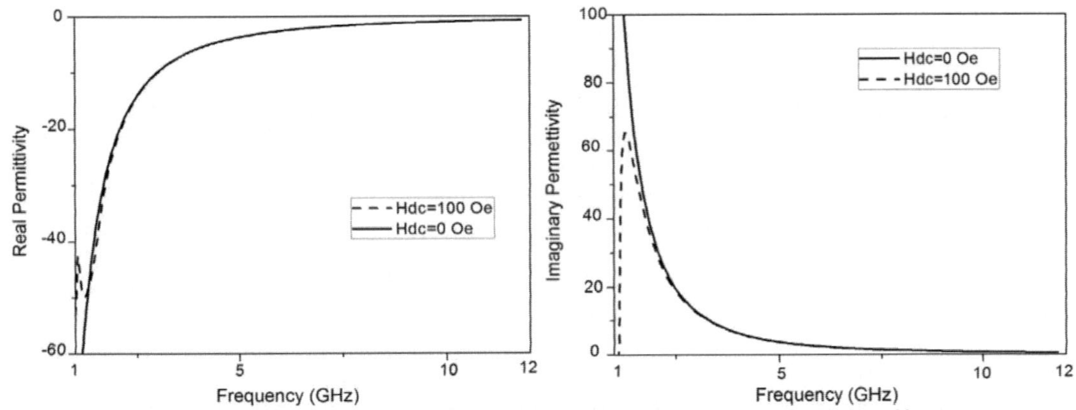

Figure 2. Variations of the real part and the imaginary part of the effective dielectric permittivity depending on the frequency for CoMnSiB

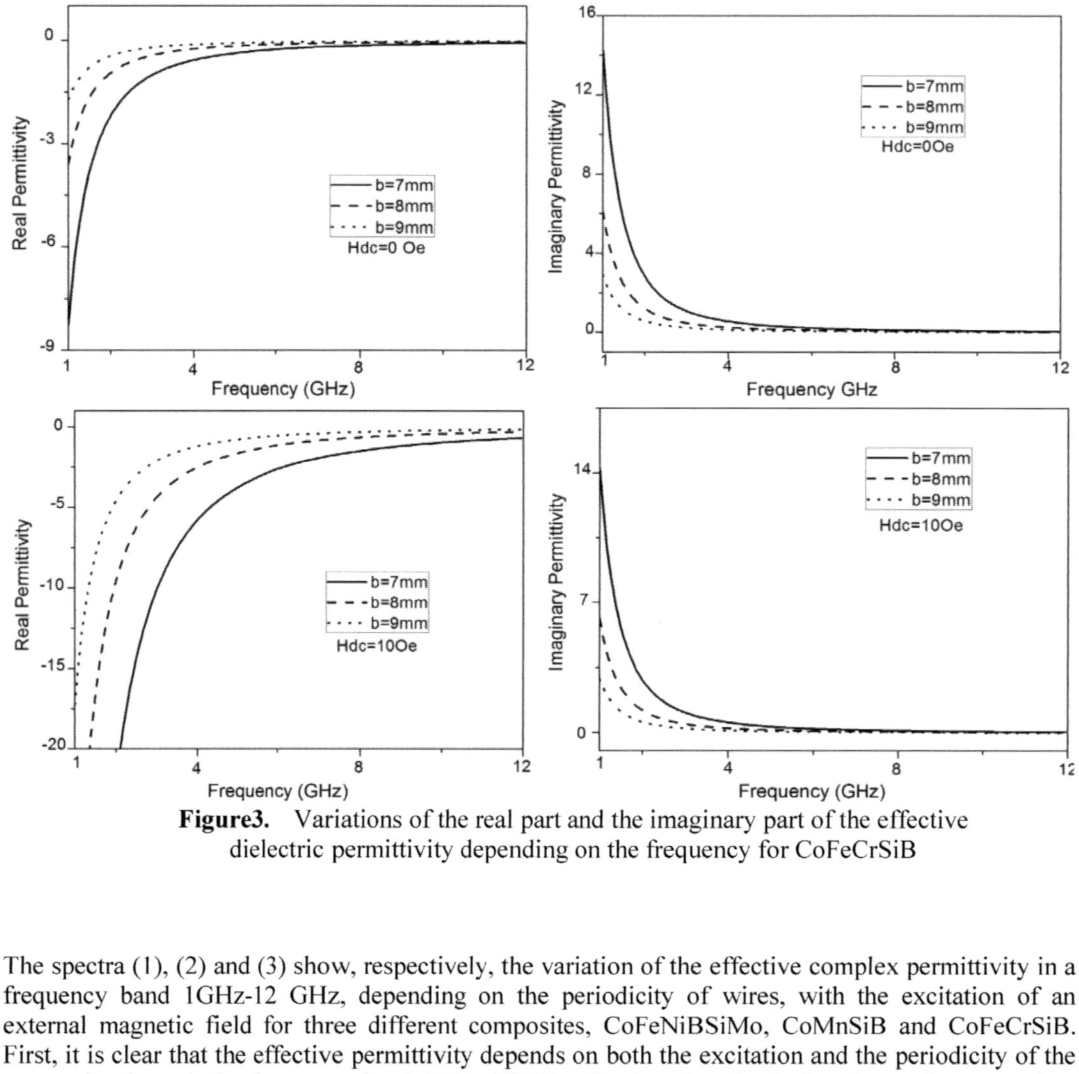

Figure3. Variations of the real part and the imaginary part of the effective dielectric permittivity depending on the frequency for CoFeCrSiB

The spectra (1), (2) and (3) show, respectively, the variation of the effective complex permittivity in a frequency band 1GHz-12 GHz, depending on the periodicity of wires, with the excitation of an external magnetic field for three different composites, CoFeNiBSiMo, CoMnSiB and CoFeCrSiB. First, it is clear that the effective permittivity depends on both the excitation and the periodicity of the composite through the frequency band 1GHz-12GHz. On the other hand, the plasma frequency is a function of the periodicity of the test composite. It can also be seen that, with the increasing wire periodicity from 7mm to 9 mm, the frequency dispersion of effective permittivity on the magnetic field is remarkably a damped manner. However, a decrease of wire periodicity will increase the plasma frequency and hence the skin effect. If the skin effect is too strong, the field effect will be rather weak. Therefore, the wire diameter may also need to be decreased to compensate the decrease of skin depth. Thereby, there exists an optimum value of wire periodicity matching the diameter for the microwave tunable properties.

3.2. The effective permeability

According to the Landau–Lifshitz–Gilbert equation (8), the permeability is strongly dependent on the anisotropy field of the wires. Considering the significant stress impact on the anisotropy field of wires [1], [5] and anisotropy angle, the permeability (or impedance) spectra can be regulated by the variation

IOP Conf. Series: Materials Science and Engineering **186** (2017) 012012 doi:10.1088/1757-899X/186/1/012012

of internal stress due to that of geometry or glass-removal; and external stress [8] as shown in equation (10). The resonance frequency can be regulated by the parameter of the microwires and thickness of the absorber (number of wires).

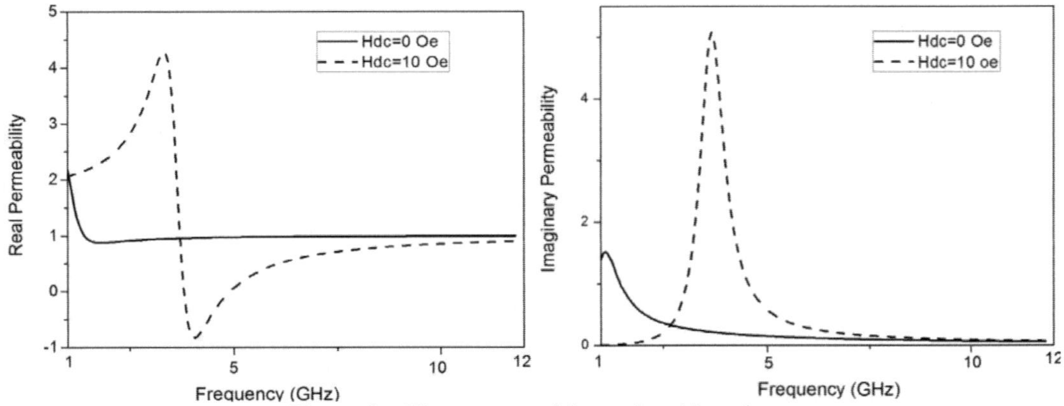

Figure4. The spectra of the real and imaginary part of the effective permeability for CoFeNiBSiMo

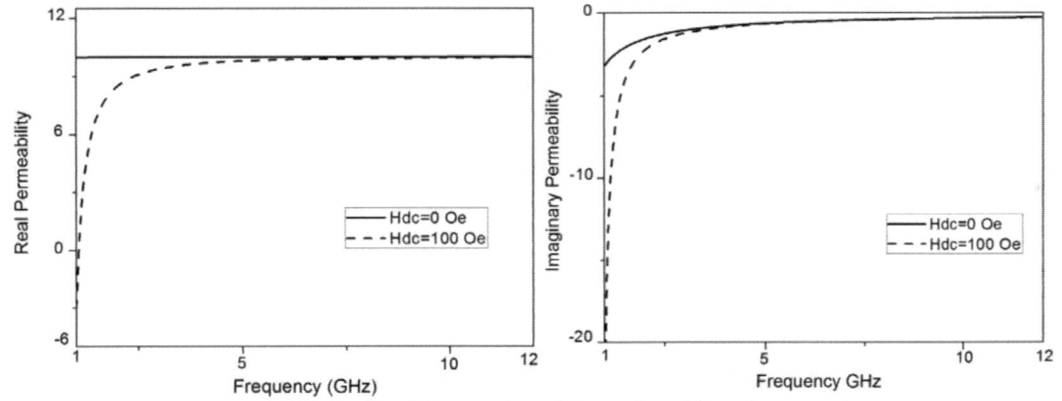

Figure5. The spectra of the real and imaginary part of the effective permeability for CoMnSiB

3.3. The reflection loss

The field dependence of the effective permittivity of the composite is caused by the field dependence of the surface impedance matrix ζ_{xx}, which determines the losses inside the inclusions. The figure (6) show respectively, the typical relationship between reflection loss, absorption and measuring frequency of the composites containing the wires CoFeNiBSiMo, with the thickness of absorber d=2,1mm. It shows the frequency dependence of the reflection loss taken at a magnetic field for the composite with b = 7 mm, and 9 mm. It can be seen that the shape of the curves varies remarkably as the wire periodicity increases from 3 mm to 9 mm. The increase of microwire periodicity, the minimum reflection loss decreases firstly and then increases, and shifts towards lower frequency region. It is seen that the plots for Hdc=10 Oe circumferential magnetization have a resonance behavior with a resonance frequency of 3,75GHz, an absorption peak, and a negative real part of the permittivity in a wide frequency band past the resonance. In the presence of the magnetic field larger than the anisotropy field which saturates the magnetization along the axis and increases the wire surface impedance, the dispersion of effective permittivity becomes a relaxation type with a gradual decrease in the imaginary part, which is always positive and with very broad losses [2].

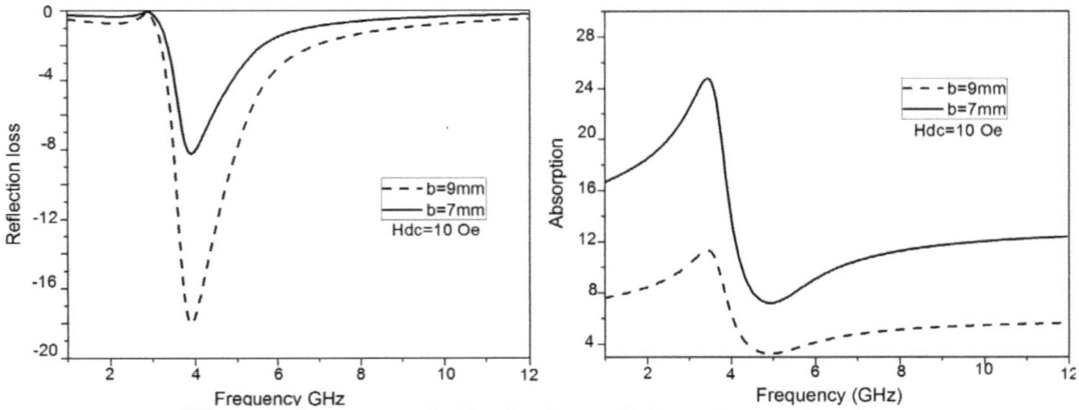

Figure6. The spectra of reflection loss and absorption as a function of frequency for CoFeNiBSiMo

4. Conclusions

In This document, the effects from the parameters of the wire on the dielectric and magnetic response depend on the nature of the ferromagnetic microwires composite, continuous in the frequency range from 1 GHz to 12 GHz, using a calculation of the effective permittivity according to the impedance of the wire surface. The real and imaginary parts of the effective permittivity show great variations with the frequencies and the wire parameters, also in the domain due to the field dependency of the impedance of the wire that controls the losses in the dielectric response. Thus, the composite wire has a Plasmon dispersion of the effective permittivity with negative values of its real part lower than the plasma frequency is in the gigahertz range for the wire spacing of about 1 cm and a diameter wire a few microns.

References

[1] F.X. Qin., and H.X. Peng, 1995 *Ferromagnetic microwires enabled multifunctional composite materials*". Progress in Materials Science 58, pp. 183–259, 2013

[2] S.I. Bozhevolnyi., I.I. Smolyaninov and A.V. Zayats,. *Near-field microscopy of surface-plasmon polaritons: Localization and internal interface imaging*, Physical. Review B 51, pp.17916-17924.

[3] A. N. Lagarkov, and A. K. Sarychev, 1996 *Electromagnetic properties of composites containing elongated conducting inclusions*, Physical Review B 53, 6318.

[4] A.N. Lagarkov, S.M. Matytsin, K.N. Rozanov and A.K. Sarychev, 1998 *Dielectric properties of fiber-filled composites*, Journal of Applied Physics 84, pp. 3806–3814.

[5] L.V. Panina, S.I. Sandacci and D.P. Makhnovskiy, 2005 *Stress effect on magnetoimpedance in amorphous wires at gigahertz frequencies and application to stress-tunable microwave composite materials*, Journal of Applied Physics 97:013701–6.

[6] Makhnovskiy D.P., L.V. Panina, C. Garcia, A.P. Zhukov and J. Gonzalez, 2006 *Experimental demonstration of tunable scattering spectra at microwave frequencies in composite media containing CoFeCrSiB glass-coated amorphous ferromagnetic wires and comparison with theory*, Physical Review B 74, 064205.

[7] L.D. Landau and E.M. Lifshitz, 1975 *Electrodynamics of continuous Media, In: Course of*

theoretical physics – Pergamon international library of science, technology, engineering and social studies. Oxford: Pergamon Press.

[8] F.X. Qin, H.X. J. Peng, and L.C. Qin, 2010 *Ferromagnetic microwires enabled polymer composites for sensing applications*, Compos Part A: Applied Science and Manufacturing 41, pp. 1823–1828.

[9] Z. Zhang, C. Wang, Y. Zhang and J. Xie, 2010 *Microwave absorbing properties of composites filled with glass-coated Fe69Co10Si8B13 amorphous microwire*, Materials Science and Engineering B 175, pp. 233–237.

IOP Conf. Series: Materials Science and Engineering **186** (2017) 012013 doi:10.1088/1757-899X/186/1/012013

Synthesis and characterization of vanadiumoxidecatalysts supported on copper orthophosphates

M Ouchabi[1], M Baalala[1], A Elaissi[1], I Loulidi And M Bensitel[1]

[1] Laboratoire de Catalyse et Corrosion des Matériaux, Université ChouaibDoukkali, Faculté des Sciences d'El Jadida, BP. 20, El Jadida 24000, Morocco.

Email: m.ouchabi@yahoo.fr

Abstract. Synthesis of a pure copper orthophosphate **(CuP)** prepared by Coprecipitation, and **CuP** modified by impregnation of vanadium (2-12 wt % of V_2O_5) have been carried out. The solids obtained were investigated as synthesized or after calcination by various physico-chemical techniques such as X-Ray Diffraction (XRD), Infrared Spectroscopy (IR), Thermogravimetric analysis (TGA), and differential thermal analysis (DTA). The results revealed that the solids **V/CuP** consisted of copper orthophosphate $Cu_3(PO_4)_2$ as major phases, together with V_2O_5 as minor phase. The diffraction lines of V_2O_5 increase by increasing the vanadium content.**Key word-** Copper orthophosphate **CuP**, Vanadium catalysts supported on copper orthophosphates **V/CuP**, XRD, DTA-TGA.

1.Introduction

Metal phosphates may be classified by their phosphate block units (PO_4^{3-}, HPO_4^{2-}, H_2PO^{4-}, $P_2O_7^{4-}$ and $P_4O_{12}^{4-}$) and attract increasing interest for environmental and technological fields [1]. In environmental field, the formation of metal phosphates help remove phosphate from waste waters, while their dissociations help regulate the slow release of fertilizers in acidic soils [2]. In technological field, the interest of metal phosphates is mainly focused on areas such as laser host [3], ceramic [4], electric [5], magnetic [6], fertilizer [7], and catalytic [8] processes because of their valuable physical–chemical properties and reactivity. Consequently, metal phosphates have become a hot research topic in academic (material and chemical sciences) and industrial fields in the recent years [9–11]. Metal phosphate group, is a promising candidate for application in the fields of catalysis, electrics, ion exchange and conductors [12–17]. Because of their acidity and porosity, layered materials represent a vast class of intercalating compounds with useful chemical and thermal properties, which have been extensively used as heterogeneous catalysts [8,18]. Furthermore, The catalysts based on vanadium oxide are used in several industrial processes for production of important chemicals such as sulphuric acid, maleic andphthalic anhydride, and for the reduction of environmental pollution e.g. nitrogenoxides from flue gas of power plants [19-21]. Vanadium oxide catalysts have been tested in a number of heterogeneous catalytic oxidation reactions such as: a) Selective oxidation of alkanes to olefins (oxidative dehydrogenation reaction) [22-24] and to aldehydes and acids [25-27]. b) Selective oxidation of olefins to unsaturated aldehydes and acids, nitriles, dienes, and organic acids [28-30].
The most of vanadia based catalysts consist of a vanadium oxide phase deposited on the surface or in pores of an oxide support. Initially, the support was considered as inert substance that provided a high surface area to carry the active metal oxide component or to improve the mechanical strength of the catalyst material. Deposition of a metal oxide on the surface of another oxide was proposed to improve the catalytic activity of the active metal oxide phase due to a gain in surface area and mechanical

strength[20-25]. However, numerous studies in the last decade indicated that the activity and selectivity of supported metal oxide catalysts are significantly affected by the properties of the support oxide material, and that the dispersed phase has a particular structure and properties different from that of the bulk oxide.
The attempt of this work is to prepare the pure copper orthophosphate **CuP** by coprecipitation, and **CuP** modified by impregnation of vanadium oxide (2-12% of V_2O_5). Additionally, the studied powders were characterized by thermogravimetry–differential thermogravimetry (TG/DTG), X-ray diffraction (XRD), Fourier transform infrared (FTIR) and UV-Visible. The data obtained will be important for further studies of the compound.

2. Experimental

Copper orthophosphate **CuP** were synthesized upon evaporation to dryness from aqueous solutions of $Cu(NO_3)_2.3H_2O$ and $(NH_4)_2HPO_4$. The resulting solids were dried for 12h at 120°C and were finally calcined in air at 450°C for 10h and at 650°C for 10h.
A series of vanadium catalysts V/CuP were prepared using impregnation method, over an CuP support. Amass of metavanadate d'ammonium NH_4VO_3 calculated to yield the desired percentage of vanadium on the support, was dissolved in a predetermined volume of distiled water.
The impregnation catalysts **V/CuP** were dried at 120°C for 12h and calcined under air flow at 450°C during 10h and at 650°C for 10h.

3. Characterization techniques

The X-ray diffraction (XRD) analysis was performed for the fresh catalyst using Bruker-eco D8 Advance diffract meter with Cu-Kα radiation (λ = 1.5418 Å). The XRD patterns were scanned from 5 to 70 θ.
TGA–DTA analysis has been recorded with a TA Instruments balance, model SDT 2966 (TGA–DTA). It allows obtaining simultaneous DTA and TGA diagrams under similar experimental conditions. Experiments were performed under an N_2 flow, from 298 K to 1073 K using a heating rate of 5°C/min.
Diffuse reflectance infrared spectra for the synthesized solids were recorded from 4000 to 400 cm^{-1} on a Nicolet 460 spectrophotometer.
UV–Visible spectra in diffuse reflectance mode were recorded by Shimadzu UV-2400PC Series spectrometer, from 200 to 1000 nm.

4. Results and Discussion

The TG/DTA curves of CuP are shown in Figure 1. The TG curve of CuP shows the three weight loss steps in the range of 50–670°C. The weight loss steps in the TG curve were observed in ranges of 50–150°C and 150–320°C. The corresponding observed weight losses were 4.372% and 41.524% by mass, which correspond to 0.44 molof water and 4.15 mol of ammonium nitrate and water, respectively. The first stage of the weight loss was related to the loss of moisture because it is not stable over 100°C. While the second stage was due to the loss of water molecules from overlapping reaction the loss of coordination water (dehydration reaction) [31-32]. These two stages appear in the corresponding DTAcurve as two endothermic peaks at 130 and 290°C.
The third stage of the weight loss was related to the start of dehydroxylation of libenthite$Cu_2(PO_4)$ (OH) which leads to an orthophosphate $Cu_3(PO_4)_2$, corresponding an one endothermic peaks at 330°C. The endothermic peak appeared at 605°C corresponds to the crystallization of Orthophosphate $Cu_3(PO_4)_2$.
The TGA curve of the solid V(10%)/CuP dried at 110 ° C (Figure 2) has two mass loss between room temperature and 670°C. These weight losses are due to the removal of water and ammonia from the structure metavanadate. The endothermic peak appeared at 605°C corresponds to the crystallization of Orthophosphate $Cu_3(PO_4)_2$. While the other endothermic peak at 645°C can be attributed to

crystallization of the V_2O_5phase.The formations of $Cu_3(PO_4)_2$and V_2O_5oxide will be confirmed by XRD and FTIR.

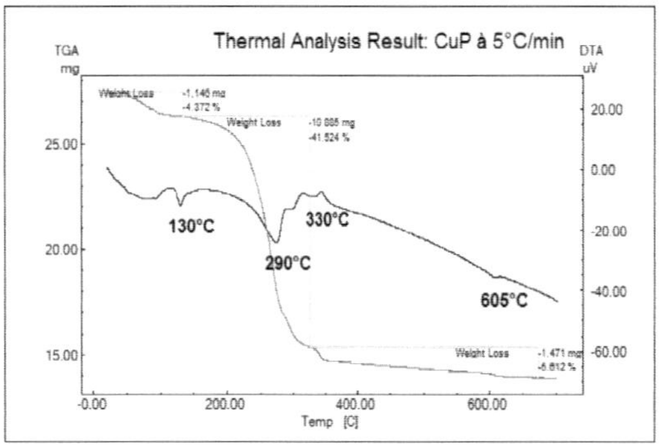

Figure 1.TGA-TDA curves for pure solid CuP

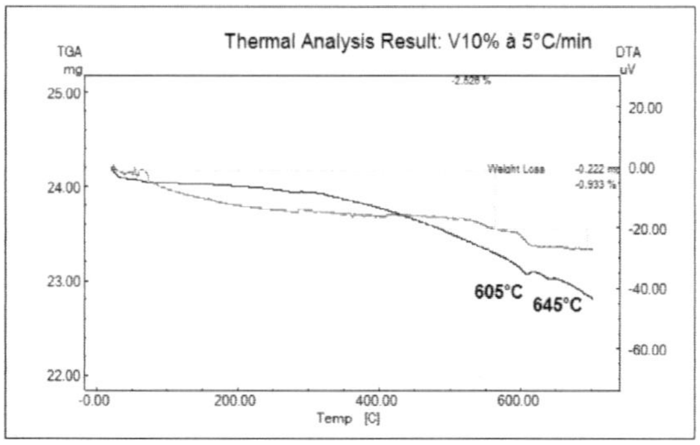

Figure 2.TGA-TDA curves for solid V(12%)/CuP

XRD pattern of the vanadium oxide catalysts supported on copper orthophosphate (V/CuP) calcined at 450°C and 650°C are shown in Figure 3 and figure 4, respectively.

The pure CuP solid treated at 450°C exhibits broad diffraction lines corresponding $Cu_3(PO_4)_2$(PDF Index Name : Copper Phosphate .ref cod: 01-070-0494), and also some weaker diffraction stripes ($2\Theta = 15°$) have been assigned to libethenite $Cu_2(PO_4)(OH)$(PDF Index Name : Copper Phosphate .ref cod: 01-083-1557). Progressively with the increase in temperature at 650°C (Figure 4), diffraction lines of the copper orthophosphate sharpen indicating a better crystallization of $Cu_3(PO_4)_2$(PDF Index Name :Copper Phosphate .ref cod: 01-070-0494).

The solids impregnated with vanadium treated at 450°C lead not only to the previous phase identified for $Cu_3(PO_4)_2$ but also to a new crystalline phase V_2O_5(PDF Index Name :Vanadium Oxide ; ref cod:01-072-0433).At high temperature treatment (650°C), another structure of vanadiumoxide$V_2O_5$is clearly identified (PDF Index Name: VanadiumOxide; ref cod:00-053-0538).This suggests that vanadium oxide is well dispersed over the surface of the copper orthophosphate.

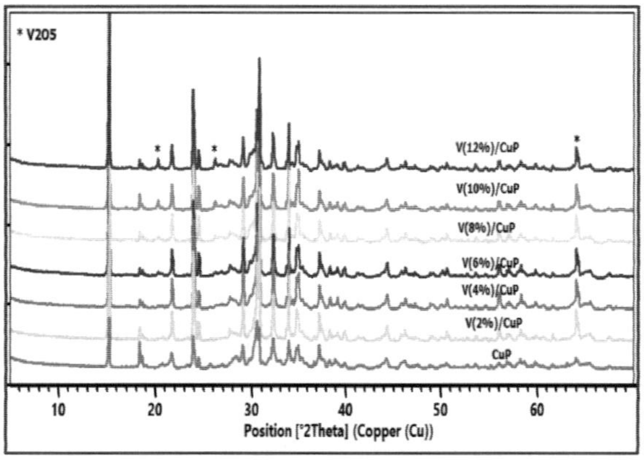

Figure 3.XRD patterns of V/CuP (450°C).

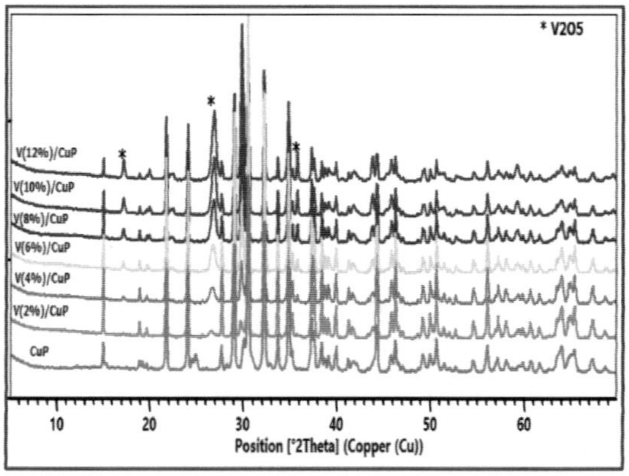

Figure 4.XRD patterns of V/CuP(650°C).

(Vanadium Oxide ; ref cod: 01-072-0433)		(Vanadium Oxide ; ref cod:00-053-0538)	
Crystal system:	Orthorhombic	Crystal system:	Orthorhombic
Space group:	Pmnm	Space group:	P E
Space group number:	59		
a (Å):	11,5100	a (Å):	7,1000
b (Å):	4,3690	b (Å):	6,2940
c (Å):	3,5630	c (Å):	3,5780
Alpha (°):	90,0000	Alpha (°):	90,0000
Beta (°):	90,0000	Beta (°):	90,0000
Gamma (°):	90,0000	Gamma (°):	90,0000

The FTIR spectra of the synthesized V/CuP are shown in Figure 5.For characteristic vibration of CuPcalcined at 650°C, FTIR bands are assigned according to the literature [33, 34] based on the fundamental vibrating unit PO_4^{3-} anion. One of the most noteworthy features of the spectrum is the presence of the strong bands at 1120, 965-935, 608-527 and 406 cm^{-1}. These bands can be assigned to the $\upsilon_{as}(PO_4^{3-})$, $\upsilon_s(PO_4^{3-})$, $\delta_s(PO_4^{3-})$ and δ_s(POP), respectively. The characteristic bands of groups(PO_4^{3-}) refine, divide, which reflects a structural evolution of crystalline compound.
For FT-IR spectra of V/CuP andcomparing with the CuP samples, it is clear that the vibrational bands of vanadate are observed in the regions 800 cm^{-1}, 652 cm^{-1}, 1024 cm^{-1} and 452cm^{-1} and 459 cm^{-1}; These bands can be assigned to the υ_{as}(V-O-V), δ_{as}(V-O-V), to the symmetric stretching V = O bond and the vibration of O-V = O respectively [35-38]. In addition When the V_2O_5 content increased the bands intensity of V/CuP samples increased because the carrier surface was being covered by the active components V_2O_5 oxide.

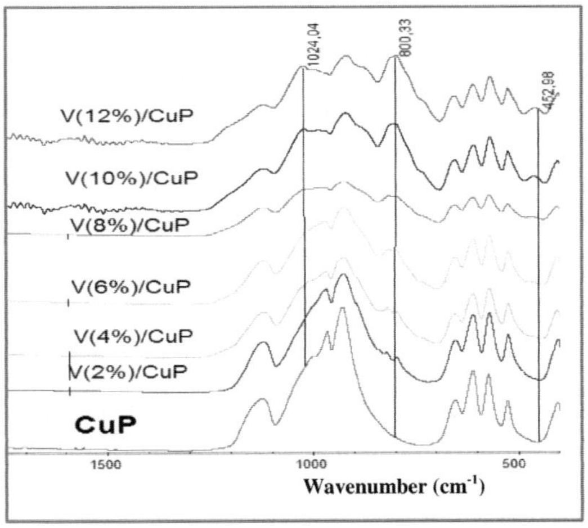

Figure 5.FTIR spectra of V/CuP solids

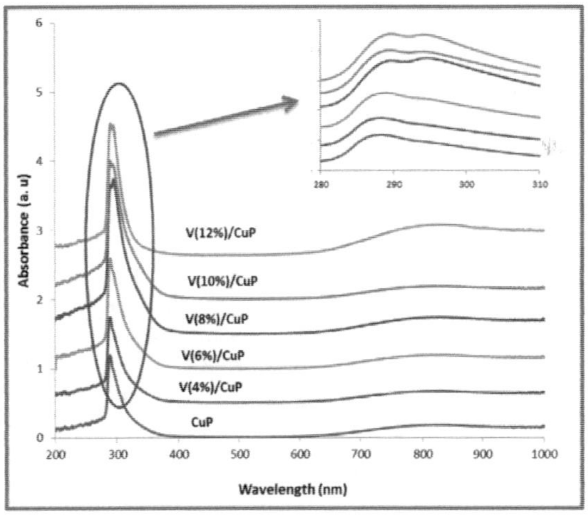

Figure 6. UV–Vis spectra of V/CuP solids.

The UV–Vis spectra of the V/CuP samples are compared in Figure6. All spectra of V/CuP samples exhibit two major CTLM (charge transfer ligand-metal) bands centered at about 289 and 294 nm of P–O, Cu–O and V-O. In addition, the V/CuP samples exhibit a small and broad visible absorption band with a maximum at 800 nm corresponding to d–d transitions ($d^{10}3d^{9}4s^{1}\rightarrow$of Cu^{2+}. This last Cu^{2+} is theonlyknownionhavingthe $3d^9$ configuration [39-43]. Theenergydiagramford^9 system inoctahedral symmetry is just the inverse of that for octahedral d^1 system. It iswellknownthat the spectraofallsolidscontainingcopper ionshaveatleast wide and unsymmetrical broad visible band centered near 800nm [41; 44]. A comparison with the spectra of aqueous and crystalline cupric

complexes shows that the Cu^{2+} ions must be present in approximately octahedral coordination [42; 44; 45]. For the catalysts supported vanadium oxide catalysts, also the strongbands between 200 and 300 nm originating from O^{2-}, V^{5+}charge transfer transitions of low coordinated V^{5+}centers according to previous assignments [46–48].
No appreciable changes of coordination of $Cu^{2+}$and V^{5+}ions composition are detected from UV–visspectra. This is in accordance with results obtained from XRD and IR.

5. Conclusions

The In this work we have reported the synthesis of a pure copper orthophosphate (CuP) and CuP modified by introduction of vanadium (2-12 wt % of V_2O_5) through an easy, cost- and timesaving route, in comparison with previous synthesis of other metal orthophosphates. The XRD and FTIR results confirmed the formation of $Cu_3(PO_4)_2$product and we have demonstrated also the formation of the V_2O_5 on the surface of support. The solids obtained are necessary for theoretical study, applications development, and catalysis reaction.
In perspective, these catalysts V/CuP will be used for the sulfuric acid production and oxidative dehydrogenation of lower alkanes. The Influence of acid–base properties of vanadium catalysts supported on copper orthophosphates in this reaction will be studied by Catalytic decomposition of Isopropyl Alcohol.

References

[1] Averbuch-Pouchot M.T., Durif A., 1996 Topics in Phosphate Chemistry, World Scientific Publishing Co. Pte. Ltd., Singapore,
[2] Silber A., Bar-Yosef B., Levkovitch I., Kautzky L., Minz D., 2008;Kinetics and mechanism of pH-dependent Mn(II) reactions in plant-growth medium, Soil Biology and Biochemistry **401** p2787–2795.
[3] Jouini A., Gaˆcon J.C., Ferid M., Trabelsi-Ayadi M., 2003; Luminescence andscintillation properties of praseodymium poly and diphosphates, OpticMaterials **24**p175–180.
[4] Hsiang H.I., Mei L.T., Lin Y.H., Formation and growth of manganesephosphate passivation layers for NTC ceramics, 2009Journal of Alloys Compounds **48**p723–728.
[5] Adam L., Guesdon A., Raveau B., Unique charge ordering of manganese in a new mixed valent phosphate K_3Mn_3IIMnIII$(PO_4)(H_{0.5}PO_4)2(HPO_4)_2$; 2009;Journal of Solid State Chemistry 182 p2338–2343.
[6] Feng D., Wang C., Cheng W., Li G., Tian S., Liao F., Xiong M., Lin J., 2009 Synthesis, crystal structure, and magnetic properties of $K_4Mn_3(HPO_4)_4(H_2PO_4)_2$, Solid State Sciences, 11 p845–851.
[7] Yuan A., Wu L., Bai S., Ma Z., Huang Tong Z., Standard molar enthalpies of formation for ammonium/3d-transition metal phosphatesNH_4MPO_4, H_2O (M = Mn^{2+}, Co^{2+}, Ni^{2+}, Cu^{2+}), 2008; Journal of Chemistryand Engineering Data 53 p1066–1070.
[8] Jibril B.Y., Al-Zahrani S.M., Abasaeed A.E., Propane oxidative dehydrogenation over metal pyrophosphates catalysts, 2001 Catalysis Letters 74 p145–148.
[9] Boonchom B., Baitahe R., Joungmunkong Z., Vittayakorn N., Grass blade-like microparticles $MnPO_4$, H_2O prepared by a simple precipitation at room temperature, 2010 Powder Technology 203 p310–314.
[10] Qiu G., Gao Z., Yin H., Feng X., Tan W., Liu F., Synthesis of $MnPO_4$, H_2O by refluxing process at atmospheric pressure, 2010 Solid State Sciences 12 p808–813.
[11] Ping Ong S., Jain A., Hautier G., Kang B., Ceder G., Thermal stabilities of delithiated olivine MPO_4 (M = Fe, Mn) cathodes investigated using first principles calculations, 2010 Electrochemistry Communications 12 p427–430.
[12] Boonchom B., 2009 Kinetic and thermodynamic studies of $MgHPO_4$, $3H_2O$ by non-isothermal decomposition data, Journal of Thermal Analysis andCalorimetry 98 p863–871.
[13] Boonchom B., Danvirutai C., 2008 A simple synthesis and thermal decomposition kinetics of $MnHPO_4$, H_2O rod-like microparticles obtained by spontaneous precipitation route, Journal

of Optoelectronics and Advance Materials 10 p492–499.

[14] Brandova D.´, Trojan M., Arnold M., Paulik F., Paulik J., 1988 Mechanism of the dehydration of $CoHPO_4$, $1.5H_2O$, Journal of Thermal Analysis 34 p673–678.

[15] Brandova D., Trojan M., Arnold M., Paulik F., Paulik J., 1988 Mechanism of the dehydration of $CuHPO_4$, H_2O, Journal of Thermal Analysis 34 p1449–1454.

[16] Rosa S., Lundager Madsen H.E., 2010 Influence of some foreign metal ions oncrystal kinetics of brushite $CaHPO_4$, $2H_2O$, Journal of Crystal Growth 45 p2983–2988.

[17] Brandova´ D. Trojan M., Paulik F., Paulik J., 1987 Mechanism of the dehydration of $ZnHPO_4$, H2O, Journal of Thermal Analysis 32 p1923–1928.

[18] Aramendı´a M.A., Borau V., Jime´nez C., Marinas J.M., Romero F.J.,1999 Synthesis and characterization of magnesium phosphates and their catalytic properties in the conversion of 2-hexanol, Journal of Colloid Interface Sciences 217 p288–298.

[19] Trifiro F., Grzybowska B., 1997Vanadia catalysts for selective oxidation of hydrocarbons and their derivatives and references therein special issue of Appl. Catal. A Gen. 157.

[20] Bond G.C., Tahir S.F., 1991Influence of phosphorus and potassium additives on the properties of vanadia/titania, catalystsCatal Today. V 10, Issue 3, 30 September 1991, P 393–395.

[21] Deo G., Wachs I.E., Haber J., 1994; Supported vanadium oxide catalysts: Molecular structural characterization and reactivityproperties;Crit. Rev. Surf. Chem. 4 p141.

[22] Khodakov A., Olthof B., Bell A.T., Iglesia E., 1999;Structure and Catalytic Properties of Supported Vanadium Oxides:Support Effects on Oxidative Dehydrogenation Reactions; J. Catal. 181 p205.

[23] Chen K., Khodakov A., Yang J., Bell A.T., Iglesia E., 1999; Isotopic tracer and kinetic studies of oxidative dehydrogenation pathways on vanadium oxide catalysts; J. Catal. 186 p325.

[24] Concepcion P.,. Lopez Nieto J.M, Perez-ParienteJ.,1995 ; Oxidative dehydrogenation of propane on $VAPO^{-5}$, $V_2O_5/ALPO_4^{-5}$ and V_2O_5/MgO catalysts. Nature of selective sites ; J. Mol. Catal. A: Chem. 97p173.

[25] Banares M.A., Alemany L.J., Granados M.L., Faraldos M., Fierro J.L.G., 1997; Partial oxidation of methane to formaldehyde on silica-supported transition metal oxide catalysts; Catal. Today 33 p73.

[26] Parmaliana A., Fursteri F., Mezzapica A., Scurrell M.S., Giordano N., 1993; Journal of Catalysis; A Basic Approach to Evaluate Methane Partial Oxidation Catalysts; Volume 143, Issue 1, September 1993, Pages 262-274

[27] Herman R.G., Sun Q., Shi C., Klier K., Wang C.B., Hu H., Wachs I.E., Bhasin M.M., 1997; development of active oxide catalysts for the direct oxidation of methane to formaldehyde; Catal. Today 37 p1-14.

[28] Vitry D., Morikawa Y., Dubois J.L., Ueda W., (2003); Mo-V-Te-(Nb)-O mixed metal oxides prepared by hydrothermal synthesis for catalytic selective oxidations of propane and propene to acrylic acid; Appl. Catal. A, 251p 411.

[29] Stobbe-Kreemers A.W., Seede M., Makkee M., Scholten J.J.F., (1995); heteropolyanions as redox components in heterogeneous wacker oxidation catalysts; journal of catalysis issue: 2, V 154, pages: 175 – 186.

[30] Lemonidou A.A., 2001; n-Butane Oxidative dehydrogenation Kinetics over Vanadia Magnesia Catalyst; Appl. Catal. A, 216 p277.

[31] Sadiq M., Bensitel M., Lamonier C., Leglise J., 2008 Solid State Sciences 10 p434-437.

[32] Boonchom B., Vittayakorn N., 2012; Simple fabrication of polyhedral grain-like microparticle $Cu_{0.5}Zn_{0.5}HPO_4{\cdot}H_2O$ and porous structure $CuZnP_2O_7$Ceramics International, 38 p411-415.

[33] Onmura N., Tagawa T., Goto S., 1998 Appl. Catal A: Gen166 p321.

[34] Abadzhieva N., Tzokov P., Uzunow I., Minkov V., Klissurski D., V. Rives;methanol oxidation to formaldehyde on bismuth phosphate-based catalysts;reaction Kinetics and catalysis letters. issue: 2, volume: 53, year: 1994, pages: 413 - 418

[35] Abello L., Husson E., Repelin Y., Lucazeau G., (1983); Vibrational spectra and valence force

field of crystalline V_2O_5; Spectrochim. Acta 39A p641.
[36] P.Ettireddy.Reddy and S.Rajender.Varma, 2004; Preparation, characterization, and activity of Al_2O_3-supported V_2O_5 catalysts; Journal of catalysis catal.221 p93-101.
[37] Frederickson L.D and Hausen D.M., 1963; Infrared Spectra-Structure Correlation Study of Vanadium-Oxygen Compounds;Anal.Chem.35(7) p818.
[38] Agunaou M., 1991., Evolution des proprietes physico-chimiques des oxydes $Bi_xGd_{1-x}VO_4$, catalyseurs d'hydrogenation du dioxyde de carbone., thèse de doctorat, Université de Lille,
[39] Cotton F.A., Wilkinson G., Murrillo C.A., Bochmann M., 1999.Advanced Inorganic Chemistry, 6th ed., John-Wiley-Interscience, New York,
[40] Bates T., in: J.D. Mackenzie (Ed.)1962;Modern Aspects of the Vitreous State, Butterworths, London 2 p195.
[41] Bamford C.R., 1977; Color Generation and Control in Glass;Elsevier Science Publisher, Amsterdam,
[42] Jorgensen C.K., 1957; Recent progress in ligand field theory;Acta Chem. Scand. 11 p73.
[43] Duran A., Fernandez Navarro J.M., 1985 ; The colouring of glass by Cu^{2+} ions; Phys. Chem. Glasses 26 p126.
[44] Cable M., XiangZ.D., 1989 ; Cuprous-cupric equilibrium in soda-lime-silica glasses melted in air;Phys. Chem. Glasses 30 p237.
[45] Morsi M.M., Metwalli E.S., Mohamed A.A., 1999; Modified chemical method for determination of Cu^{+} ions in some Cu containing phosphate glasses and effect of composition on the Cu^{2+}/Cu(tot) ratio; Phys. Chem. Glasses 40 p314.
[46] Wachs I.E., 1996,Raman and IR studies of surface metal oxide species on oxide supports: Supported metal oxide catalysts, Catal. Today 27 p437.
[47] Busca G., 2002; Differentiation of mono-oxo and polyoxo and of monomeric and polymeric vanadate, molybdate and tungstate species in metal oxide catalysts by IR and Raman spectroscopy; Raman Spectrosc 33 p348.
[48] Banares M.A., Wachs I.E., 2002; Molecular structures of supported metal oxides catalysts under different environnements;J. Raman Spectrosc. 33 p359.

IOP Conf. Series: Materials Science and Engineering **186** (2017) 012014 doi:10.1088/1757-899X/186/1/012014

Statistical Mechanics of Confined Biological Materials

R El Kinani[1], M Benhamou[1] and H Kaïdi[2]

[1]Equipe Matériaux Avancés et Applications ENSAM, Moulay Ismail University, P.O. Box 15290, Al Mansour, Meknes, Morocoo
[2]CRMEF, P.O.Box 255, Meknes, Morocco

Email: m.benhamou@ensam-umi.ac.ma, radouane_kin@yahoo.fr

Abstract. In this work, we propose a model to study the Statistical Mechanics of a confined bilayer-membrane that fluctuates between two interactive flat substrates. From the scaling laws point of view, the bilayer-membranes and strings are very similar. Therefore, it is sufficient to consider only the problem of a string. We assume that the bilayer-membrane (or string) interact with the substrate via a Double Morse potential that reproduces well the characteristics of the real interaction. We show that the Statistical Mechanic of the string can be adequately described by the Schrödinger equation approach that we solve exactly using the Bethe method. Finally, from the exact value of the energy of the ground state, we extract the expression of the free energy density as well as the specific heat.

1. Introduction

Cell membranes are crucial in life, because they separate the cell from its environment, and then act as a selective barrier of matter. Descriptive details of the structural organization of bilayer-membranes and basic functions can be found in Refs. [1-8].Cell membranes are formed by a bilayer of phospholipids combined with a variety of proteins and cholesterol. Particularly, the latter provide the fluidity of the bilayer. A phospholipid is an amphiphilic molecule that has a polar hydrophilic head attached to two hydrophobic chains carbonic.

Since their discovery, the phospholipid bilayers have been the subject of numerous studies. They are, in fact, two-dimensional model systems, and are particularly interesting from a fundamental point of view. Fluctuations, phase transitions or shape transitions, membrane fusion and vesiculation are all exciting complex phenomena for the physicist [9-11].On the other hand, the "lipid bilayer" is the basic building block of all living organisms, since it constitutes the frame of the cell wall. The cell membrane is a complex system composed of a mixture of phospholipids, forming a fluid bilayer in which are inserted a number of functional molecules.

In addition, the phospholipids bilayers are also of interest to biologists and biophysicists. This is due, in large part, by the growing attention they have been the subject in recent decades. With the DNA molecule, one can even consider that they are the cause of the rise of “biophysics”. Today, physicists are increasingly called upon to assist in the understanding of life. In this perspective, researchers are trying to work on model systems with the closest possible properties of real membranes. Within the context of the fundamental or biomimetic studies, three types of systems are mainly used: the lamellar phases, vesicles and single bilayer supported on a solid substrate.

Very often, the biological membranes are confined in areas with geometric edges. As a first result, these edges interact with the confined bilayer-membranes and then alter its thermodynamic properties.

In this work, we specifically study these thermodynamic properties, using the approach of the Schrödinger equation.
The remaining of presentation proceeds as follows. Section 2 is devoted to a description of the used model. In Section 3, we present the thermodynamic study of the considered system. Finally, some concluding remarks are drawn in the last section.

2. Description of the model

A bilayer-membrane generally fluctuates near other neighboring bilayer-membranes. Consider then a bilayer membrane confined between two interactive parallel substrates separated by a distance 2L.Since bio-membranes and the strings are governed by similar scaling laws, we simply consider the problem of a confined string in the same geometry (Figure. 1).The string is located at a distance L + z of the substrate (1), and at a distance L - z of the substrate (2).We note that the interaction felt by a point (a) on the confined string is similar to the interaction felt by the hydrogen atom in the O-H-O molecule that is part of the chromic acid (CrO_2H).

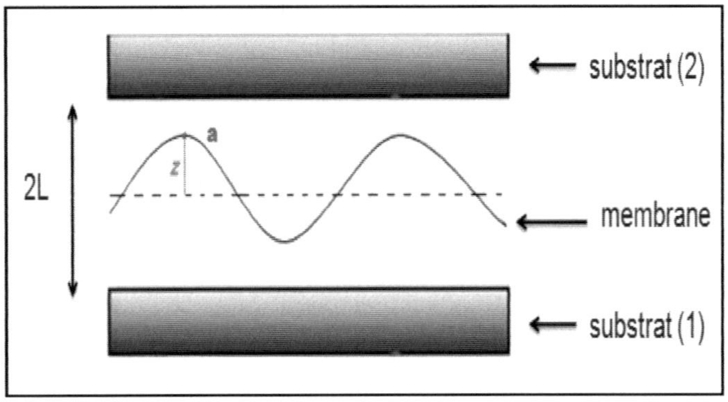

Figure 1. String confined between two substrate

First, we recall the one-dimensional expression of the standard Morse potential [11-13](Figure. 2), for a bilayer-membrane or (string) fluctuating in the vicinity of a substrate,

$$V(z) = D\left[e^{-2\alpha(z-z_m)} - 2e^{-\alpha(z-z_m)}\right] \tag{1}$$

Therefore, this interaction potential depends on three kinds of parameters (D, z_m, α).The parameter $D > 0$represents the potential depth. The second is the minimum potential, while the inverse of the latter, $\alpha > 0$, defines the potential-range. We recall that such a potential has been used in the past for the description of the vibration motion of diatomic molecules.

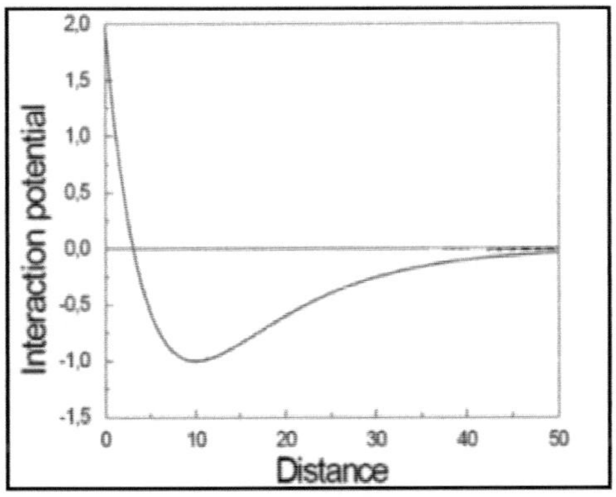

Figure 2. Standard Morse potentiel versus distance

In the case of confined bilayer-membrane or strings, the adopted potential is a Double Morse potential (DMP) (Fig. 3), given by the following expression

$$V(z) = D\left[\frac{A^2}{2}\cosh(2\alpha z) - 2A\cosh(\alpha z)\right], \qquad (2)$$

with$A = 2e^{-\alpha(L-z_m)}$. Here $(D, \alpha)$are the usual parameters of the standard Morse potential, and z is the displacement of a point on the string between the two substrates. We note that the DMP,$V(z)$, admits two minima, if and only if $A < 1$. These minima values are located at

$$z = z_{\pm} = \alpha^{-1}\ln\zeta\,,\ \zeta = \left(1 \pm \sqrt{1 - A^2}\right)/A\,. \qquad (3)$$

The condition$ln\zeta \geq 0$ leads to determining a critical value, L_c, of separation L, namely
$L_c = \alpha^{-1}\ln(2) + z_m$(4)

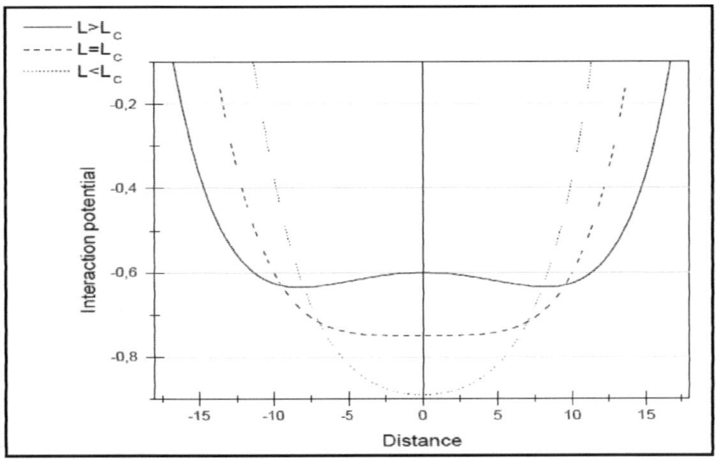

Figure 3. Reduced DMP, V(z)/2D, versus distance

3. hermodynamic study

Without details, we show that the one-dimensional Schrödinger equation describing the system, is given as follows

$$\varphi''(z) + \frac{2\sigma}{(k_B T)^2}\left[E - D\left(\frac{A^2}{2}\cosh(2\alpha z) - 2A\cosh(\alpha z)\right)\right]\varphi(z) = 0 \quad (5)$$

Here,σis the tension of the string, $T$is the absolute temperature, and k_B is the Boltzmann's constant.In this equation, $E$is energy and $\varphi(z)$ is the associated wave-function which satisfies the condition: $\varphi(\pm\infty) = 0$. We find that the solution of this equation is given by

$$\varphi(u) = e^{-Au\sqrt{\gamma}}\phi(u)\,, u = \cosh(\alpha z) \quad (6)$$

where the function ϕ solves the following differential equation,

$$(u^2 - 1)\phi''(u) + (-2A\sqrt{\gamma}u^2 + u + 2A\sqrt{\gamma})\phi'(u) + Au(2\gamma - \sqrt{\gamma})\phi(u) = \left(\frac{\gamma A^2}{2} - \xi\right)\phi(u) \quad (7)$$

with the notations

$$\gamma = \frac{2\sigma D}{\alpha^2(k_B T)^2}, \xi = \frac{2\sigma E}{\alpha^2(k_B T)^2} \quad (8)$$

We show that this equation is almost exactly soluble, with

$$\phi(u) = \prod_{i=1}^{n}(u - u_i) \quad (9)$$

Where the u_i, are the roots of certain known polynomials. So, the exact values of the energies associated with these wave functions are given by:

$$\xi_n = 2A\sqrt{\gamma}\sum_{i=1}^{n} z_i + \frac{\gamma A^2}{2} - n^2 \quad (10)$$

with $\sqrt{\gamma} = (n + 1/2)$

The value n = 0 corresponds to the ground state,

$$E_0 = \frac{(k_B T)^2\alpha^2 A^2\gamma}{4\sigma}, \varphi_0 \sim e^{-Au\sqrt{\gamma}} \quad (11)$$

The free energy density

$$f = E_0 = \frac{(k_B T)^2\alpha^2 A^2\gamma}{4\sigma} \quad (12)$$

This energy we extract the value of the specific heat of the string,

$$C = -T\frac{\partial^2 f}{\partial T^2} = \frac{2}{\sigma}\left(\frac{k_B\alpha A}{4}\right)^2 T \quad (13)$$

Then the specific heat varies linearly as a function of temperature.

4. Conclusion

In conclusion, this model has allowed us to determine the thermodynamics of contained bio-membrane, using the approach of the Schrödinger equation, by adopting a Morse Double potential. The exact solution of this equation gave us free energy, from which we extracted the specific heat.

Finally, we plan to extend this study to the case of a pair of bilayer membranes and to study the effect of confinement on their relocation transition.

References

[1] Bretscher MS and Munro S 1993*Sci.***261** 5126 1280–81
[2] Gauthier NC Rossier OM, Mathur A, Hone JC and Sheetz MP 2009*Mol. Biol. Cell.* **20**3261–72
[3] Edidin M1997*Curr. Opin. Struct. Biol.***7** 528–32
[4] Hackenbrock CR1981*Trends Biochem Sci.***6**151–54
[5] Tanford C 1980 The *Hydrophobic Effect: Formation of Micelles and Biological Membranes*(New York Wiley-Interscience)
[6] Vance D E and Vance J E 1996*Biochemistry of lipids, lipoproteins and membranes* (Elsivier) pp 1-33
[7] ZhangFLee G M and JacobsonK1993*Bioessays***15** 9 579–88
[8] Pope CA III ThunMJ Namboodiri MM Dockery DWEvans JSSpeizer FE and Heath CW 1995*Am. J. Respir Crit. Care Med.***151** 669–74
[9] LipowskyR 1993*Biophys. J.* **64**1133
[10] LipowskyR 1991*Nature***349** 475–81
[11] Asfaw M2008. *Physica* A **387**3526–36
[12] Morse P M1929*Phys. Rev.***34** 57
[13] Huber K P and Hezberg G1950*Molecular Spectra and Molecular Structure: I. Spectra of Diatomic Molecules* (New York Springer Science)

IOP Conf. Series: Materials Science and Engineering **186** (2017) 012015 doi:10.1088/1757-899X/186/1/012015

Stark effect of shallow donor impurities in HgS Inhomogeneous Quantum Dots

S M'zerd[1], K Rahmani[2], S Janati[1], Y Chrafih[2] ,I Zorkani[1] and A Jorio[1].

[1] Solid Group of Nanomaterials and Renewable Energies, Laboratory of Solid State Physics, Faculty of Sciences Dhar el Mahraz, University Sidi Mohamed Ben Abdellah, Atlas Fez, Morocco
[2] LIRST, FP, USMS, Beni Mellal, Morocco

E-mail: safae.mzerd@gmail.com

Abstract. Using the variational method, within the effective mass approximation, we have calculated the binding energy and the Polarizability of a shallow donor confined to move in [CdS/HgS/CdS] Inhomogeneous Quantum Dots, in the presence of a uniform electric field. We consider an infinitely deep well and we present our results as function of the size of the well and for several values of the electric field strength: (i) The Polarizability decreases when the electric field increases. We find that the Polarizability it is more influenced by the quantum confinement than by the electric field. (ii) The binding energy depends on the inner and the outer radius of the IQD, decrease when the electric field increases and depends strongly on the donor position, (iii) We have demonstrated the existence of a critical value of radius ratio which can be used to distinguish the tree dimension confinement from the spherical surface confinement.

1. Introduction

In the past ten years, studies on the impurity states in semiconductors nanostructures, such as quantum well (QW), quantum well wire (QWW) and quantum dots (QD), have become subjects of extensive investigations both in basic and applied researches [1-5]. Recently, a new class of quantum dots called Quantum-Dots Quantum-well (QDQW) or Inhomogeneous quantum dots (IQD), have been studied both theoretically and experimentally [6-9]. IQD are composed of two semiconductor materials one of which, with the smaller band gap, is embedded between a core and outer shell of the material with the larger band gap. El Ghazi et al [10] have used the variational approach within the framework of the effective mass approximation to calculate the ground- state Hydrogenic shallow-donor binding energy in InGaN/GaN SQD-QW as a function of the ratio of the inner and the outer radius. The effect of electric field is investigated for two confinement regimes. Rezaei et al [11] have calculated the binding energy and self-polarization of hydrogenic impurity confined in a finite confining potential square quantum well wire, under the action of external electric field and hydrostatic pressure. In a previous work, Zounoubi et al [5] have calculated the binding energy and the polarizability of shallow donor impurity in cylindrical QD under applied magnetic field. The ground electron state and the binding energy of a shallow donor impurity in a ZnS/CdSe core–shell QD under applied electric fields were calculated by Cristea et al [12] taking into account the dielectric mismatch at the dot interface.

In the present work, we use a variational method to calculate the shallow donor binding energy in IQD. We will examine the effects of the donor position, the electric field on the binding energy and polarizability. This paper is organized as follows: in section 2, we present the general formalism; we deduce the analytical values of the binding energy and polarizability as a function of the radius of IQD. The numerical results and discussions are presented in section 3.

2. General formalism

We consider a shallow donor located at the position r0 in a Inhomogeneous Quantum Dots made out of [CdS (Core)/HgS (Well)/CdS (Shell)]. In the effective mass approximation, the Hamiltonian of the system is written as:

$$H = -\nabla^2 - \frac{2}{|r - r_0|} + f(r\cos\theta - r_0) + V(r) \tag{1}$$

Where the confining potential V(r) is assumed to an infinitely deep well:

$$\mathrm{V(r)} = \begin{cases} 0, & \mathrm{R_1} < r < \mathrm{R_2} \\ \infty, & r < \mathrm{R}_1 \text{ and } \mathrm{r} > \mathrm{R_2} \end{cases} \tag{2}$$

And

$$|\vec{r} - \vec{r}_0| = \sqrt{r^2 + r_0^2 - 2 r r_0 \, Cos\theta} \tag{3}$$

R_1 and R_2 are the inner and the outer radius of the IQD respectively. We use the effective Bohr radius $a^* = \frac{\hbar^2 \varepsilon_0}{m^* e^2}$ and the effective Rydberg $R^* = \frac{m^* e^4}{2\hbar^2 \varepsilon_0^2}$ as the units of length and energy (respectively).

ε_0 is the dielectric constant, m* is the electron band effective mass and $f = \frac{e a^* F}{R^*}$ is a dimensionless measure of the electric field F. We use a variational method approach to determine the ground state, we adopt the wave function given by:

$$\psi(\vec{r}) = \psi_0(\vec{r})\,[1 + \beta f(r\cos\theta - r_0)] \tag{4}$$

Where β is a variational parameter which takes into account the presence of the electric field and $\psi_0(\vec{r})$ is the wave function in the absence of the electric field (f = 0) given by:

$$\Psi_0(\vec{r}) = \frac{\sin[K(r - R_1)]}{r} e^{-\alpha|\vec{r} - \vec{r_0}|} \tag{5}$$

α is a variational parameter and $K = \frac{\pi}{R_2 - R_1}$. The exponential factor exp-$\alpha|\vec{r} - \vec{r}_0|$ describes the Coulomb spatial interaction. The energy is obtained by the minimization with respect to the variational parameter α , The binding energy E_b of the donor is defined as the ground state energy of the system without Coulomb term minus the ground state energy in the presence of the Coulomb term:

$$E_b = E_{Sub} - E_t \tag{6}$$

The Polarizability can be calculated from the dipole moment and is defined as:

$$\alpha = \frac{P}{F} \tag{7}$$

Where the dipole moment is given as :

$$P = -e(\langle|\vec{r} - \vec{r}_0|\rangle_{f \neq 0} - \langle|\vec{r} - \vec{r}_0|\rangle_{f=0}) \tag{8}$$

3. Results and discussion

In this section we will discuss the results obtained by applying the direct variational method for calculating the binding energy of a shallow donor impurity in the absence and presence of an electric field in IQD and the Polarizability by the formula dipole moment P. Numerical calculations are applied to a Inhomogeneous Quantum Dots made out of [CdS/HgS/CdS]. The physical parameters characterizing the system are given in the table 1:

Table 1. The physical parameters characterizing the system

Material	m^*/m_0	ε_0	ε_∞	a*(A°)	R*(eV)	Eg(eV)
Cds	0,2	9,1	5,5	5,818	3,65	2,5
HgS	0,036	18,2	11,36	5,851	5,0	0,5

3.1 Absence of the electric field

In Figure 1 (a) and (b), we plot the variations of the donor binding energy E_b as a function of the ratio $\frac{R_1}{R_2}$ and for different values of the outer radius R_2. The ratio $\frac{R_1}{R_2}$ varies between 0 and 1. First, we analyze the two particular cases, $\frac{R_1}{R_2}=0$ and $\frac{R_1}{R_2} \approx 1$.In the first case ($\frac{R_1}{R_2} = 0$), the situation correspond to a HQD. In Figure 1 (a) we present the variations of the binding energy E_b of a donor placed in the center of a spherical layer, as a function of the ratio ($\frac{R_1}{R_2}$) and for different values of the outer radius R_2>0.7a*.We remark that when the ratio $\frac{R_1}{R_2}$ tend to 1, the donor binding energy tends to the limit of 4R* which represents the well-known tow dimensional donor binding energy. We also remark that the binding energy E_b presents a minimum for a critical value of the ratio $\left(\frac{R_1}{R_2}\right)_{crit}$ depending of the value of R_2. For $\frac{R_1}{R_2}$ equal to zero, E_b recovers the limit corresponding to a HQD. When the core radius R_1 increases, the donor is confined to hollow nanosphere, its confinement becomes less important and its binding energy decreases to a minimal value. For a critical value $\left(\frac{R_1}{R_2}\right)_{crit}$ there is equilibrium between the surface confinement effects and the tree dimensional confinement effects. For one value of the ratio $\frac{R_1}{R_2}$ greater than $\left(\frac{R_1}{R_2}\right)_{crit}$, the surface confinement effects becomes predominant and the binding energy increases. We also notice that the minimum of the binding energy moves toward small values of $\left(\frac{R_1}{R_2}\right)_{crit}$ when the size R_2 of the dot increases and vanishes for large values of R_2.This situation is explained by the fact that in the limit $R_2 \rightarrow \infty$, the quantum dot behavior is similar to the bulk case. When $\frac{R_1}{R_2}$ increases from 0 to 1, the binding energy increases monotonously from the value corresponding to the bulk (3D) to the value corresponding to the surface (2D). In Figure 1(b), we plot the variations of the donor binding energy E_b against the ratio $\frac{R_1}{R_2}$ for different values of the outer radius R_2<0.8 a*. We see that the binding energy decreases monotonously when the ratio $\frac{R_1}{R_2}$ increasing from 0 to 1. Indeed, when the inner radius R_1 increases, the donor binding energy moves from the HQD situation where the kinetic energy is dominating to a small spherical surface where the coulomb energy prevails. We note that all the curves tend to the values which are in good agreement with the results obtained in reference [7]. The second cases ($\frac{R_1}{R_2}$= 1) correspond to a narrow HgS well.

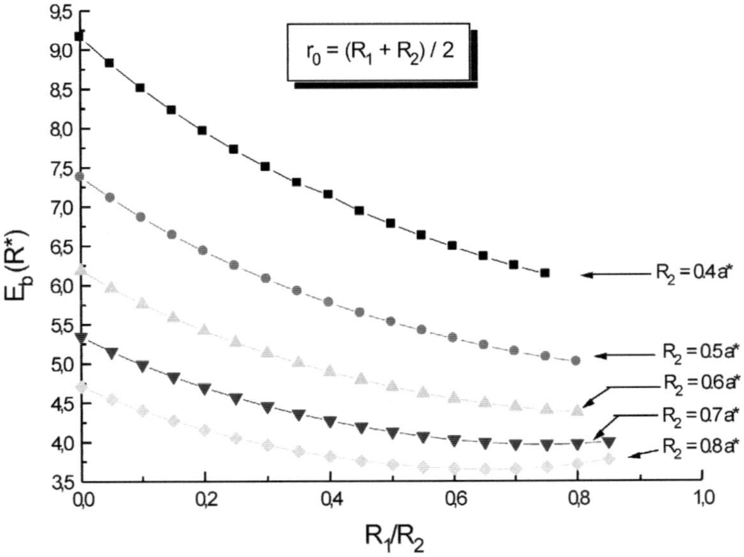

(a)

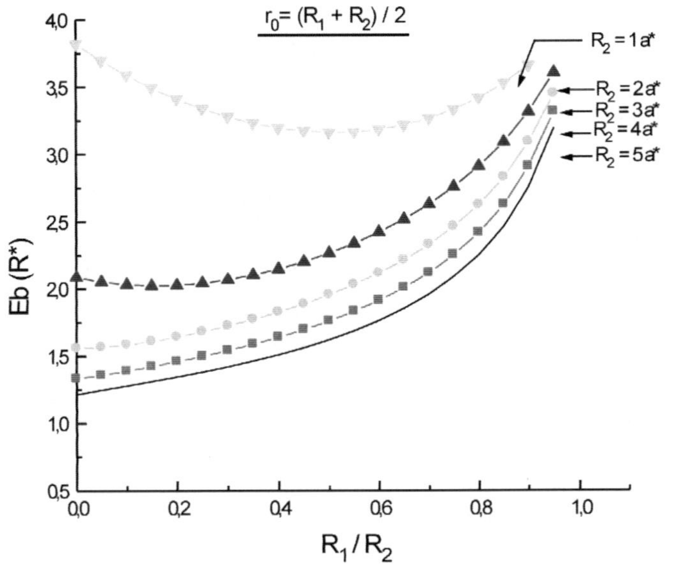

(b)

Figure 1. Variations of the binding energy E_b as a function of the ratio R_1/R_2 for different values of the outer radius of the IQD, (a) $R_2 = 0.4, 0.5, 0.6, 0.7$ and 0.8 a* and (b) $R_2 = 1, 2, 3, 4$ and 5 a*.

In Figure 2, we fix the outer radius at $R_2 = 3$ a* and we plot the variations of the donor binding energy E_b as a function of the donor position r_0 and for different values of the inner radius R_1 ($R_1 = 0.5$ a*, 1 a*, 1.5 a*, 2 a*). We remark that the binding energy presents a maximum when the donor is placed on the center of the spherical layer $r_0 = \frac{R_1+R_2}{2}$ .When the donor is placed at the center of the spherical layer, the orbital electronic presents the spherical symmetry, the energy is maximal and decreases when the donor moves toward extremities of the spherical layer. This behavior had been reported for HQD [13].

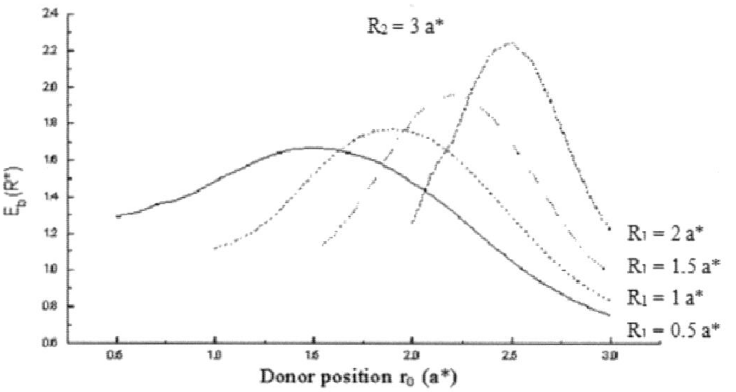

Figure 2. Variations of the binding energy E_b as a function of the donor position r_0 for different values of the inner radius of the IQD: $R_1 = 0.5$, 1, 1.5 and 2 a* where $R_2 = 3$ a*.

3.2 Presence of the electric field

In Figure 3, we present the binding energy E_b as a function of the ratio R_1/R_2 for two different confinement regimes R_2= 1 a* and R_2= 2 a* and for two several values of the electric field f = 0, 0.4 and 0.8. We remark that when the electric field increases the binding energy decreases and the effect of the electric field becomes more important when the ratio R_1/R_2 offers toward 1 or the outer radius of the IQD increases. We can explain this result by the fact that the confinement decreases.

In Figure 4, we fix the outer and the inner radius of the IQD respectively at $R_2 = 3$ a* and $R_1 = 1$ a* and we plot the binding energy E_b as a function of the donor position r_0 for two different values of the electric field f = 0 and 0.4. We notice that the electric field reduced the binding energy. Its effect is more pronounced when the impurity is placed on the center of the spherical layer ($r_0 = \frac{R_1+R_2}{2}$)and decreases when the donor moves toward extremities of the spherical layer.

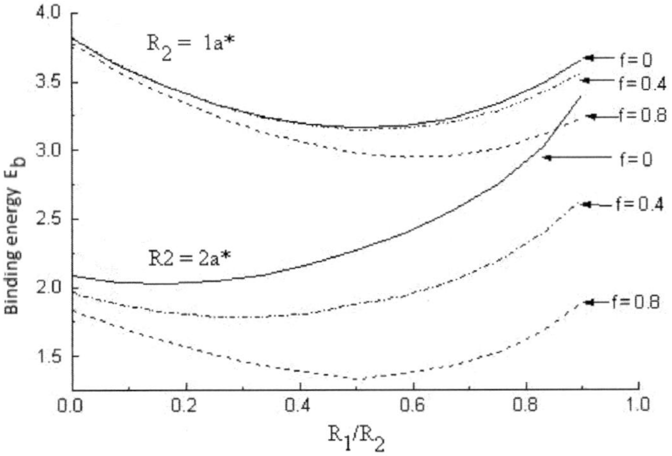

Figure 3. Variations of the binding energy E_b as a function of the ratio R_1/R_2 for two different values of the outer radius of the IQD, R_2= 1 a* and 2 a* and two values of the electric field f = 0.4 and 0.8. The solid curve corresponds to the binding energy variation in the absence of electric field.

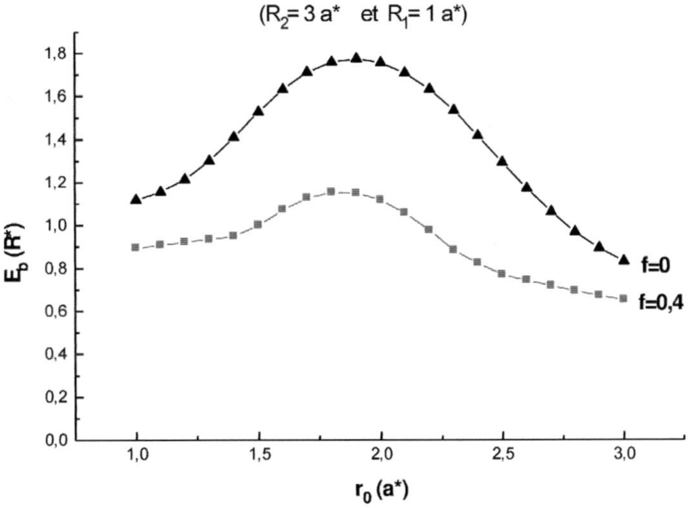

Figure 4. Variations of the binding energy as a function of the impurity position for two different values of the electric field (IQD).

We present in figure 5, the Polarizability of a donor placed in a HQD as a function of the dot radius R_2 and for different values of the electric field. We note that the Polarizability decreases with increasing the intensity of the electric field. We remark also that the Polarizability decreases with the size of the quantum dot. The results show that the Polarizability decreases as the radius of the dot decreases for both cases. For larger quantum dots, the electronic confinement is negligible and therefore the results tend to the case of HgS bulk. The Polarizability decreases when the intensity of the electric field increases.

The Polarizability of a donor as a function of the position of r_0/R_2 HQD is plotted in figure 6. We show that the Polarizability increases with the position until it reaches the report $r_0/R_2 \approx 0.5$. For $r_0/R_2 > 0.5$ Polarizability decreases with increasing the size of the dot. This implies that the confinement effect is more significant than the effect of the electric field. Its effect is more pronounced when the impurity is placed between $r_0/R_2 \approx 0.5$ to 0.7 and decreases as the donor moves to the ends of the dot. We also note that the Polarizability increases when we increase the electric field.

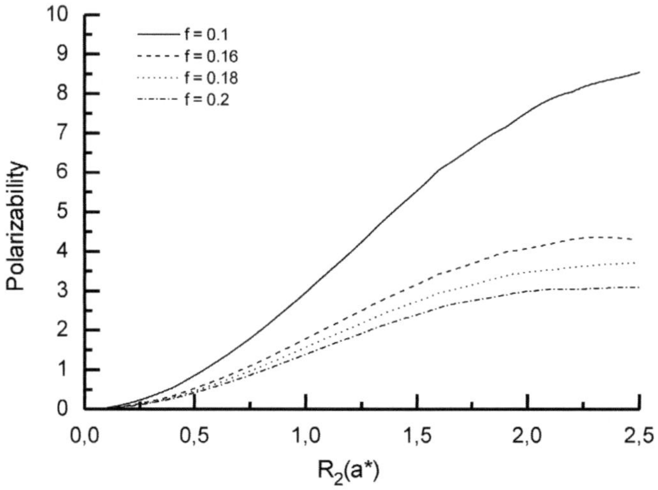

Figure 5. Polarizability of an impurity function of the radius R_2 for different values of the electric field f = 0,1, f = 0.16, f = 0 .18 et f = 0.2

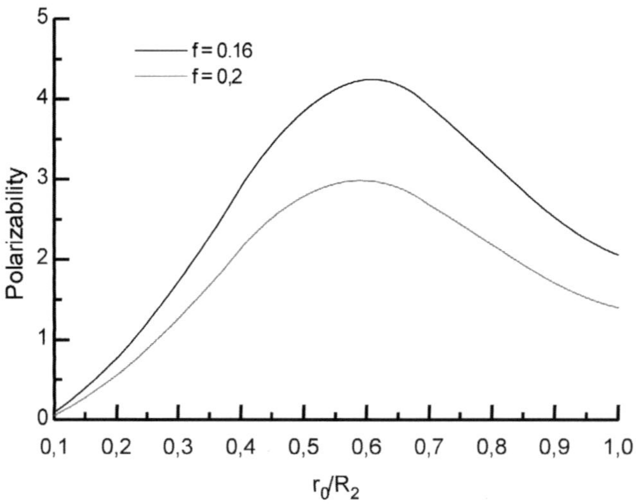

Figure 6. Polarizability of an impurity donor function deposition r_0 /R_2 for two values of the electric field.

4. Conclusion

In conclusion, we have studied the donor binding energy and the Polarizability in IQD. The calculation was performed within the effective mass approximation and using the variational method.

The results show that (i) The Polarizability decreases when the electric field increases. We find that the Polarizability it is more influenced by the quantum confinement than by the electric field. (ii) The binding energy depends strongly on the core and the shell radius, we have illustrate the existence of a critical value which may be important for the nanofabrication techniques. This value may also be used to distinguish the tree dimensional confinement from the spherical surface confinement. (iii) The binding energy depends strongly on the donor position. We demonstrated that it is maximal when the donor is placed in the center of the HQD or IQD and the effect of the electric field reduces the binding energy. The calculations of the Polarizability of a donor in Inhomogeneous quantum dots IQD are in progress.

Acknowledgments

The authors gratefully acknowledge the Abdus Salam International Centre for Theoretical Physics, Trieste-Italy, where this work was finally completed, for very kind hospitality and its support in the preparation of the results.

References

[1] El Said M and Tomak M 1990 *Physics Review B* **42** 3129-3131
[2] Silva-valencia J and Porras-Montenegro N 1997 *Journal of Applied Physics* **81** 901-904
[3] Zorkani I and Fillali L 1999 *Phy Status Solidi (b)* **215** 999-1004
[4] Zounoubi A, El Messaoudi K, Zorkani I and Jorio A 2001 *Superlatt Microstruct* **30** 189-200
[5] Zounoubi A, Zorkani I, Messaoudi K and Jorio A 2003 *Physics Letters A* **312** 220-227
[6] H Cheng-Ying 2000 *Chinese Journal of Physics* **38** 478-490
[7] Khamkhami J, Feddi E, Assaid A, Dujardin F, Stébé B and Diouri J 2001 *Phys Low Dim Struct* **9** 131-142
[8] Ying L, Jing X H, Kun M B, Qiang L J and Lin Z J 1999 Chin *Phys Lett* **16** 680- 682
[9] Chaudhuri S *Phys Res B* 1983 **28** 4480- 4487
[10] El Ghazi H, Jorio A and Zorkani I, 2013 *Physica B* **422** 47–50
[11] Rezaei G, Mousavi S and Sadeghi E 2012 *Physica B* **407** 2637–2641
[12] Cristea M and Niculescu E C 2013 *Physics Letters A* **377** 1221-1226
[13] Assaid E, Feddi E, Khaidar M, Dujardin F and Stébé B 2001 *Physica scripta* **63** 329-335

IOP Conf. Series: Materials Science and Engineering **186** (2017) 012016 doi:10.1088/1757-899X/186/1/012016

Effect of magnesium substitution on the magnetic properties of diluted magnetic spinels $Ni_{1-x}Mg_xFe_2O_4$

S Salmi[1], M El Azhari[1], A El Grini[1], A Hourmatallah[2], N Benzakour[1], K Bouslykhane[1] and A Marzouk[2]

[1] Laboratoire de Physique du Solide-USMBA-FSDM-Fès
[2] Equipe de Physique du Solide-USMBA-ENS

Email: saidsalmi@hotmail.com

Abstract. The magnetic properties of Ni-Mg ferrites with compositions $Ni_{1-x}Mg_xFe_2O_4$, have been studied using the mean field theory and high temperature series expansion theory (HTSE), extrapolated with the Padé approximants method. The nearest neighbour super-exchange interactions for intra-sites and inter-sites of the Ni-Mg ferrites in the range $0 \leq x \leq 1$ have been computed using the distribution method of magnetic cations. The transition temperature T_C is calculated as a function of Mg concentration. The critical exponent associated with the magnetic susceptibility was then deduced. The obtained results are in good agreement with experimental results and critical exponent values are consistent with those suggested by the universality hypothesis.

1. Introduction

Ni ferrite ($NiFe_2O_4$) and substituted ones are technologically important materials, which have been studied in many experimental and theoretical works [1-4]. $NiFe_2O_4$ is a typical soft ferromagnetic material crystallizing in a completely inverse spinel structure with all nickel ions located in the B-sites and ferrite ions occupying both A-sites and B-sites [5]. The compound, thus can be represented by the formula $(Fe^{3+})_A[N^{i2+}Fe^{3+}]_BO_4^{2-}$. $MgFe_2O_4$ has well known spinel structure [6]. The cations Mg^{2+} and Fe^{3+} are unequally distributed among the tetrahedral (A) and octahedral (B) lattice sites. The cations distribution in $MgFe_2O_4$ indicated that this ferrite is 86 % inverted as given by neutron scattering [7] and Mossbauer [8], for example ($Fe_{0.86}$ $Mg_{0.14}$) in A and ($Mg_{0.86}$ $Fe_{1.14}$) in B and this remained unchanged in the temperature range 90 –700K,with $T_c = 665$ K [9].

The objective of this paper is to study the magnetic behavior of spinel compounds situated between the pure inverse spinel compounds $NiFe_2O_4$ and $MgFe_2O_4$ [10].The high temperature series expansions (HTSEs) extrapolated with the padé approximants (PAs) method are given to determine the critical T_C versus x and the critical exponent associated with magnetic susceptibility γ.

2. Theoretical methodologies

2.1. Exchange interactions calculations

Magnetic exchange interactions, inter and intra-sublattices (J_{AB}, J_{BB} and J_{AA}) in $[Mg_{(0.14x)}Fe_{(1-0.14x)}]_A[Ni_{(1-x)}Mg_{(0.86x)}Fe_{(1+0.14x)}]_BO_4$ diluted spinels, with different Mg concentrations, in the range $0 \leq x \leq 1$, were calculated. A probability law based on distribution of ions in A and B sublattices, given by neutron scattering (**Table 2**) [7] has been used. The interactions $J_{AA}^{Fe^{3+}-Fe^{3+}}$, $J_{AB}^{Fe^{3+}-Ni^{2+}}$, $J_{AB}^{Fe^{3+}-Fe^{3+}}$, $J_{BB}^{Fe^{3+}-Fe^{3+}}$, $J_{BB}^{Fe^{3+}-Ni^{2+}}$, and $J_{BB}^{Ni^{2+}-Ni^{2+}}$, in the inverted pure spinel, given by Srivastava et al. [11], have also been used. The exchange interactions between B sites, between A and B sites and between A sites in the range $0 \leq x \leq 1$ are given by:

$$J_{AA}(x) = (1-0.14x)^2 J_{AA}^{Fe^{3+}-Fe^{3+}}$$

$$J_{AB}(x) = \frac{1}{2}\left[(1+0.14x)(1+0.14)J_{AB}^{Fe^{3+}-Fe^{3+}} + (1-x)(1-0.14x)J_{AB}^{Fe^{3+}-Ni^{2+}}\right]$$

$$J_{BB}(x) = \frac{1}{4}\Big[(1+0.14x)^2 J_{BB}^{Fe^{3+}-Fe^{3+}} + (1-x)(1+0.14x)J_{BB}^{Fe^{3+}-Ni^{2+}}$$

$$+(1-x)^2 J_{BB}^{Ni^{2+}-Ni^{2+}}\Big] \quad (1)$$

Table 1. Exchange interaction ion-to-ion for the pure component $NiFe_2O_4$ [12].

$J_{AA}^{Fe^{3+}-Fe^{3+}}$	$J_{AB}^{Fe^{3+}-Fe^{3+}}$	$J_{AB}^{Fe^{3+}-Ni^{2+}}$	$J_{BB}^{Fe^{3+}-Ni^{2+}}$	$J_{BB}^{Fe^{3+}-Fe^{3+}}$	$J_{BB}^{Ni^{2+}-Ni^{2+}}$
-14	-28	-27,4	-9	-10	+29.0

From **Table 1**, one can see that all exchange interactions are antiferromagnetic except J_{BB}^{Ni-Ni} which is ferromagnetic. Furthermore, when increase in the Mg content all the interactions decrease except J_{BB}^{Fe-Fe} which increases (**Table 2**). However, $J_{AB}(x)$ is stronger than the other two, which means that the interactions between A and B are higher than the interactions inside the sublattices A and B, but their value decreases with increase in Mg concentration x. This is due to the decrease in Ni concentration in B sites, and to the replacement of the magnetic ion Ni^{2+} by the non-magnetic ion Mg^{2+}.

2.2. High-temperature series expansion (HTSE)

The Hamiltonian that describes the $[Mg_{(0.14x)}Fe_{(1-0.14x)}]_A[Ni_{(1-x)}Mg_{(0.86x)}Fe_{(1+0.14x)}]_BO_4$ spinel, under an external field h_{ex} within the Heisenberg model ,may be written as follows:

$$H = -J_{AA}^{Fe-Fe}\sum_{\langle ii'\rangle}\vec{S_i}\vec{S_{i'}} - J_{BB}^{Fe-Fe}\sum_{\langle jj'\rangle}\vec{S_j}\vec{S_{j'}} - J_{BB}^{Fe-Ni}\sum_{\langle jj'\rangle}\vec{S_j}\vec{\sigma_{j'}}$$
$$-J_{BB}^{Ni-Ni}\sum_{\langle jj'\rangle}\vec{S_j}\vec{S_{j'}} - J_{AB}^{Fe-Fe}\sum_{\langle ij\rangle}\vec{S_i}\vec{S_j} - J_{AB}^{Fe-Fe}\sum_{\langle ij\rangle}\vec{S_i}\vec{\sigma_j}$$
$$-\mu_B h_{ex}\left(g^{Fe}\sum_{i\in A}S_i^z - g^{Fe}\sum_{j\in B}S_j^z - g^{Ni}\sum_{j\in B}\sigma_j^z\right) \quad (2)$$

Where $\vec{S}$ and $\vec{\sigma}$ are spin vectors of $\vec{S}^2 = S(S+1)$ and $\vec{\sigma}^2 = \sigma(\sigma+1)$ for Fe and Ni, respectively, $g^{Fe} = 2.091$ and $g^{Ni} = 2.183$ are the corresponding gyromagnetic factors. The symbol <...> denotes summation over nearest neighbours. J_{AA}, J_{BB}, and J_{AB} are the intra-and the inter-sublattice exchange interactions in ferrimagnetic spinels.

The magnetization of the system is given by:

$$M = \mu_B\left(g^{Fe}\sum_{i\in A}\langle S_i^z\rangle - g^{Fe}\sum_{j\in B}\langle S_j^z\rangle - g^{Ni}\sum_{j\in B}\langle \sigma_j^z\rangle\right) \tag{3}$$

Following the process in [12,13], we obtained the general expression of susceptibility for the collinear inverse ferrimagnetic spinel as follows.

$$\chi = \left(\frac{\mu_B^2}{3k_BT}\right)\Bigg[N_A^{Fe}g_A^2\vec{S}^2 + N_B^{Fe}g_A^2\vec{S}^2 + N_A^{Ni}g_B^2\vec{\sigma}^2 + N_A^{Fe}g_A^2\Gamma_{AA}\vec{S}^2$$
$$+\frac{N_B^{Fe}(N_B^{Fe}-1)g_A^2\Gamma_{BB}\vec{S}^2}{(N_B^{Fe}+N_B^{Ni}-1)} + 2\frac{N_B^{Fe}N_B^{Ni}g_Ag_B\Gamma_{BB}\vec{S}\vec{\sigma}}{(N_B^{Fe}+N_B^{Ni}-1)} + \frac{N_B^{Fe}(N_B^{Fe}-1)g_B^2\Gamma_{BB}\vec{\sigma}^2}{(N_B^{Fe}+N_B^{Ni}-1)}$$
$$-2\left(\frac{N_A^{Fe}N_B^{Ni}g_Ag_B\Gamma_{AB}\vec{S}\vec{\sigma}}{(N_A^{Fe}+N_A^{Ni})} + \frac{N_B^{Fe}N_B^{Fe}g_A^2\Gamma_{BB}\vec{S}^2}{(N_A^{Fe}+N_A^{Ni})}\right)\Bigg] \tag{4}$$

Where $N_{A(B)}$ are the number of magnetic ions both A and B sites, and $S = 5/2$, $\sigma = 1$ are the spin values of Fe^{3+} and Ni^{2+}, respectively. The spin correlation functions Γ_{AA}, Γ_{BB} and Γ_{AB} in terms of powers of $\beta = 1/k_BT$ (k_B is the Boltzmann constant) and mixed powers of couplings have been computed in ref. [14].

$$\Gamma_{AA} = \sum_{q=1}^{7}\sum_{m=0}^{q}\sum_{n=0}^{q-m}\sum_{p=0}^{q-m-n} a(m,n,p,q)J_{BB}^m J_{AB}^n J_{AA}^p \beta^q$$
$$\Gamma_{BB} = \sum_{q=1}^{7}\sum_{m=0}^{q}\sum_{n=0}^{q-m}\sum_{p=0}^{q-m-n} b(m,n,p,q)J_{BB}^m J_{AB}^n J_{AA}^p \beta^q$$
$$\Gamma_{BA} = \sum_{q=1}^{7}\sum_{m=0}^{q}\sum_{n=0}^{q-m}\sum_{p=0}^{q-m-n} c(m,n,p,q)J_{BB}^m J_{AB}^n J_{AA}^p \beta^q \tag{5}$$

Non-zero coefficients a(m,n,p,q), b(m,n,p,q), and c(m,n,p,q) up to order 7 in $\beta = 1/k_BT$ are given in Ref. [15].

A PA [M, N] to a magnetic susceptibility $\chi(T)$ is a rational fraction P_M / Q_N, with P_M and Q_N, polynomials of orders M and N in $\beta = 1/k_B T$ (k_B is the Boltzmann constant) such that $\chi(\beta) \approx P_M / Q_N + O(\beta^{M+N+1})$, M and N are the degrees of the P and Q polynomials, respectively. Estimates of T_C and γ for $Mg_xNi_{1-x}Fe_2O_4$ have been obtained using the PA method [16]. The simple pole corresponds to T_C and the residues to the critical exponent γ.

3. Results and discussions.

The powerful Padé approximants method has been used to estimate the critical temperature, and the critical exponent associated with magnetic susceptibility. The locations of singularities in the PA method to the HTSEs of the magnetic susceptibility determine the Curie point.

The high temperature series expansions extrapolated with the Padé approximants method [12-14] is shown to be a convenient method to provide valid estimations of the critical temperatures for real system. By applying this method to the magnetic susceptibility χ (x), we have estimated the critical temperature T_C for each dilution x in the $[Mg_{(0.14x)}Fe_{(1-0.14x)}]_A[Ni_{(1-x)}Mg_{(0.86x)}Fe_{(1+0.14x)}]_BO_4$ system. The values of T_C obtained by HTSE are in good agreement with experiment measurements. The estimated magnetic susceptibility exponents are γ =1.32 for $CoFe_2O_4$ and γ = 1.27 for $MgFe_2O_4$.
Figure 1 shows the phase diagram of the $Ni_{1-x}Mg_xFe_2O_4$ systems. It contains the ferrimagnetic and paramagnetic (PM) phases, in the range $0 \leq x \leq 1$, which cover the range between the opposite pure systems $NiFe_2O_4$ and $MgFe_2O_4$. The experimental results obtained by magnetic measurements [10] are also presented in **Figure** 1. They are in good agreement with theoretical results.

Table 2. The values of critical temperature obtained by HTSE and the exchanges integrals J_{AB}, J_{BB} and J_{AA} for the ferrites $Ni_{1-x}Mg_xFe_2O_4$ ($0 \leq x \leq 1$) with the cation distribution $[Mg_{(0.14x)}Fe_{(1-0.14x)}]_A[Ni_{(1-x)}Mg_{(0.86x)}Fe_{(1+0.14x)}]_BO_4$.

x	Distribution ion in A site [9]	Distribution ion in B site [9]	J_{BB}(K)	J_{AB}(K)	J_{AA}(K)	T_C(K) *exp.[6]*	T_C(K) HTSE
0	$(Fe_1)_A$	$(Ni_1Fe_1)_B$	------	-27.70	-14.00	860	866
0.1	$(Fe_{0.986}Mg_{0.014})_A$	$(Ni_{0.9}Mg_{0.086}Fe_{1.014})_B$	-0.80	-25.96	-13.61	----	850
0.2	$(Fe_{0.972}Mg_{0.028})_A$	$(Ni_{0.8}Mg_{0.172}Fe_{1.028})_B$	-1.70	-24.26	-13.22	831	838
0.3	$(Fe_{0.958}Mg_{0.042})_A$	$(Ni_{0.7}Mg_{0.258}Fe_{1.042})_B$	-2.44	-22.59	-12.84	----	827
0.4	$(Fe_{0.944}Mg_{0.056})_A$	$(Ni_{0.6}Mg_{0.344}Fe_{1.056})_B$	-3.02	-20.97	-12.47	809	815
0.5	$(Fe_{0.930}Mg_{0.070})_A$	$(Ni_{0.5}Mg_{0.430}Fe_{1.070})_B$	-3.45	-19.39	-12.10	----	802
0.6	$(Fe_{0.916}Mg_{0.084})_A$	$(Ni_{0.4}Mg_{0.516}Fe_{1.084})_B$	-3.72	-17.84	-11.74	786	793
0.7	$(Fe_{0.902}Mg_{0.098})_A$	$(Ni_{0.3}Mg_{0.602}Fe_{1.098})_B$	-3.84	-16.33	-11.39	----	774
0.8	$(Fe_{0.888}Mg_{0.112})_A$	$(Ni_{0.2}Mg_{0.688}Fe_{1.112})_B$	-3.80	-14.86	-11.03	759	763
0.9	$(Fe_{0.874}Mg_{0.126})_A$	$(Ni_{0.1}Mg_{0.744}Fe_{1.126})_B$	-3.60	-13.43	-10.69	----	738
1	$(Fe_{0.86}Mg_{0.14})_A$	$(Mg_{0.86}Fe_{1.14})_B$	-3.24	-12.04	-10.35	719	725

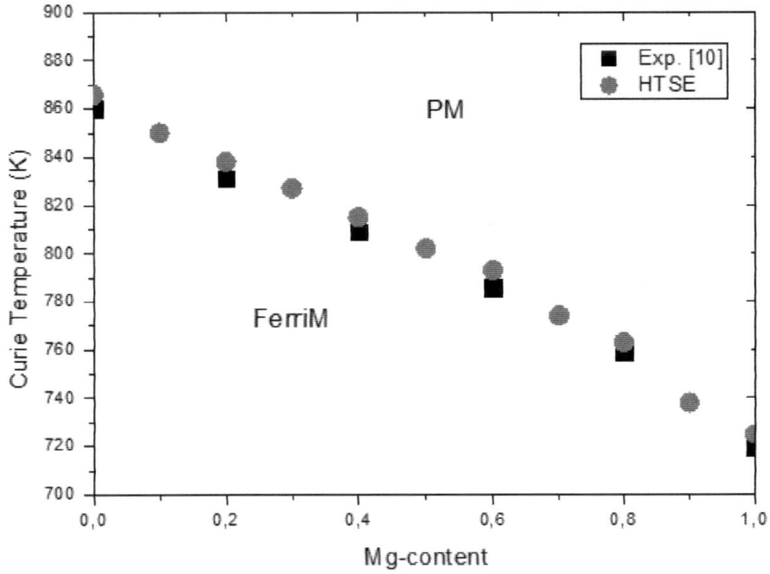

IOP Conf. Series: Materials Science and Engineering **186** (2017) 012016 doi:10.1088/1757-899X/186/1/012016

Figure. 1 Magnetic phase diagram for the $Ni_{1-x}Mg_xFe_2O_4$ ferrite with $[Mg_{(0.14x)}Fe_{(1-0.14x)}]_A[Ni_{(1-x)}Mg_{(0.86x)}Fe_{(1+0.14x)}]_BO_4$ cations distribution

4. Conclusions

The general expression of exchanges integrals J_{AB}, J_{BB} and J_{AA} of the cation distribution $[Mg_{(0.14x)}Fe_{(1-0.14x)}]_A[Ni_{(1-x)}Mg_{(0.86x)}Fe_{(1+0.14x)}]_BO_4$ are given. The magnetic properties of spinel ferrites depend on the magnetic interaction and cation distribution in the two sublattices i.e. tetrahedral (A) and octahedral (B) lattice sites.

The HTSEs combined with the Padé approximants methods are applied to the $[Mg_{(0.14x)}Fe_{(1-0.14x)}]_A[Ni_{(1-x)}Mg_{(0.86x)}Fe_{(1+0.14x)}]_BO_4$ system, to obtained the critical temperature for different values of x The value of critical exponent associated with magnetic susceptibility has been estimated.

References

[1] Ponpandian N, Balaya P, Narayanasamy A, *J. Phys.: Condens. Matter.* **14** (2002) 3221.

[2] Torres L, Zazo M, Flores A G, Raposo V, Inigues J, *J. Appl. Phys.* **79** (1996) 5422.

[3] George M, John A M, Nair S S, Joy P A, Anantharaman M R, *J. Magn. Magn.Mater.* **302** (2006) 190.

[4] Bhasker D. Ravindre, P. Vijaya Reddy, *J. Magn. Magn. Mater.* **263** (2003) 127 – 133.

[5] Chinnasamy C N, Narayanasamy A, Ponpandian N, Chattopadhyay K, Shinoda K, Jeyadevan B, Tohji K, Nakatsuka K, Furubayashi T, and Nakatani I, *Phys. Rev.* B **63** (2001) 184108.

[6] Barth T W F, Posnjak E , *Z. Kristallogr.* **82** (1932) 325.

[7] Murthy N S S, Rao L M, Begum R J, Natera M G and Youssef S I, *J. Phys. Colloq.* **32** (1971) 318 – 319.

[8] Nagakomi F, da Silva S W, Grag V K, Oliveira A C, Morais P C, Franco Jr A, *J. Solid State Chem.* **182** (2009) 2423.

[9] Ahmed M A , EL Khawlani A A, *J. Magn. Magn. Mater.* **321** (2009) 1959.

[10] Mahbubul Haque Md , Kazi Hanium Maria, Shamima Choudhury, Mahabub Alam Bhuiyan and.Hakim M A, *Journal of Ceramic Processing Research.* **14** (2013) 82–86.

[11] Srivastava C M, Srinivasan G, Nanadikar N G, *Phys. Rev.* B **19** (1979) 499.

[12] Stanley H E, *Phys. Rev.* **158** (1967) 537.

[13] Hamedoun M, Houssa M, Benzakour N, Hourmatallah A, *J. Phys. : Condens. Matter* **10** (1998) 3611.

[14] Masrour R, Hamedoun M, Benyoussef A, *Phys. Lett.* A **373** (2009) 2071–2074.

[15] Hamedoun M, Bakrim H, Bouslykhane K, Hourmatallah A, Benzakour N, Filali A, Masrour R, *J. Phys: Condens. Matter* **20** (2008) 125216.

[16] Baker G.A, P. Graves-Morris (Eds.), *Padé Approximants*, Addison-Wesley, London, 1981.

Form effect on the Diamagnetic Susceptibility of a magneto-donor in Quantum Dot

S Janati.Edrissi[1], S M'Zred[1], Y Chrafih[2], K Rahmani[2], I Zorkani[1], A Mmadi[1] and A Jorio[1]

[1]Groupe des Nanomatériaux et Energies Renouvelables, L P S,
Faculté des Sciences Dhar Mehraz, BP 1796, Fès-Morocco
[2]Faculté polydisciplinaire-Beni Mellal-Morocco

E-mail: janati.edrissi.sanae@gmail.com

Abstract: The diamagnetic susceptibility of a shallow donor confined to move in Quantum Dots 'QD' in the presence of a magnetic field is theoretically investigated. The numerical calculations are performed in the effective mass approximation, using a variational method. We describe the effect of the quantum confinement by an infinite deep potential. The form effect is studied for the Spherical Quantum Dot 'SQD' and Cylindrical Quantum Dot 'CQD'. The results for these two forms of structures show that the diamagnetic susceptibility and the binding energy increase with the magnetic field. There are more pronounced for larger dot. We remark that for a zero magnetic field, the binding energy and the diamagnetic susceptibility are decreasing functions of the quantum dot radius.

1. Introduction

In the recent years, it became possible to fabricate high quality semiconductor quantum dots (QDs) with good electronic and optical properties due to the development of experimental crystal growth techniques [1-3]. These structural quantum dots have attracted a great deal of attention from academic and technological viewpoints [4-6]. The physical properties are improved by the reduction of the dimensional quantum confinement of carriers, they drastically changes the electronic structure in QDs from that of bulk semiconductors. In the past few years, several theoretical [7] and experimental [8] investigations have been reported on the behavior of shallow hydrogenic donor impurity in quantum dot. The effects of applied magnetic fields on the physical properties are given in the reference [9-11]. Recently, Ulas et al [12] have reported the study of electric field and geometrical effects on a shallow donors binding energy in a quantum well wire. They have obtained that the donor binding energy is sensitive to the interplay of the electric field and geometrical effects. Zounoubi et al [13] have studied the influence of magnetic fields on the binding energy and polarizability of a shallow donor impurity placed at the center of a cylindrical quantum dot (CQD). They have showed that the magnetic field increases the binding energy and strongly reduces the polarizability. For higher field strength and large dot, the magnetic field effects are predominant. Ulas et al [14] have also calculated electric field and geometrical effects on the self polarization in GaAs–(Ga-Al)As quantum well wires of square, rectangular and cylindrical cross-sections. They have found that the self-polarization effect outside of the center depends on both the geometrical form of the wire and the impurity position in the same structure. Several properties of the donors such as the polarizability [15] and diamagnetic susceptibility [16-19] are yet to be obtained experimentally. Recently, several theoretical works have been reported on the diamagnetic susceptibility of impurity in low dimensional systems. Peter et al [20] have computed and compared the susceptibility for a hydrogenic donor in a spherical confinement, harmonic oscillator-like and rectangular well-like potentials for a finite QDs. They observed a strong influence of the shape of confining potential and geometry of the dot on the susceptibility. Rahmani et al [21] have studied the diamagnetic susceptibility of a confined donor in Ga1-xAlxAs-GaAs Inhomogenous Quantum Dot without a magnetic field. They found that the diamagnetic susceptibility depend strongly on the core and the shell radius and presents a minimum for a critical value of the ratio R1/ R2 depending on the value of the outer radius. Kilicarslan et al [22] have investigated the effects of the magnetic field and the dielectric screening on the diamagnetic susceptibility of a donor in a QW with anisotropic effective mass. Koksal et al [23] have studied the magnetic field effects on the diamagnetic susceptibility and binding energy of a hydrogenic impurity in a QWW by taking into account spatially dependent screening. They show that the diamagnetic susceptibility is more important for donors in QWW over a large range of wire dimensions. Kilcarslan et al [24] have studied the magnetic field effects on the diamagnetic susceptibility in a GaxIn1-x NyAs1-y/GaAs QW and found that the diamagnetic susceptibility and binding energy of the magneto-donor in the GaxIn1-x NyAs1-y/GaAs QW increases with Nitrogen mole fraction.. El Ghazi et al [25] have studied the dependence of the binding energy as function of external magnetic field and donor's position in GaN | (Ga,In)N | GaN spherical quantum dot–quantum well (SQDQW) Their results show that the magnetic field effect is more marked in large layer than in thin layer and it is more pronounced in the spherical layer center than in its extremities. A. Mmadi et al [26,27] have calculated the effects of an applied magnetic field on the diamagnetic susceptibility of a shallow donor confined to move in a spherical homogeneous Quantum Dots "HQD''and in Cylindrical Quantum Dot "CQD''. They have found that the diamagnetic susceptibility increases with the magnetic field and is more important especialy for larger structure of quantum dot.

The present paper is organized as follows: the model and the calculation method for calculating the ground-state binding energy and diamagnetic susceptibility of a shallow donor in QD are presented in Section2. The numerical results and discussions are shown for a Spherical QD of radius Rs and Cylindrical QD of radius Rc and length Hc in Section3.

2. Model and calculation

We consider a donor impurity located at the center of the spherical quantum dot "SQD" of raduis Rs and the cylindrical quantum dot "CQD" with radius Rc and length Hc in the presence an applied a magnetic field $\vec{B}$ one along the z-direction.

2.1 Spherical Quantum Dot

The Hamiltonian that describes the problem of a hydrogenic impurity located in a spherical QD at the centre of the dot in the effective-mass approximation and in spherical coordinates can be expressed as [9]:

$$H = -\nabla^2 - \frac{2}{r} + \frac{1}{4}\gamma^2 r^2 \sin^2\theta + V(r) \tag{1}$$

where the first term is the kinetic energy of the electron of the impurity, the second term correspond to the potential energy of the impurity the third and the fourth terms correspond to the effect of magnetic field. The last term $V(r)$ is the infinite confinement potential given by:

$$V(r) = \begin{cases} 0 & r < R_s \\ \infty & r > R_s \end{cases} \tag{2}$$

We use the effective Bohr radius $a^* = \frac{\hbar^2 \varepsilon_0}{m^* e^2}$ and the effective Rydberg $R^* = \frac{m^* e^4}{2\hbar^2 \varepsilon_0^2}$ as the units of length and energy. Furthermore, we introduce the dimensionless parameter $\gamma = \frac{\hbar\omega_c}{R^*}$ and

$\omega_c = \frac{eB}{m^* c}$ characterizing the strength of the magnetic field and the effective cyclotron frequency respectively.

We use a variational method approach to determine the ground state binding energy; we adopt the wave function given by [9]:

$$\psi_0(r) = \begin{cases} \psi = \frac{Sin(Kr)}{r} \exp - \lambda r & for\ r \le R_s \\ 0 & for\ r > R_s \end{cases} \quad (3)$$

$K = \frac{\pi}{R_s}$ and λ is a variational parameter. The exponential factor $\exp - \lambda r$ describes the Coulomb spatial interaction. The corresponding energy is obtained by minimization with respect to the variational parameter λ:

The binding energy of the donor impurity located at the center of a Spherical Quantum Dot is given by

$$E_b = E_{Sub} - \langle H \rangle_{\min} \quad (4)$$

Where $\langle H \rangle_{\min}$ is the minimum of the expectation value of the Hamiltonian obtained by varying the variational parameter λ.

2.2 Cylindrical Quantum Dot

In this section, we consider the case cylindrical quantum dot (CQD) with radius R and length H, with a donor impurity placed at the center of the dot. In the effective mass approximation, the Hamiltonian can be written in cylindrical coordinates and reduced units as [26]:

$$H = -\nabla^2 - \frac{2}{\sqrt{\rho^2 + z^2}} + \frac{\gamma^2}{4} \rho^2 + \gamma L_z + V(\rho, z) \quad (5)$$

Where ρ and z are the electron coordinates in the plane perpendicular and along the cylinder axis respectively. Lz is the z component of the angular momentum operator. $V(\rho, z)$ is the confining potential given by.

$$V(\rho, z) = \begin{cases} 0 & for\ \rho < R_c\ and\ |z| < \frac{H_c}{2} \\ \infty & for\ \rho > R_c\ and\ |z| > \frac{H_c}{2} \end{cases} \quad (6)$$

Since the Schrödinger equation cannot be solved exactly, we follow the Hass variational method. We choose the wave function for the impurity ground-state as [14]:

$$\psi(\rho, z) = \begin{cases} N J_0\left(\theta_0 \frac{\rho}{R}\right) \cos\left(\frac{\pi z}{H}\right) \exp\left(-\left(\frac{\rho^2}{8\alpha^2} + \frac{z^2}{8\beta^2}\right)\right) & for\ \rho \le R\ and\ |z| \le \frac{H_c}{2} \\ 0 & for\ \rho > R\ and\ |z| > \frac{H_c}{2} \end{cases} \quad (7)$$

where J_0 is the Bessel function of zero order; $\theta_0 = 2.40482$ is its first zero, α and β are variational parameters and N is the normalization constant. The corresponding energy is obtained by minimization with respect to the variation parameters α and β.

The binding energy of the donor impurity located at the center of a Cylindrical Quantum Dot is given by

$$E_b = E_{Sub} - \langle H \rangle_{\min} \quad (8)$$

Where $\langle H \rangle_{min}$ is the minimum of the expectation value of the Hamiltonian obtained by varying the variational parameters α and β.

The Schrodinger equation is solved variationally to find the ground state wave function, which has been used in the computation of diamagnetic susceptibility χ_{dia} of the hydrogenic donor given as [27]:

$$\chi_{dia} = -\frac{e^2}{6m^* \varepsilon_0 c^2} < (\vec{r})^2 > \tag{9}$$

The final results on the diamagnetic susceptibility are obtained by numerical minimization of the energy expression with respect to the parameters λ for SQD and α and β for CQD.

3. Numerical results and Discussion

(For the GaAs material, we have: the effective Bohr radius a * = 98.6 ° A, the effective Rydberg R * = 5.85 MeV, the effective mass m * = 0.067 me and finds dielectric = 12.5)

In this section, the diamagnetic susceptibility of a hydrogenic impurity confined in a quantum dot with and without applied magnetic field is calculated numerically. Using a variational procedure in the effective mass approximation. Numerical applications are for a typical GaAs SQD and CQD. The binding energy versus the quantum dot radius for different magnetic fields (γ = 0, 0.5 and 1) is presented in Figure1. From this figure, we remark that in the strong spatial confinement case (Rs< 2a*) and (Rc<1.5a*), the binding energy is relatively insensitive to the magnetic field and is identical to the zero magnetic field case. This explains that the main contribution to the binding energy is the electron spatial confinement energy and that the electron spatial confinement prevails over the magnetic field confinement. For the week spatial confinement (Rs>2a*) and (Rc>1.5a*), the different magnetic field curves tend to deviate from each other and reach steady values as the dot radii increase. Also, we can see that without of a magnetic field, the binding energy tends to the value of the bulk semiconductor (Eb→1R*), while for a given value of the magnetic field, the binding energy is larger than for the case without a magnetic field. The physical meaning of this is that increasing the strength of the magnetic field shrinks the electron wave function and decreases the cyclotron radius for the electron relative to the quantum radius and confines the electron closer to the on-center impurity.

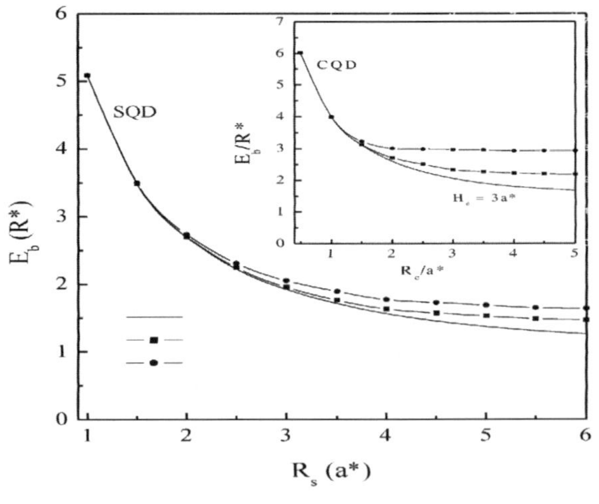

Figure.1 Variation of donor binding energy as a function of Rs raduis of a SQD and as a function of

Rc radius and H= 3a* of a CQD for three values of magnetic field (γ = 0, 0.5 and 1)

We have reported in figure2-a and figure2-b the diamagnetic susceptibility χ_{dia} of a donor as a function of the spherical radius Rs and as a function of the cylindrical radius Rc with different values of the lengths (Hc=0.6a*, 1a*, 3a* and 20a*) for (γ = 0). We consider the case of a donor impurity placed at the center of this two structures. From figure 2-a,we can see that the diamagnetic susceptibility χ_{dia} decreases as the radius Rs increases. We remarque also that diamagnetic susceptibility decreases strongly and eventually converges to the bulk limit value (−1.1a.u) [16, 18] for ($R_s \to \infty$). For figure2-b, we remark that the diamagnetic susceptibility χ_{dia} decreases when the radius dot Rc increases for each value H. our results show that for strong axial confinement (Hc=1a*) and weak radial confinement (Rc >>1a*), the diamagnetic susceptibility χ_{dia} decreases and tends to the quantum well value (-0.2a.u) [19]. Nevertheless, for weak axial confinement (Hc>>1a*) and weak radial confinement (Rc>1a*), we see that the diamagnetic susceptibility decreases with the increase of CQD length and approaches to the three dimensional value (-1.1a.u) which correspond to the bulk limit case (see Ref[16, 18]). It should be noted that the diamagnetic susceptibility χ_{dia} is more sensitive for large dimensional in these two cases. Our results are in perfect agreement with previous calculations without the magnetic field.

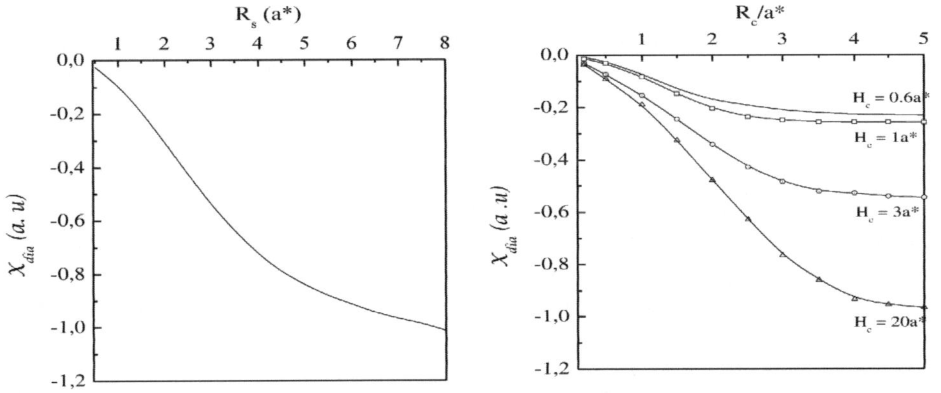

Figure.2 Variations of the diamagnetic susceptibility χ_{dia} as a function of Rs raduis of a SQD :(a) and as a function of Rc radius and with several values of the length (H=0.6a*, 1a*, 3a* and 20a*) of a CQD: (b)

The diamagnetic susceptibility of a shallow donor as a function of a SQD and a CQD dimension obtained by using a variational method with an infinite barrier are presented in table1 and table2. From Table1, we conclude that the diamagnetic susceptibility χ_{dia} values decreases with increasing spherical radius. In the bulk limit χ_{dia} should approach (-1.1a.u). We obtain (χ_{dia} = -1.1012 a.u) for Rs = 11a*. As in table2, It is easily observed that the variation of the diamagnetic susceptibility

χ_{dia} increases with cylindrical size and approaching the bulk value (-1.0807a.u) for Rc =11a* and Hc=11a*. These two form of quantum dot gives good results of the diamgnetic susceptibility when the quantum dot size becomes very large, a comparison has been reported in Ref [28].

Table1. Diamagnetic susceptibility of a Spherical Quantum Dot

SQD	RS =1a*	RS=3a*	RS=5a*	RS=7a*	RS=9a*	RS=11a*
Xdia(a.u)	-0.0928	-0.5408	-0.9684	-0.9684	-1 .0416	-1.1012

Table 2. Diamagnetic susceptibility of a Cylindrical Quantum Dot

CQD	Rc =1a* Hc=1a*	Rc=3a* Hc=3a*	Rc=5a* Hc=5a*	Rc=7a* Hc=7a*	Rc= 9a* Hc= 9a*	Rc= 11a* Hc= 11a*
Xdia(a.u)	-0.0827	-0.4848	-0.7955	-0.9411	-0.9869	-1.0807

In order to investigate the magnetic field and the form effect, we display in figure3-a and figure3-b the diamagnetic susceptibility χ_{dia} as a function of the dot radius Rs and Rc of SQD and CQD respectively width several values of magnetic field (γ = 0, 0.5 and 1). The donor is placed in the centre of the dot. There is a competition between the geometric confinement and the magnetic confinement. From figure3-a, the magnetic field effect on the diamagnetic susceptibility is not remarkable for small dot (Rs <2a*), the magnetic field effect becomes important for large QD (Rs $\geq$ 2a*). Also as expected the diamagnetic susceptibility increases with magnetic field due to compression of the electron wave function with the magnetic field. It's important to note that the diamagnetic susceptibility χ_{dia} decreases when the dot radius Rs increases. For figure3-b, the effects of the magnetic field on diamagnetic susceptibility χ_{dia} as a function of the radius Rc for different values of the length Hc (Hc = 1a* and Hc=3a*) has been plotted. We notice that for strong radial confinement (Rc<1.5a*), the magnetic field effect on the diamagnetic susceptibility is not remarkable. The diamagnetic susceptibility increases with the magnetic field. This increase is due to a shrinking of the charge distribution when an external magnetic field is applied. Furthermore, for given values of Rc and γ, the diamagnetic susceptibility increases when the length of the dot decreases which reflects the increasing confinement. These results explain that in the presence of the magnetic field, the diamagnetic susceptibility χ_{dia} remain to be constant over a large dot [25, 32]. Also, the diamagnetic susceptibility are found to be almost identical for dot of spherical and cylindrical if the dot dimensions taken to be comparable.

Now, by comparing figure4-a and figure4-b, we plotted the variation of the diamagnetic susceptibility as a function of the magnetic field strength for fixed QD geometries. This figure reflects correctly the effect of the magnetic field, which confines more the electron and increases the diamagnetic susceptibility. From figure4-a, we took three values of a spherical dot radius Rs (Rs = 1a*, 1.25a* and 2a*). We notice that for quantum dots quite small geometrical dimensions (Rs = 1a* and 1.25a*), the diamagnetic susceptibility is relatively insensitive to the magnetic field and is identical to the zero magnetic field case. For the quantum dots which is the geometrical dimension wide enough (Rs= 2a*), the diamagnetic susceptibility χ_{dia} increases when the magnetic field

IOP Conf. Series: Materials Science and Engineering **186** (2017) 012017 doi:10.1088/1757-899X/186/1/012017

increases and reaches a limit value when the magnetic field becomes very strong. For figure4-b, we plot the variation of the donor diamagnetic susceptibility χ_{dia} for three different radius values of cylindrical quantum dot (Rc=1a*, 2a*and Rc = 3a*) and H = 3a. We can remark that the diamagnetic susceptibility χ_{dia} increases as CQD radius Rc decreases. The diamagnetic susceptibility is totally insensitive to the increase of the magnetic field for small radius confinement (Rc=1a*). For large radius confinement case (Rc>2a*), the variation of the diamagnetic susceptibility is much more pronounced due to the stronger confinement effect of the magnetic field. From the two forms of the dot, it is clearly seen that the diamagnetic susceptibility decrease as the QD radius increases or the magnetic field strength decreases. As observed that the magnetic field effect and geometry effect, influence directly on the diamagnetic susceptibility behavior.

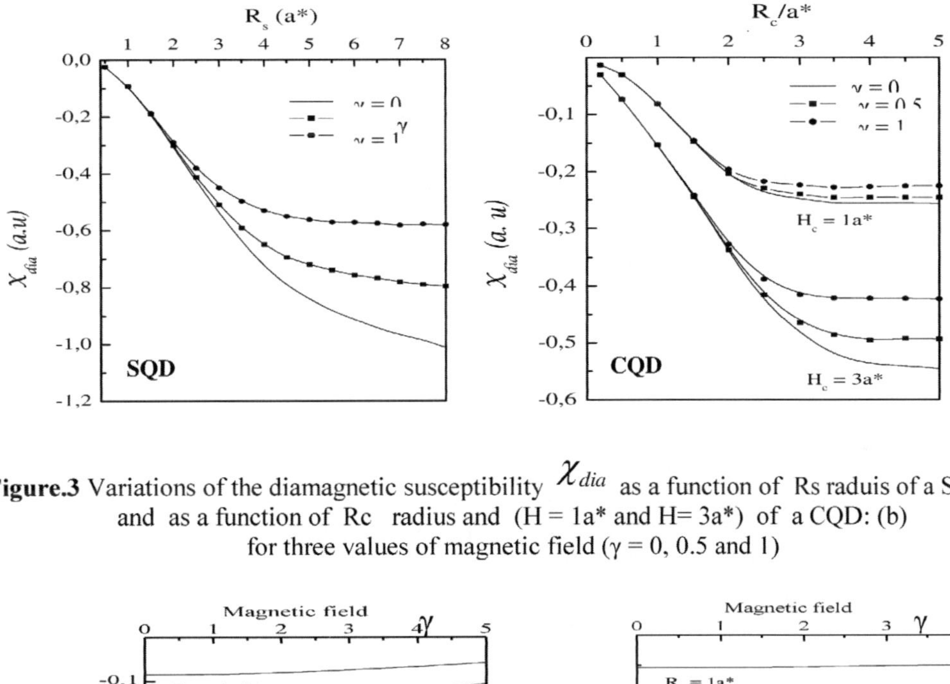

Figure.3 Variations of the diamagnetic susceptibility χ_{dia} as a function of Rs raduis of a SQD: (a) and as a function of Rc radius and (H = 1a* and H= 3a*) of a CQD: (b) for three values of magnetic field (γ = 0, 0.5 and 1)

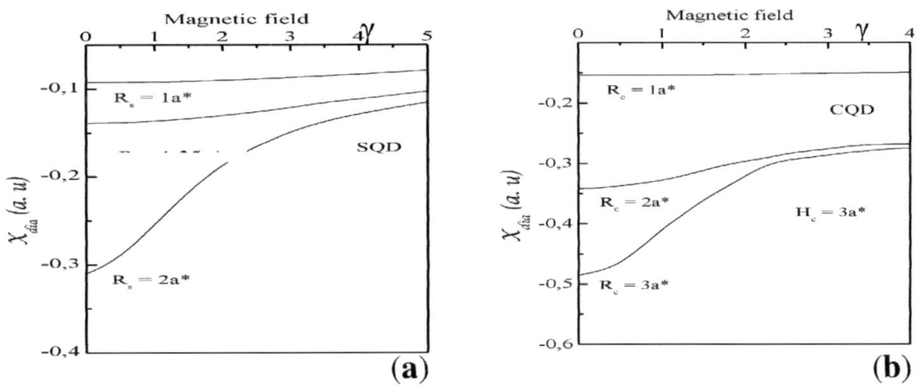

Figure.4 Variation of the diamagnetic susceptibility χdia function of magnetic field with three

values RS (RS= 1a *,1.25a * and 2a *) of a SQD: (a) and with three values radius (Rc=1a*, 2a* and 3a*) and Hc = 3a* of a CQD: (b)

4. Conclusion

In the present work, we have reported the form and magnetic field effects on the diamagnetic susceptibility of a hydrogenic donor placed in spherical and cylindrical quantum dot by using the effective mass approximation. The diamagnetic susceptibility is dramatically dependent on the size of the dot and increases as the radius of the dot decreases. Under a magnetic field, additional increases for diamagnetic susceptibility are presented for nanostructure over a large range of QD dimensions.

References

[1] D. Bimberg, V.A. Shchukin, N.N. Ledentsov, A. Krost, F. Heinrichsdorff, Applied Surface Science 130 (1998) 713
[2] B.A. Joyce, D.D. Vvedensky, Mater. Sci. Eng. Reports 46 (2004) 127.
[3] A. Babinski, M. Potemski, Peter C.M. Christianen C. R. Physique 14 (2013) 121
[4] G. Vastola, Y.-W. Zhanga, Vivek B. Shenoy, Current Opinion in Solid State and Materials Science 16 (2012) 64
[5] L. Leea, W-C. Fanb, K-F. Chienb, A-J. Tzoub, Wu-Ching Choub, J. Cr. Growth 378 (2013) 222
[6] G. Biasiol, S.Heun. Phys Rep 500 (2011) 117.
[7] Gh. Safarpour, A.Zamani, M.A.Izadi, H.Ganjipour, J.Lumin.147 (2014) 295
[8] L. Lu, W. Xie, S. Liang, Current Applied Physics 11 (2011) 1302
[9] C. Dane, H.Akbas, S.Minez, A.Guleroglu, Physica E 42 (2010)1901
[10] J. Eymery, G. Biasiol, E. Kapon, T. Ogino, C. R. Physique 6 (2005) 105.
[11] X. Zhigang, Z. Jiqian, H. Fengai, J. Appl. Phys. 79 (1996) 9181
[12] M. Ulas, H. Akbas, M. Tomakt, Tr. J. of Physics 22 (1998) 369
[13] A. Zounoubi, I. Zorkani, K. El Messaoudi, A. Jorio, Phys. Lett. A 312 (2003) 220
[14] M. Ulas, I. Erdogan, E. Cicek, S. Senturk Dalgı, Physica E 25 (2005) 515
[15] K. El Messaoudi, A. Zounoubi, I. Zorkani, A. Jorio, Phys. Stat. Sol. (b) 233 (2002) 270
[16] A.J. Jasper, D. Reuben K. Jayakumar Phys. Status Solidi (b) 243 (2006) 4020
[17] M. Revathi , Chang KyooYoo, A. John Peter, Physica E 43 (2010) 322
[18] P. Nithiananthi and K. Jayakumar, Solid Stat. Commun. 138 (2006) 305
[19] E. Kasapoglu, F. Ungan, H. Sari, I. Sökmen, Superlatt. Microstruct. 46 (2009) 817
[20] A.J. Peter, J. Ebenezar, J. Sci. Res. 1 (2009) 200.
[21] K. Rahmani, I. Zorkani, A. Jorio, Phys.Scr.83 (2011) 035701
[22] E. Kilicarslan, S. Sakiroglu, M. Koksal, H. Sari, I. Sokmen, Physica E 42 (2010) 1531.
[23] M. Koksal, E. Kilicarslan, H. Sari, I. Sokmen, Physica B 404 (2009) 3850.
[24] E. Kilicarslan, S. Sakiroglu, E. Kasapoglu, H. Sari, I. Sökmen, Superlatt. Microstruct. 48 (2010) 305.
[25] Haddou ElGhazi , AnouarJorio , Izeddine Zorkani .Physica B 427 (2013) 106–109
[26] A. Mmadi, K. Rahmani, I.Zorkani, A. Jorio, Superlatt. Microstruct.57 (2013) 27
[27] A. Mmadi, I.Zorkani, K. Rahmani, A. Jorio, The African Reviw of Physics 8 (2013) 219
[28] C. Kittel, Introduction to Solid State Physics, seventh ed., Wiley, New York, 1998.
[29] S. Rajashabala, K. Navaneethakrishnan, Superlatt. Microstruct.43 (2008) 247–261
[30] N. Arunachalama, C. Yoob, A. John Peter, Superlatt. Microstruct. 49 (2011) 43

IOP Conf. Series: Materials Science and Engineering **186** (2017) 012018 doi:10.1088/1757-899X/186/1/012018

Complex impedance, dielectric relaxation and electrical conductivity studies of $Ba_{1-x}Sr_xTiO_3$ ceramics

A Elbasset[1,2], S Sayouri S[2], F Abdi[1], T Lamcharfi [1,2] and L Mrharrab[2]

[1] LSSC Department of Electrical Engineering, Faculty of Science and Technology (FST),Universit Sidi Mohamed Ben Abdellah, Road Immouzer, Fes, Morocco
[2] LPTA, Physics Department, Faculty of Science DM, B.P.1796, Fes-Atlas Morocco

Email: elbasset.abdelhalim@gmail.com

Abstract. In this work, we prepared series of $Ba_{1-x}Sr_xTiO_3$ (BS_xT) powders, with different strontium concentrations (x = 0, 0.025, 0.75, 0.10, 0.125 and 0.15), by the sol-gel method. The variation of structure in the Ba1-xSrxTiO3 system was analyzed using XRD and Raman techniques. The field dependence of dielectric relaxation and conductivity was measured over a wide frequency range from room temperature to 400 °C. The activation energy, calculated from the thermal variation of the conductivity for different frequencies, showed that the Sr has significant effects on the properties of $BaTiO_3$. Relaxation times extracted using the imaginary part of the complex impedance (Z''(ω)) and the modulus (M''(ω)) were also found to follow the Arrhenius law and showed an anomaly around the phase transition temperature.
Key words. BST, Conductivity, Activation energy, Complex impedance (Z'(ω) and Z''(ω)), Relaxation times

1. Introduction

Ferroelectric materials of the ABO_3 perovskite type crystal structure have been widely used in microelectronic industries due to their high dielectric constant and low leakage current [1–6]. The most important examples of such ferroelectric materials are $BaTiO_3$ (BT) and $SrTiO_3$ (ST). $BaTiO_3$ is a ferroelectric for which the origin of ferroelectricity derives from the displacement of ions relative to each other, It is well known also that the temperature of transition from ferro-to-paraelectric state, Tc, and the dielectric properties of $BaTiO_3$ can be systematically changed by chemical substitution of barium and/or titanium by a wide variety of isovalent and aliovalent dopants [7–9] while $SrTiO_3$ is in paraelectric state of polarization. A well known solid solution of both $BaTiO_3$ and $SrTiO_3$ is barium strontium titanate ($Ba_{1-x}S\ r_xTiO_3$ or BSxT), has been considered to be an important material for tunable microwave devices such as phase shifters, tunable filters, delay lines, tunable oscillators, induced piezoelectric transducer etc. This because of its high dielectric constant, large electric field tunabilities, relatively low dielectric loss, variable Curie temperature (from 243 to 127 °C) depending on the composition of strontium, large polarizations, large permittivity and large induced strains that are achievable [10–13].

On the other hand, it is known that all the above mentioned properties are basically influenced by the temperature and the frequency in perovskite structure-type materials. Therefore, the temperature dependence of the dielectric permittivity for some frequencies has been widely studied and reported as an important feature in order to determine the temperatures and character of the phase transition in ferroelectric materials [14–16]. To the best of our knowledge, no works focusing on the study of the dielectric properties in the BSxT system in a very wide frequency interval, and for different temperatures of measure, have been reported in the literature, but only for few frequencies. Therefore the objective of the present work is to investigate the phenomenon of relaxation of the BSxT ferroelectric ceramics in a wide temperature and frequency ranges.

2. Preparation

Solutions of barium acetate, Strontium carbonate and the ground Ti are mixed in stoichiometric proportions, depending on the chemical formulation $Ba_{1-x}Sr_xTiO_3$ under agitation for 5 min. The destabilization of this solution is assured by evaporation of the solvent in an oven at the temperature of 80 °C for 96 h.The different steps in the preparation of the $Ba_{0.0975}S_{0.025}TiO_3$ powders are similar to those we have in our previous work [17]. Raw powder, after grinding, is calcined in air at 1000 °C for 4 h, in a programmable oven. Phase identification of the samples was performed using X-ray diffraction (Cu K ray, λ = 1.5418 Å) and Raman spectrum. These compositions were then pressed into pellets of 12 mm diameter and a thickness of 1 mm under a uniaxially pressure equal to 10 tons/cm^2. The pelletized samples were finally sintered at 1100 °C for 8 h in a programmable furnace. Silver paste was added to both faces of the disks, These compositions were then pressed into pellets of 12 mm diameter and a thickness of 1 mm under a uniaxially pressure equal to 10 tons/cm^2. The pelletized samples were finally sintered at 1100 °C for 8 h in a programmable furnace. Silver paste was added to both faces of the disks, which were then fired at 60 °C as electrodes.

3. Results and discussion

3.1. X-ray diffraction studies

Fig. 1 shows the X-ray diffraction patterns obtained on the powders of BSxT (x = 0, 0.025, 0.05, 0.075, 0.1, 0.125, and 0.15) calcined at 1000 °C for 4h. In this figure it is clear that all BSxT powders crystallize in the perovskite phase without presence of secondary phases. Zoom in on the peak (111) (Fig. 2) shows a shift in the position of the latter to larger angles, which highlights the incorporation and effect of Sr on BT, these results are consistent with the literature [18-21]. On the other hand the intensity of the (111) peak also diminishes with the mole fraction Ba/Sr, due to the difference of atomic numbers between Ba and Sr [21]. In the Fig. 3, and for low Sr content (from 0 to 0.075), we note the presence of (200) and (002) peaks around 2θ = 44.58°, which are characteristic of the quadratic phase [22], With the increase in the rate of Sr, the two peaks tend to merge, and cannot practically be distinguished for x = 10%.Which indicates a decrease in the tetragonality of the unit cell and thus a gradual transition from tetragonal to pseudocubic phase.

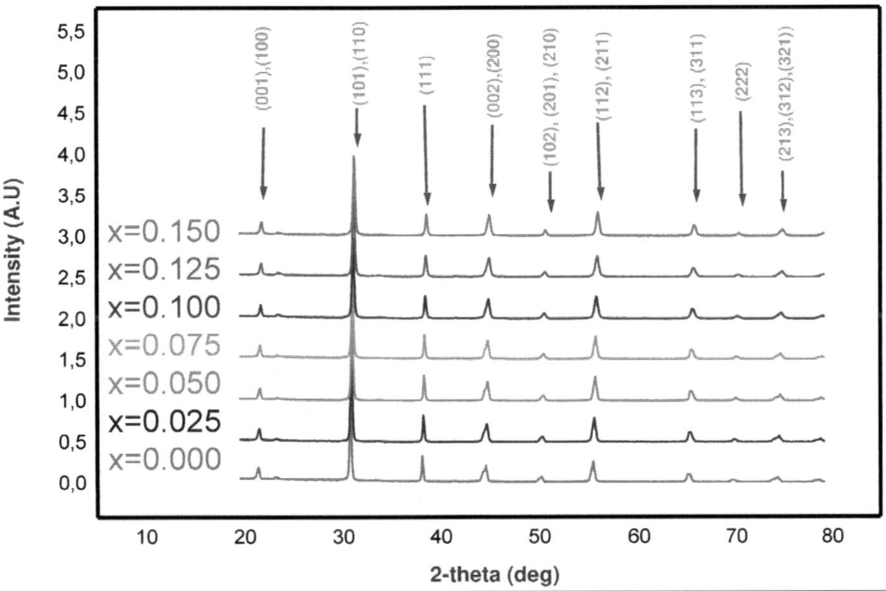

Figure 1. Evolution of the XRD spectra of the $Ba_{1-x}Sr_xTiO_3$ powder, calcined at 1000°C.

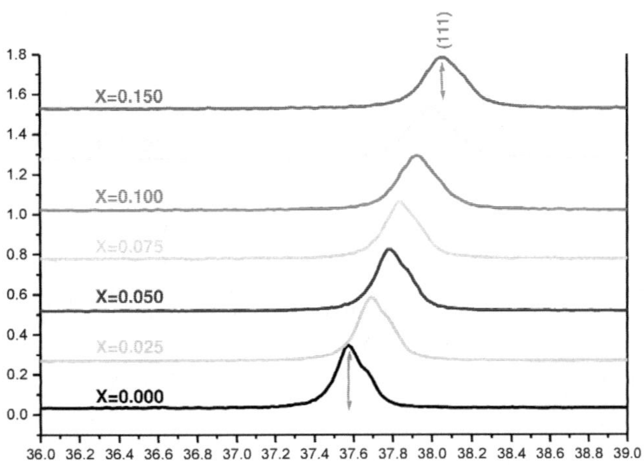

Figure 2. The zoom peak (111) in the range 36°< 2θ < 39°

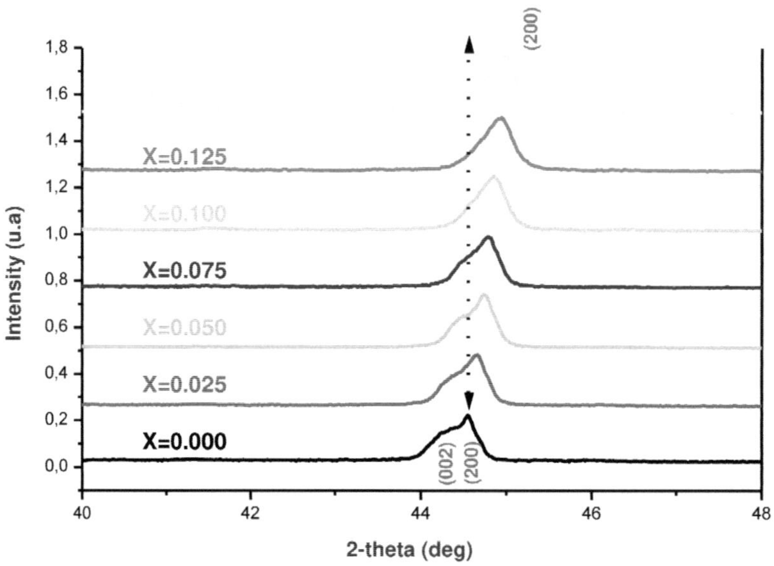

Figure 3. Zoom peak (200) in the range 40°< 2θ < 48°

Fig.4 shows the evolution of the Raman spectra at room temperature of the solid solution $Ba_{1-x}Sr_xTiO_3$ as a function of x, in the frequency range 100-1000 cm^{-1}. On the Raman spectrum of the pure sample (x = 0), we observe a strong and asymmetrical wide band due to the recovery of A(TO1), A(LO1) and A(TO2) modes between 200 cm^{-1} and 280 cm^{-1}. It is also observed a narrow band around 305 cm^{-1} associated with the modes B1 and E(TO3 + LO2), and a broad and asymmetric band at around 520 cm $^{-1}$ associated with A1(TO3) and E(TO4) modes, and another wide band (weak intensity) at around 720 cm^{-1} associated with A1(LO3) and E(LO4) modes /Furthermore, the peak intensity at 305 cm^{-1}, assigned to B1+E(TO3 + LO2) modes, decreases as the temperature increases and disappears when the phase becomes cubic [23-25]. When Sr concentration increases (Fig.4), the three peaks at around 200 cm^{-1}, 230 cm^{-1} and 285 cm^{-1} in BT tend to merge into a broad band located between 190 cm^{-1} and 290 cm^{-1} in BSxT. We also note the decrease in intensity of modes E(TO3 + LO2)/B1 and E(TO4)/A1 (TO3) and a widening of the band E(LO4)/A1(LO3). These are the result of shifting of Raman frequencies of the modes, but also of the presence of a local disorder of the phase which is due to the presence of Sr. The E(TO3) mode around 305 cm^{-1} tends to expand and decrease in intensity indicating a tendency towards pseudo-cubic phase. These results are consistent with those found by Shou-Yi Kuo, Wen-Yi Liao, and Wen-Feng Hsieh [20].

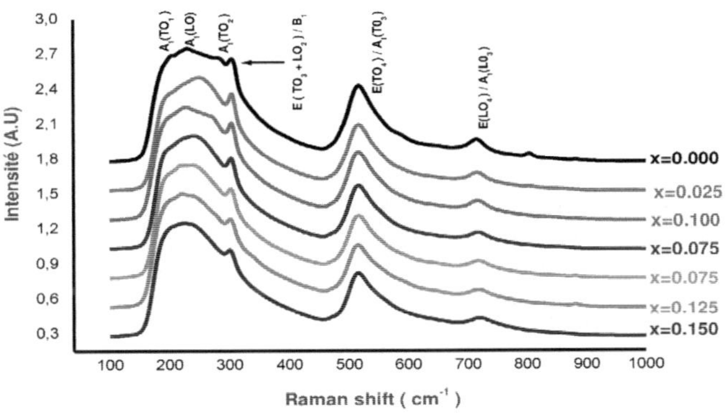

Figure 4. Raman spectra of $Ba_{1-x}Sr_xTiO_3$

3.2. Relaxation frequency fr for BSxT compositions

The thermal variation of the relaxation frequency, fr, for BSxT samples is shown in the curves of Fig.5. The results obtained show that $BaTiO_3$ relaxation frequency decreases with the temperature to reach its minimum value near Tc then increases. A possible mechanism responsible of this behaviour is based on the fact that the Ti^{4+} ion is displaced from the center of the oxygen octahedron and can occupy different potential wells (double potential well model). The relaxation process is then associated with Ti^{4+} ions cooperative jumps between the different wells. This movement would be consistent about a length Ic correspondent to that of a correlation chain [26-27]. The observed minimum can be explained by the growing Number of Ti^{4+} ions contributing to the cooperative movement along the correlation chains when the temperature approaches and tends towards Tc. Thus, the inertia of the chain increases (Ic increases) and the relaxation frequency decreases to a minimum value in the vicinity of the temperature of transition. On the other hand when the strontium content increases, the minimum of the curve fr = f (T) becomes more broad (diffuse). This study shows that the relaxation frequency value fr, at Tc (Fig.6) increases when x increases from 0 to 0.15, this can be explained by the substitution of Sr in $BaTiO_3$ which leads to a deformation of BO6 octahedra, probably involving shorter correlation chains and therefore leads to an increase of the relaxation frequency, which may be related to the nature of A-O bond (A: Ba, Sr) and / or to the polarisability of atoms.

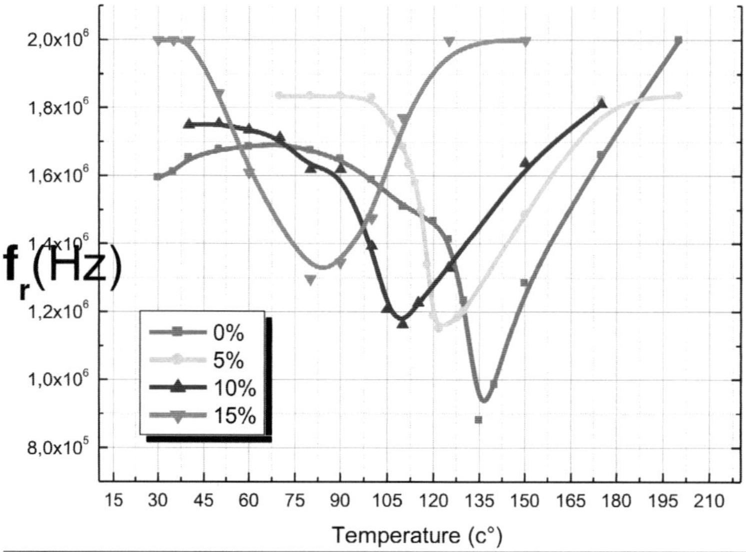

Figure 5. Thermal variation of fr as a function of the temperature of BSxT

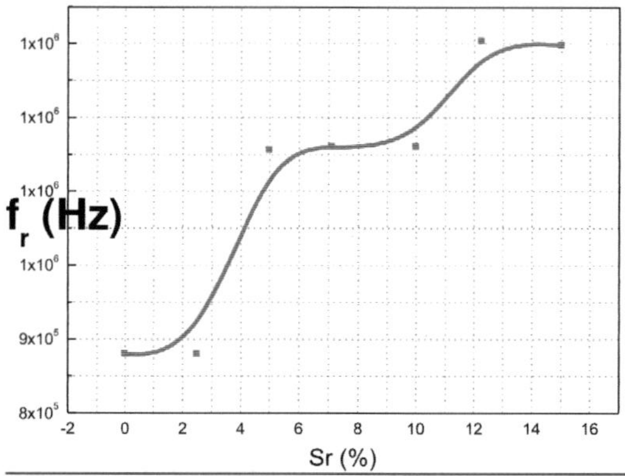

Figure 6. Evolution of relaxation frequency fr of BSxT

3.3. Conductivity

To study the transport mechanism in BTSx, we plotted the thermal variation of the conductivity for different frequencies. The conductivity was calculated using the following relation:

$$\sigma = \omega \varepsilon_0 \, \varepsilon'' = \omega \, \varepsilon_0 \varepsilon' \tan(\delta) \quad (1)$$

Fig.7 shows that the conductivity of $BS_{2.5}T$ passes through a maximum at the temperature Tm of ferroelectric-to-paraelectric transition. In addition, it increases as a function of frequency from $3.95.10^{-6}$ for 0.1 kHz to 4.10^{-6} for 100 kHz in the vicinity of T_m. The increase becomes significant when the frequency exceeds 5.10^5 Hz. This behaviour can be related directly to the relaxation phenomenon which results in a leakage current due to the energy dissipation. The shape of the curves recorded on the other BSxT compounds ($x \neq 0.025$) are similar to that of $BS_{2.5}T$, and they differ from each other only by the position of the transition temperature T_m. In order to ensure the reproducibility of these results, the experiments have also been conducted in cooling mode. On the fig.8 we present the thermal evolution, on heating and cooling, of the conductivity of BSxT compounds recorded for the frequency 100 kHz. The results show that the temperature of ferroelectric-to- paraelectric transition in cooling process is lower than that corresponding to the heating process.

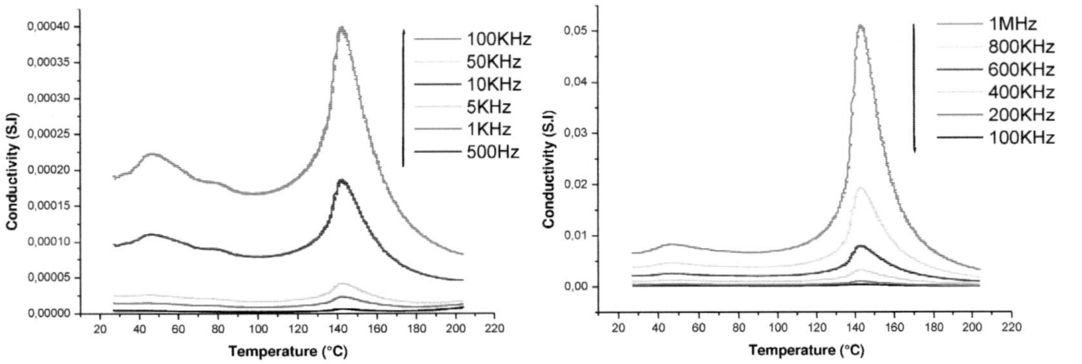

Figure 7. *Conductivity of $BS_{0.025}T$ as a function of temperature*

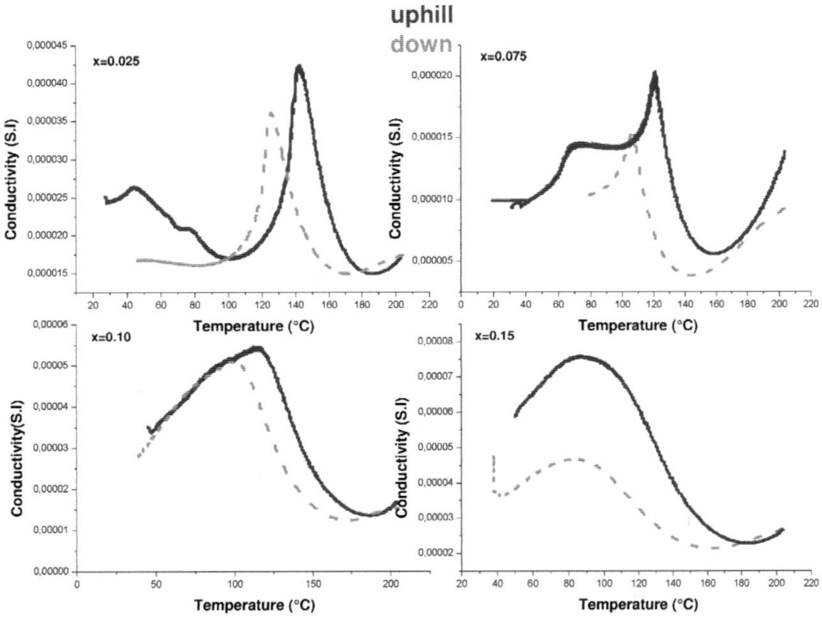

Figure 8. Evolution of the conductivity as a function of the temperature, uphill and down of BSxT

On the other hand, it is noted that maximum values are lower during the cooling process. When the temperature increases, and below Tm, the conductivity follows the Arrhenius law:

$$\sigma = \sigma_0 e^{\frac{-E_a}{K_b T}} \qquad (3)$$

With the activation energy Ea, T the absolute temperature in Kelvin and k_B is Boltzmann's constant (k_B= 1.3806.10-23 J/K).

The fig.9 shows activation energy variation as a function of the frequency. The parameter Ea shows an increase for low frequencies (f <100 kHz) followed by an almost linear increase with frequency above 100 kHz. The corresponding recorded values remain low confirming that the B $S_{0.025}$Tis a conventional ferroelectric material [28-29].

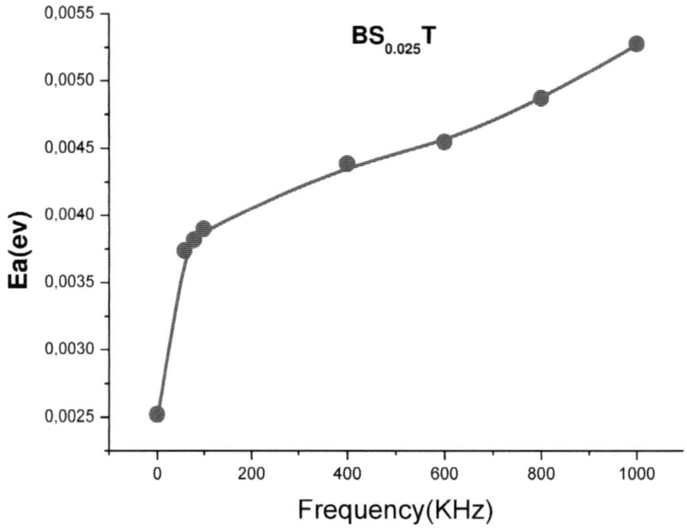

Figure 9. Activation energy (Ea) as a function of frequency of $BS_{0.025}T$

3.4. The Complex Impedance

The permittivity is not the only possible representation of the response of a material to an alternating current. Additional derived quantities can be used as: The admittance Y (ω): Y (ω) = 1 / Z (ω), Modulo, M (ω): M (ω) = iωC0 Z (ω) = iωC0 / Y (ω). Where C_0 is the capacity of the empty cell.
From these relations one can find the real part Z 'and Z imaginary "impedance.
$Z' = \varepsilon'' / 2.pi.f\, \varepsilon_0 * (\varepsilon'^2+\varepsilon''^2)$ and $Z'' = \varepsilon' / 2.pi.f\, \varepsilon 0 * (\varepsilon''^2+\varepsilon'^2)$
The results of the complex impedance can be represented in different ways:
Argand diagram (Z "= f (Z ')), Z" and Z' as functiuns the frequency. Argand diagrams of these samples showed a single semi semicircle in the frequency range considered (Fig.10), which means that the electric material interface and the domain walls are not contributing to the electrical response [30].

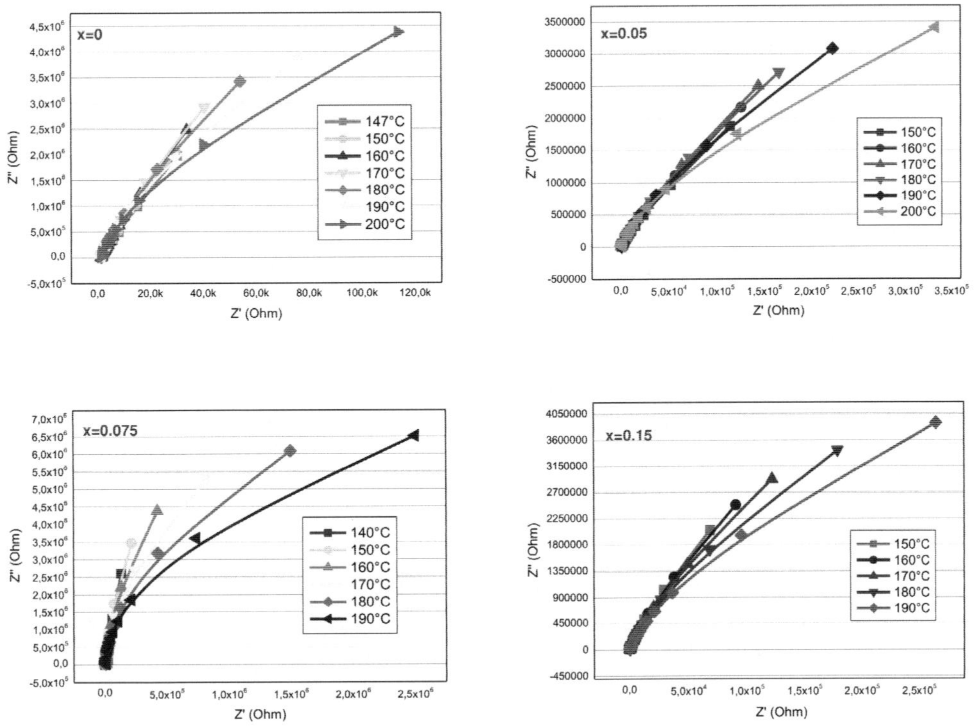

Figure 10. Argand diagram (Z "= f (Z ')) of different samples BSxT at different temperatures.

However, it is possible in the case of a ceramic not to observe separately the second arc representing this mentioned contribution. It can either be masked by the existing semi-circle arc (deformed) or be completely absent from the diagram [31] which is the case of ZrO_2 (Y_2O_3 + MgO) and Bi_2O_3: Er_2O_3 compounds. Or, since we do not have enough frequency values, it was not possible to observe the second arc. It is well known that the ceramic microstructure also plays a key role in the electrical behaviour of $BaTiO_3$ doped with a donor [32].

Fig.11 shows the variation of the real part of the impedance (Z') as a function of the frequency at different temperatures and compositions. It shows a decrease as a function of the frequency in the low frequency region followed by a saturation region in the high frequency region. This indicates the presence of different type of polarizations in the material at low frequency.

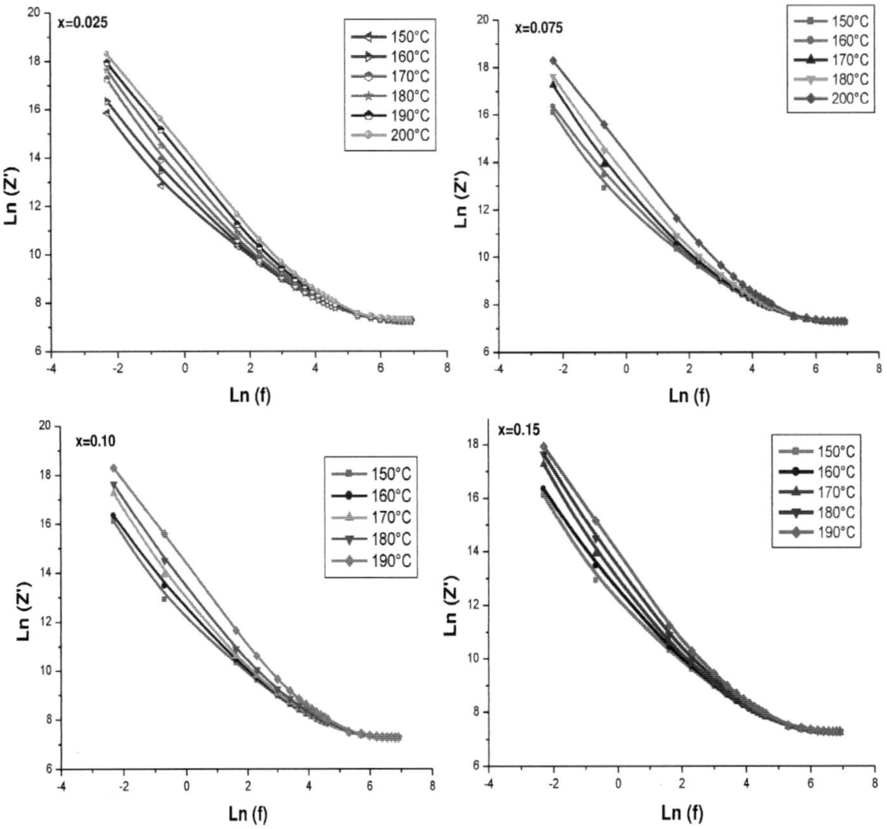

Figure 11. Variation of ln (Z ') as a function of the frequency of BTZx

The dielectric data were used to evaluate the relaxation time (τ) of the electrical phenomena in the material using the relation: $\tau = \dfrac{1}{\omega_{\max}} = \dfrac{1}{2.\pi.f_{\max}}$ (4)

Where f_{max} is the relaxation frequency. The nature of the variation of τ with temperature of all the compounds is shown in Fig.12. The curves appear to follow the Arrhenious relation which is given by:

$$\tau = \tau_0 \exp\frac{E_\tau}{KT} \quad (5)$$

Where τ_0 is a pre-exponential factor, Eτ is the activation energy, k is the Boltzmaan constant and T is the absolute temperature. Activation energy values evaluated from the slope of log (τ) vs. $10^3\ T^{-1}$ curve are 3.18, 1.51, 1.14, 0.81 eV for x =0, 0.05, 0.10, 0.15 respectively. It is clear that the value of activation energy decreases with an increase in Sr concentration. The small value of activation energy may be due to the ionized oxygen vacancies.

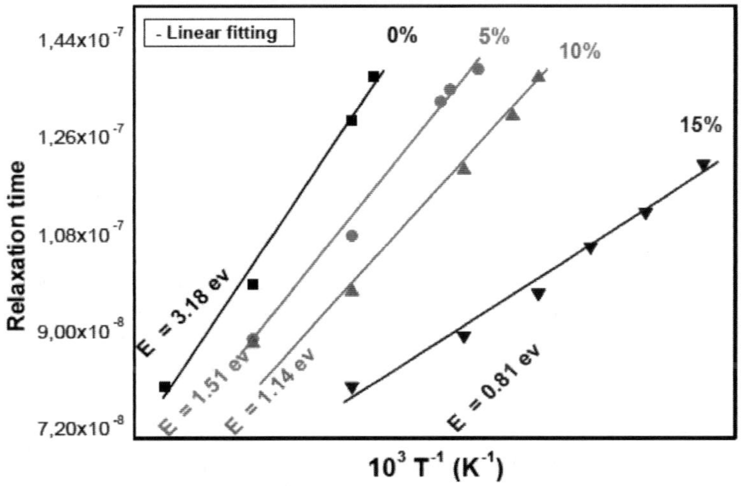

Figure 12. Variation of relaxation time with inverse of temperature of BSxT

4. Conclusion

In this work, we have successfully prepared a series of $Ba_{1-x}Sr_xTiO_3$ samples. The phase formation of the samples was confirmed by XRD technique and spectroscopy Raman. The frequency dependent electrical data were used to study the conductivity and to estimate the activation energy (Ea). The results showed that the relaxation frequency value fr, at Tc increases when x increases from 0 to 0.15, which behaviour may be explained by the fact that the substitution of Sr in $BaTiO_3$ leads to a deformation of the BO6 octahedron, which probably implies shorter correlation chains and can be linked, with the AO binding nature (A: Ba, Sr) and/or polarizability of atoms. Argand diagrams are represented by a single semi semicircle which means that the electro material interface and the domain walls are not contributing to the electrical response, however the behaviour of the conductivity may be directly related to the relaxation phenomenon which results in a leakage current due to the energy dissipation.

Reference

[1] Scott JF, Auraujo CAP, Melnick BM, McMillan LD, Zuleeg R 1991 Quantitative measurement of spacecharge effects in lead zirconatetitanate memories *J. Appl. Phys.* **70** 382

[2] Lines ME, Glass AM 2001 Principles and applications of ferroelectrics and related materials *Oxford Classic Texts in the Physical Sciences* 1977 680.

[3] Bunting LN, Shelton GA, Creamer AS 1947 Properties of barium strontium titanate dielectrics *J. Am. Ceram. Soc.* **30** 114

[4] Horikawa T, Mikami N, Ito H, Ohno Y, Makita T, Sato K 1994 $(Ba_{0.75}Sr_{0.25})TiO_3$ Films for 256 Mbit DRAM *IEICE Trans. Electron E77-C* **3** 385

[5] Zhung J, Li GS, Gao XC, Guo XB, Huang YH, Shi ZZ, Lu YYWJ 1991 Preparation and photorefractive properties of barium strontium titanate $(Ba_{1-x}Sr_xTiO_3)$ *Opt. Commun.* **82** 69

[6] Li W, Xu Z, Chu R, Fu P, Hao J 2009 Structure and electrical properties of $BaTiO_3$ prepared by sol gel process *J. Alloys Compd.* **482** 137

[7] Sengupta LC and Sengupta S 1999 Breakthrough advances in low loss, tunable dielectric materials *Mater. Res. Innov. 2* **5** 278

[8] Liang XF, Meng ZG, and Wu WB 2004 Effect of Acceptor and Donor Dopants on the Dielectric and Tunable Properties of Barium Strontiu Titanate *J. Am. Ceram. Soc* **87** 2218

[9] Feteira A, Sinclair DC, Reaney IM, Somiya Y, and Lanagan MT 2004 $BaTiO_3$-Based Ceramics for Tunable Microwave Applications *J. Am. Ceram. Soc* **87** 1082
[10] RoutS K, Rout SK, Bera J 2004 in: A.P. Tandon (Ed.), Ferroelectrics and Dielectrics, Allied Publishers Pvt. Ltd., New Delhi pp.37
[11] Fu C, Yang C, Chen H, Wang Y, Hu L 2005 Microstructure and dielectric properties of $Ba_1Sr_{1-x}TiO_3$ ceramics *Mater. Sci. Eng. B.* **119** 185
[12] Thakur O. P, Prakash C, Agarwal DK 2002 Dielectric behavior of $Ba_{0.95}Sr_{0.05}TiO_3$ ceramics sintered by microwave *Mater. Sci. Eng. B.* **96** 221
[13] Alexandru HV, Berbecaru C, Ioachim A, Nedelcu L, Dutu A 2006 BST solid solutions, temperature evolution of the ferroelectric transitions *Applied Surface Science* **253** 354
[14] Kim SW, Choi HI, Lee MH, Park JS, Kim DJ, Do D, Kim MH, Song TK, Kim WJ 2013 Electrical properties and phase of $BaTiO_3$ –$SrTiO_3$ solid solution *Ceramics International* **39** S487
[15] Horchidan N, Ianculescu CA, Curecheriu LP, Tudorache F, Musteata V, Stoleriu S, Dragan N, Crisan D, Tascu S, Mitoseriu L 2011 Preparation and characterization of barium titanate stannate solid solutions *J. Alloys Compd.* **509** 4731
[16] Darwish AGA, Badr Y, El Shaarawy M, Shash NMH, Battisha IK 2010 Influence of the Nd^{3+} ions content on the FTIR and the visible up conversion luminescence properties of nano-structure $BaTiO_3$ prepared by sol-gel technique *J. Alloys Compd* **489** 451
[17] A. Elbasset, S. Sayouri 2015 Thermal treatment effect on piezoelectric and pyroelectric properties of BaTiO3 ceramics. Adv. Mater. Lett **6(11)** 999-1003
[18]D.H. cruz. Thèse présentée à la Faculté des études supérieures de l'Université Laval 2004
[19] X. Wei, G. Xu, Z. Ren, Y. Wang, G. Shen, G. Han 2008 *Journal of Crystal Growth* **310** 4132
[20] S.Y. Kuo, W.Y. Liao, and W.F. Hsieh 2001 *Physical Review B* **64** 224103
[21] B. Hou, Y. Xu, D. Wu, Y. Sun 2006 *Powder Technology* **170** 26
[22] J. Zhang, J. Zhai, X. Chou, J. Shao, X. Lu, X. Yao 2009 Acta Materialia **57** 4491
[23] J. Di Domenico, H. Wample and S. Porto 1968 *Phys. Rev.***174** 522
[24] A. Scalabrin, A.S. Chaves, D.S. Shim, S.P.S. Porto, R.P. Buman 1977 *Status Solid B* **79** 731
[25] A.S. Chaves, R.S. Katiyar, S.P.S. Porto 1974 *Phys. Rev.B.***10** 3522
[26] R.Come lambert and Guinier 1968 *Solid State Comm.* **6** 715
[27] M. Maglion, R. Bohmer, A. Loid and U.T. Hochli 1989 *Phys. Rev. B* **40** 11441
[28] S. Cole, R. H. Cole 1951 *J. Chim. Phys.* **19** 1484
[29] A. Smolenskii et al. 1984 Ferroelectrics and Related Materials *Gordon and Breach New York* 589
[30] R. James, S. Priya, Uchino, Jpn. *J. Appl. Phys.* **41** 5272
[31]J. R. Macdonald, John Wiley and sons (1987).
[32]M.H. Chan, M.P. Harmer, D.M. Smyth 1986 *J. Am. Ceram. Soc.* **69** 507

Structural study of emulsions stabilized by charged nanoparticles

S El-moudny[1], M Badia[2] and M Benhamou[1]

[1]Research Team Advanced Materials and Applications, ENSAM, Moulay Ismail University, B. O Box 1529, Al Mansour 500000, Meknes, Morocco
[2]Mechanical Department, DFST, Royal Air School, P. O. Box 40002, Menara, Marrakech, Morocco

Email: elmoudny.soukaina@gmail.com

Abstract. The aim is a quantitative study of the structure of the Pickering emulsionsthat are stabilized by charged nanoparticles. We assume that the interaction potential between droplets is of Sogami type. First, we compute the structural properties using the *Integral Equation Method* with the so-called *hybridized-mean spherical approximation* (HMSA). Finally, the validity of this method is tested by confronting the results of the correlation functions found by HMSA integral equation with those from Molecular Dynamics (MD) simulations.

1. Introduction

Very often, an emulsion presents in the form of oil droplets in water or water droplets in oil, but in both cases the droplets need to be stabilized to prevent them from re-coalescing, because when two droplets come into contact, they naturally coalesce to minimize the surface energy. Emulsion droplets stabilization is often achieved through the addition of amphiphilic molecules such as surfactants, which act by decreasing the interfacial tension between the phases. In addition to molecular surfactants, solid particles having intermediate wetting properties can also be used to stabilize emulsions. In fact, the solid particles form a shell protecting the emulsion droplets against coalescence. Emulsions stabilized by solid particles are known as Pickering emulsions [1, 2],in which emulsifiers irreversibly adsorb at the oil–water interface and require a much higher energy for desorption as compared to conventional surfactants [3]. In addition to the high stability of Pickering emulsions, the stabilization of emulsions by solid particles gives the possibility to control the size of the droplet of the emulsion, because the solid stabilizing particles are necessarily smaller than the emulsion droplets. Solid particles of nanometric size allow the stabilization of droplets as small as few micrometers diameter. Micron-sized solid particles can stabilize larger droplets, the diameter of which possibly reaching few millimeters [4].

The Pickering emulsions have a complex structure which is present in the form of a core with discrete spherical particles arranged on a (spherical) surface. Therefore, simple models that may have a crude resemblance to the structures are often considered. The most frequently used model for these systems is to assume a core–shell structure, where the shell has a homogenous density and thickness [5, 6]. This approach is not strictly appropriate but may often be valid if the preparing conditions (such as wettability, charge, concentration, shape and size of particles, etc.) are right.

In this work, using the model of core-shell supra-colloids that we assume interacting via Sogami and Ise potential, we investigate the structural properties of Pickering emulsions, stabilized by charged nanoparticles, i.e. formed by micrometer-sized droplets, through the use of the so-called integral equation method. We validate the use of this method by MD simulation.

This paper is organized as follows. In Sec. 2, we present the useful theoretical backgrounds. Results and discussion are the aim of Sec. 3. Finally, some concluding remarks are drawn in the last section.

2. Theoretical backgrounds

2.1. Interaction Potential

We consider a system of supra-colloids of diameter σ, interacting via the Sogami and Ise potential which has usual screened Coulomb repulsion and also an attractive part leading to a secondary minimum [7, 8], defined as

$$U_s(r) = \frac{Z^2e^2}{2\varepsilon}\left[\frac{\sin(\kappa\sigma/2)}{(\kappa\sigma/2)}\right]^2\left[\frac{2+k\sigma\coth(\kappa\sigma/2)}{r} - \kappa\right]\exp(-\kappa r), \quad (1)$$

wherer is the inter-particle distance, and κ is the inverse screening length that is given by

$$\kappa^2 = \frac{4\pi}{\varepsilon k_B T}(\rho Z e^2 + C_s). \quad (2)$$

Here ρ is the number density of droplets, each having a surface charge Ze , C_s is the salt concentration, ε is the relative dielectric constant of water and $k_B$is the Boltzmann's constant. The absolute temperatureT is fixed to the value 298°K. Then, the chosen potential depends on seven parameters:$(Z, \sigma, \kappa, \varepsilon, \rho, C_s, T)$. The above potential exhibits a minimum point for

$$r_m = \frac{\frac{\kappa\sigma}{2}\coth\left(\frac{\kappa\sigma}{2}\right)+1+\left[\left(\frac{\kappa\sigma}{2}\coth\left(\frac{\kappa\sigma}{2}\right)+1\right)\left(\frac{\kappa\sigma}{2}\coth\left(\frac{\kappa\sigma}{2}\right)+3\right)\right]^{1/2}}{\kappa}. \quad (3)$$

2.2. Theory and simulation details

The static description of the homogenous classical liquids, whose particles are taken to interact through an effective pair potential, can be obtained from the exact Ornstein-Zernike (OZ) [9] equation which represents the starting point of the Integral Equation Theory (IET).

$$h(r) = c(r) + \gamma(r) = c(r) + \rho\int c(r')h(|r - r'|)dr' \quad (4)$$

The total correlation function $h(r)$ is a sum of two contributions. The first contribution is the direct correlation function, $c(r)$, which represents the correlation between a particle of a pair with its closest neighbor separated by a distance r. The second contribution is the indirect correlation function, $\gamma(r)$, which represents the correlation between the selected particles of the pair with the rest of the fluid constituents. To solve equation (4), an approximation (closure) relation which relates the correlation functionc(r) and $h(r)$to the pair-potential should be added. In this study, we will use HMSA,

$$g^{HMSA}(r) = \exp(-\beta u_o(r))\left\{1 + \frac{\exp\{f(r)[g(r)-1-c(r)-\beta u_1(r)]\}-1}{f(r)}\right\}, \quad (5)$$

where$u_o(r)$ and $u_1(r)$are respectively the short-range and the long-range contributions of the potential according to Weeks *et al.* [10],

$$u(r) = u_o(r) + u_1(r), \quad (6)$$

where

$$u_o(r) = \begin{cases} u(r) - u(r_m), & r < r_m \\ 0, & r > r_m, \end{cases} \quad (7)$$

and

$$u_1(r) = \begin{cases} u(r_m), & r < r_m \\ u(r), & r > r_m. \end{cases} \quad (8)$$

Here,r_m denotes the position of the minimum of the potential, and$f(r)$ is the mixing-function, proposed by Bretonnet and Jakse [11], which ensures the requirement of the thermodynamic consistency, in calculating the isothermal compressibility by two different routes.

Within the framework of MD method, the equations of motion are solved in the microcanonical ensemble using the velocity Verlet algorithm [12] with the thermostat of Berendsen [13], in order to keep the temperature constant. Periodic boundary conditions are applied to remove the surface effects and simulate an infinite system. In the following, we will use dimensionless units, where the length unit is σ, time in units of $\sigma\sqrt{m/\epsilon}$, where ε is the depth of the interaction potential and m is the droplet mass, $k_B T$ is the energy unit, and $L_0 = \sqrt[3]{6V/\pi N} \times N$ is the box size, where N is the number of droplets and V is the volume of simulation box (in periodic conditions). MD simulations where carried out with 1728 particles and the (reduced) time step for the velocity Verlet algorithm is chosen to be 0.01.

3. Results and discussions

We start by comparing the theoretical results for the correlation functions $\mathbf{g(r)}$ calculated from HMSA approximation and MD simulation. From the figure 1(a)-(b) we particularly note that the peaks in the correlation functions from molecular dynamics simulations are in good agreement with the corresponding values calculated from the integralequation. Thus,HMSA solution of OZ relation provides a very accurate description of our system.

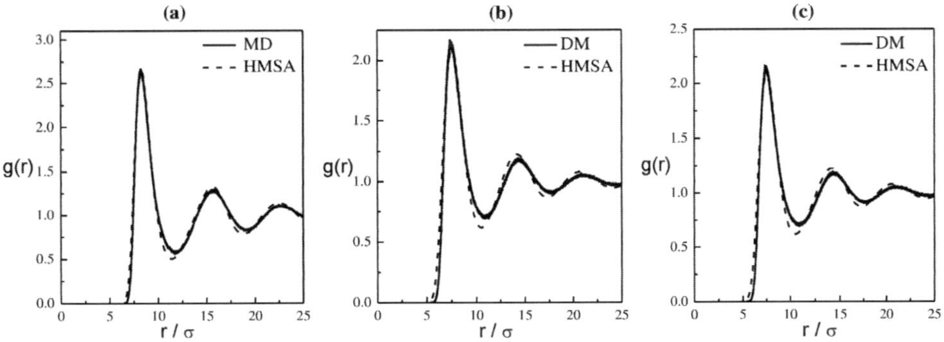

Figure 1.Comparison for the correlation function calculated from HMSA approximation and from MD simulation, (a) at $\boldsymbol{Z} = \mathbf{2000}$ and $\boldsymbol{\rho}^* = \mathbf{0.0020}$ and $\boldsymbol{\sigma} = \mathbf{20000}$Å, (b) at $\boldsymbol{Z} = \mathbf{2000}$ and $\boldsymbol{\rho}^* = \mathbf{0.0025}$ and$\boldsymbol{\sigma} = \mathbf{20000}$Å, (c) at $\mathbf{Z} = \mathbf{2000}$and $\boldsymbol{\rho}^* = \mathbf{0.0025}and\boldsymbol{\sigma} = \mathbf{30000}$Å.

3.1. The density effect

In this section, we study the effect of the increasing density of the supra-colloids, at fixed $\boldsymbol{\sigma} = \mathbf{20000}$Å and $\mathbf{Z} = \mathbf{2000}$. Figure 2shows the correlation function for different densities obtained by the HMSA closure. It is clear from this figure, on the one hand, that as the density of supra-colloids increases, the height of the main peak of $\boldsymbol{g(r)}$ steadily increases (1.33 for $\boldsymbol{\rho}^* = \mathbf{0.0010}$ to 2.09 for $\boldsymbol{\rho}^* = \mathbf{0.0025}$) and the peak positions steadily decreases in $\boldsymbol{r}$.Such a result indicates that, when the number density is increased, the local density around a droplet, taken as origin, also increases. The decreasing peak position in the structure factor (figure 3) with the increasing density of supra-colloids is also consistent with this feature. On the other hand, the height of the second peak increases at increasing density and is absent at$\boldsymbol{\rho}^* = \mathbf{0.0010}$. Thus, a local structural order is established, represented by the first and second peak, and the local density increases (height of the main peak), when the supra-colloid density increases.

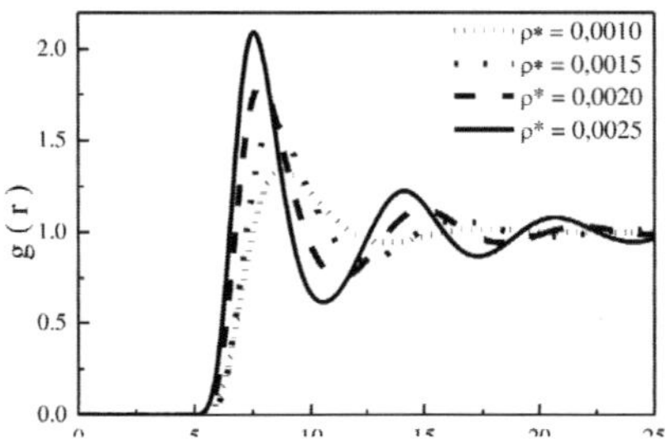

Figure 2. Variation of the correlation functionwith the reduced density ($\rho^* = \rho\sigma^3$) of supra-colloids, at fixed charge ($Z = 2000$) and fixed size ($\sigma = 20000$Å).

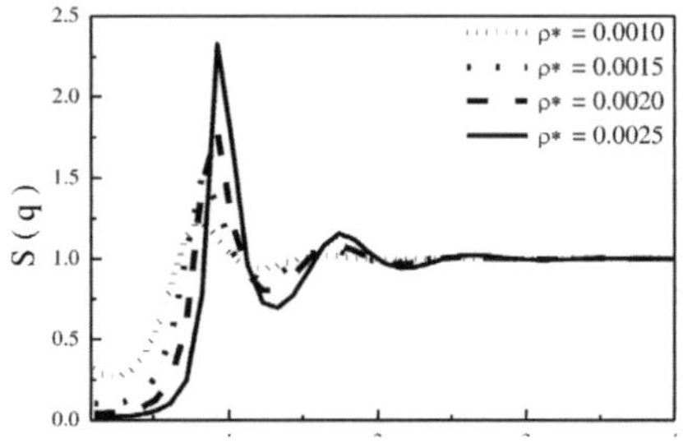

Figure 3. Variation of the factor structure with the reduced density of supra-colloids at fixed charge ($Z = 2000$) and fixed size ($\sigma = 20000$Å).

3.2. The charge effect

The charge of the supra-colloids is varied from 500 to 2000 at fixed $\boldsymbol{\sigma} = \mathbf{20000}$**Å** and $\boldsymbol{\rho}^* = \mathbf{0.0020}$. The effective pair-potential is shown in figure 4. It explains that as the charge of particles is increased, the potential depth is diminished. Thus, the effective interaction between the supra-colloids becomes less and less attractive as the charge increases and finally it becomes repulsive. This tendency is confirmed in figure 5 and figure 6 that represent the correlation function and structure factor, respectively. In fact, we remark that the position of the principal peak of the correlation function is shifted towards the higher values by an increasing of the charge. In contrast, the main peak of the factor structure is shifted towards smaller distances as the charge of particles increases.

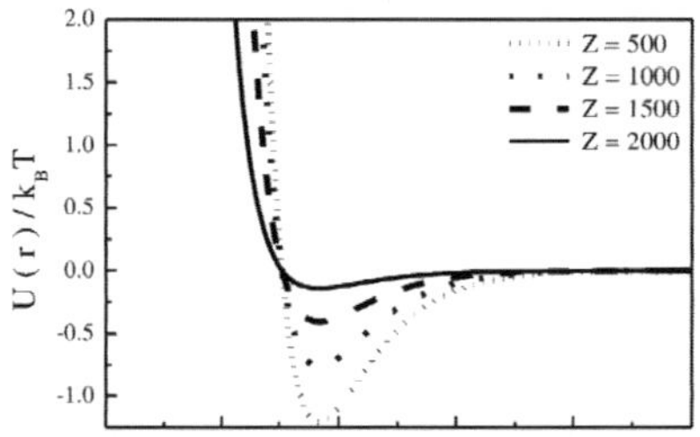

Figure 4. Variation of the effective pair-potential with the charge of supra-colloids, at fixed density ($\rho^* = 0.0020$) and fixed size ($\sigma = 20000$Å).

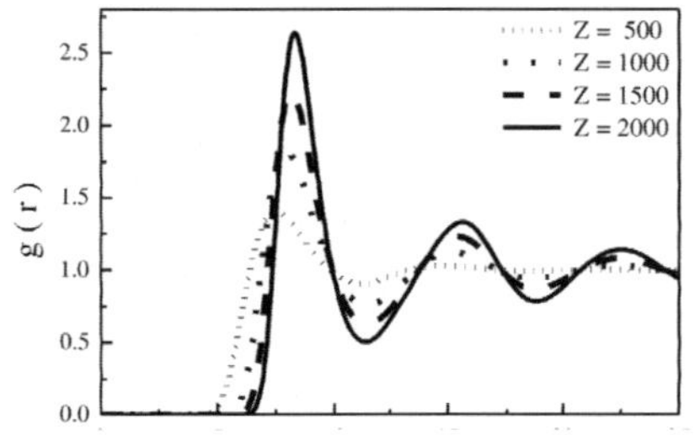

Figure 5. Variation of the correlation function with the charge of supra-colloids at fixed density ($\boldsymbol{\rho}^* = \mathbf{0.0020}$) and fixed size ($\boldsymbol{\sigma} = \mathbf{20000}$**Å**).

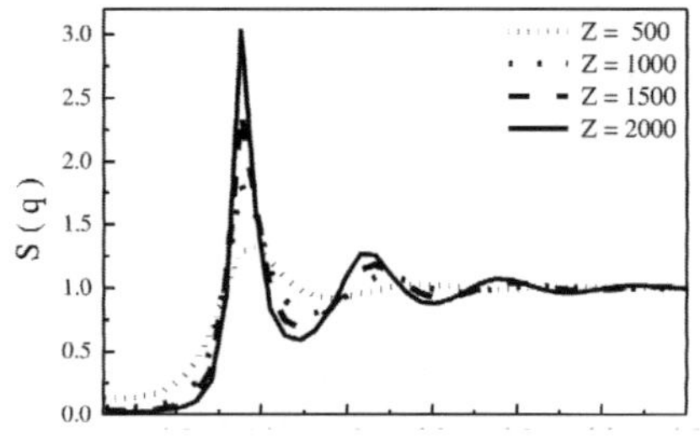

Figure 6. Variation of the factor structure with the charge of supra-colloids at fixed density ($\boldsymbol{\rho}^* = \mathbf{0.0020}$) and fixed size ($\boldsymbol{\sigma} = \mathbf{20000}$**Å**).

3.3. The size effect

In this section, the effect of the variation of the size of the supra-colloids on the structure is studied at $\rho^* = 0.0020$ and $Z = 2000$. The variation of the effective pair-potential (figure 7) is very much similar to that of the charge variation, i.e. as the size increases; the effective interaction between the supra-colloids becomes less and less attractive. The figure 8 shows that the height of the principal peak of the correlation-function decreases progressively, and its position is shifted to higher values, as the droplet size is increased. This means that the size of the local ordered region made of first neighbors (around some droplet) increases with increasing droplet diameter.This result is confirmed by the variation of the factor structure shown in figure 9.

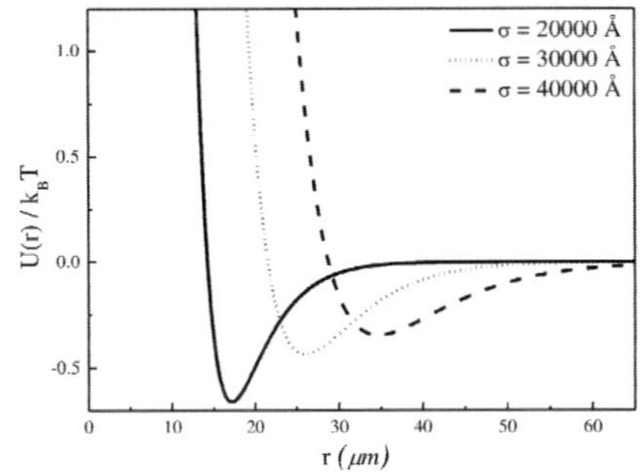

Figure 7. Variation of the effective pair-potential with the size of supra-colloids at fixed density ($\rho^* = 0.0020$) and fixed charge ($Z = 2000$).

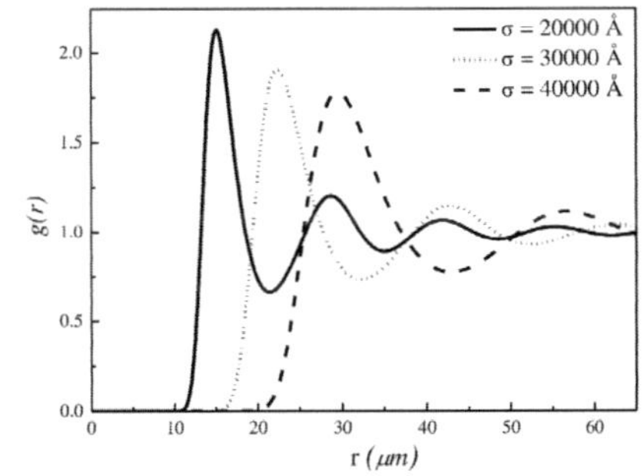

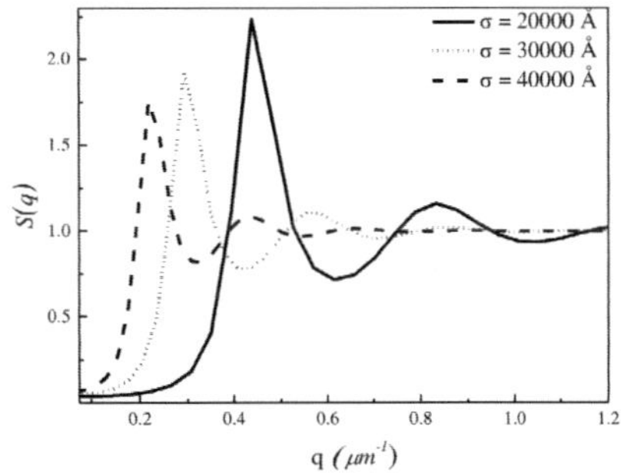

Figure 8. Variation of the correlation function with the size of supra-colloids at fixed density ($\rho^* = 0.0020$) and fixed charge ($Z = 2000$).

Figure 9. Variation of the factor structure with the size of supra-colloids, at fixed density$\boldsymbol{\rho}^* = \mathbf{0.0020}$and fixed charge (Z=2000).

4. Conclusion

The structure of Pickering emulsions, using the model of supra-colloids, is studied through of the use of the integral equation method. We find a systematic variation in correlation functions, the structure factors and the effective pair-potentials as the density, the charge and the size of supra-colloid are varied.

References

[1] Pickering S U 1907*J. Chem. Soc.* Trans. **91**2001–21
[2] Ramsden W 1903 *Proc. R. Soc.* Vol 72 (London) 156–64

[3] BinksB P2002 *Curr. Opin. Colloid Interface Sci.***7(1)**21–41
[4] Chevalier Yand BolzingerM A 2013 *Colloids Surf.* A**439** 23–34
[5] Larson-Smith K, JacksonA and Pozzo D C 2010 *J. Colloid Interface Sci.***343(1)**36–41
[6] VerritoV J and KilpatrickP K2008 *Langmuir*,**24(22)**, 12807–22
[7] SogamiI 1983 *Phys. Lett.* A,**96(4)**199–203
[8] SogamiI and IseN 1984 *J. Chem. Phys***81(12)** 6320–32
[9] Ornstein L Sand Zernike F1914*In Proc. Akad. Sci.* Vol 17 *(Amsterdam)*p 793
[10] WeeksJ D, ChandlerD and AndersenH C 1971*J. Chem. Phys.* **54(12)** 5237–47
[11] BretonnetJ L and Jakse N 1992 *Phy.s Rev.*B**46(9)** 5717
[12] Allen M P and Tildesley D J 1989 *Computer Simulation of Liquids* (Oxford University Press)
[13] BerendsenH J, PostmaJ P M, van GunsterenW F, DinolaA R H J and HaakJ R1984 *J.chem. Phys.***81(8)** 3684-90

IOP Conf. Series: Materials Science and Engineering **186** (2017) 012020 doi:10.1088/1757-899X/186/1/012020

Spherical Brownian diffusion of particles on liquid-liquid interfaces of Pickering emulsions

S El-moudny[1], M Badia[2] and M Benhamou[1]

[1]Research Team Advanced Materials and Applications, ENSAM, Moulay Ismail University, B. O Box 1529, Al Mansour 500000, Meknes, Morocco
[2]Mechanical Department, DFST, Royal Air School, P. O. Box 40002, Menara, Marrakech, Morocco

Email: elmoudny.soukaina@gmail.com

Abstract. We investigate the Brownian diffusion of particles that are confined in liquid/liquid interface in Pickering emulsions. These particles are subject to a spherical diffusion. Using the Green's function techniques, we *exactly* derive the diffusion laws.

1. Introduction

Pickering emulsions are emulsions that are stabilized by solid particles [1, 2]. In fact, the solid particles form a spherical shell and impede coalescence when two droplets approach each other. Pickering emulsifiers irreversibly adsorb at the oil–water interface and require a much higher energy for desorption ($\approx 10^6 - 10^8 k_B T$) as compared to the conventional surfactants ($\approx 7\, k_B T$). The emulsifier-free character of Pickering emulsions makes them attractive regarding applications where the surfactants have detrimental effects, in particular, when contacted with living matter for health and body care applications [3-4].

Pickering emulsions have been the focus of considerable research in the past decade due to their properties such as high stability with respect to the coalescence, as well as due to advances in nanotechnology that allows us to create and characterize the nano-scale structures in new ways. In fact, the colloidal assembly of solid particles within Pickering emulsions can be used as templates for advanced materials such as Janus colloids, composite particles, and colloidosomes by analogy with liposomes. In addition to increasing possibilities for practical applications, Pickering emulsions also provide a new and convenient experimental model system for investigating spherical diffusion, i.e. Brownian motion of solid particles at the liquid-liquid interfaces [5-7].

Diffusion on the sphere is a problem that arises in several contexts. Diffusion is an important mode of transport of biological substances (lipids and proteins) on the cell walls which are curved surfaces. Spherical diffusion also crops up in surface smoothening in computer graphics and global migration patterns of marine mammals. While the planar diffusion has been studied extensively both analytically and numerically, there have been fewer analytical studies of diffusion on spherical surfaces.

In this work, using Pickering emulsions as template, we establish the diffusion laws of particles, confined on the surface of a sphere, undergoing Brownian motion.

This paper is organized as follows. In Sec. 1, we present the useful theoretical backgrounds. Results and discussion are the aim of Sec. 2. Finally, some concluding remarks are drawn in the last section.

2. Theory

In this section, we compute the Green's function that is solution of the spherical diffusion equation enabling us the derivation of the diffusion laws

When modeling the diffusion phenomena in an infinite medium, one often solves the diffusion equation for the probability density, $u(x,t)$, which is subject to given initial condition: $u(x,t=0)=f(x)$.

A one dimensional linear diffusion equation with constant coefficient is defined as [8]

$$\Delta u(x,t) = \frac{1}{D}\frac{\partial}{\partial t}u(x,t)\,, \tag{1}$$

where the parameter D is called the diffusion coefficient.The spherical diffusion equation is given by

$$\Delta_{S^2} u(\theta,\varphi,t) = \frac{1}{D}\frac{\partial}{\partial t}u(\theta,\varphi,t)\,. \tag{2}$$

Here, Δ_{S^2}accounts for the spherical Laplace operator defined as

$$\Delta_{S^2} = \frac{1}{\sin\theta}\frac{\partial}{\partial\theta}\left(\sin\theta\frac{\partial}{\partial\theta}\right) + \frac{1}{(\sin\theta)^2}\frac{\partial^2}{\partial\varphi^2}\,. \tag{3}$$

In obtaining the solutions to the Laplace's equation in spherical coordinates, it is traditional to introduce the spherical harmonics,$Y_l^m(\theta,\varphi)$,

$$Y_l^m(\theta,\varphi) = N_{lm}\,P_l^m(\cos\theta)e^{im\varphi}\,, \tag{4}$$

with the normalization factor

$$N_{lm} = \sqrt{\frac{(2l+1)}{4\pi}\frac{(l-m)!}{(l+m)!}}, \tag{5}$$

and the associated Legendre functions

$$P_l^m(x) = \frac{(-1)^m(1-x)^{m/2}}{2^l l!}\frac{d^{l+m}}{dx^{l+m}}(x^2-1)^l\,. \tag{6}$$

For solving the diffusion equation, we use the standard separation of variables,

$$u_{lm}(\theta,\varphi,t) = Y_l^m(\theta,\varphi)\,T(t)\,. \tag{7}$$

Inserting this decomposition into the diffusion equation and using the fact that the spherical harmonics are eigen-functions of the spherical Laplace operator yields

$$u_{lm}(\theta,\varphi,t) = Y_l^m(\theta,\varphi)e^{-l(l+1)kt}\,. \tag{8}$$

In order to obtain the Green's function, we will impose the following initial condition

$$u_{lm}(\theta,\varphi,0) = \delta_{S^2}(\theta,\varphi)\,, \tag{9}$$

where the symbol $\delta_{S^2}(\theta,\varphi)$is the spherical Dirac function defined by [9]

$$\delta_{S^2} = \sum_{l\in N} \frac{2l+1}{4\pi} P_l(\cos\theta) . \quad (10)$$

Thus, we obtain for the Green's function the final result

$$G(\theta,\varphi,t) = \sum_{l\in \mathbf{N}} \sqrt{\frac{2l+1}{4\pi}} u_{l0}(\theta,\varphi,t), \quad (11)$$

with

$$u_{l0}(\theta,\varphi,t) = Y_l^0(\theta,\varphi) e^{-l(l+1)kt}. \quad (12)$$

3. Results and discussion

The statistical properties of a Brownian particle moving on the liquid-liquid interface, are well described by the Green's function (propagator), $G(\theta,\varphi,t)$, which represents the probability density of finding a particle at point r at time t, together with the initial condition: $G(r,r',0) = \delta(r - r')$. Once the propagator is known, all the statistical properties can be evaluated. For instance, the mean-square displacement (MSD) is obtained from

$$\langle r^2 \rangle = R^2 \sum_{l=0}^{\infty} (2l+1) e^{-l(l+1)t/\tau} \int_0^{\pi} d\theta \sin\theta (1 - \cos\theta) P_l(\cos\theta), \quad (13)$$

where$\tau = R^2/D$ is the relaxation time, with the particle radius R.

From the well-known results of integrals [10], we find

$$\langle r^2 \rangle = 2R^2 \left(1 - e^{-2t/\tau}\right). \quad (14)$$

Let us comment the above obtained result. Firstly, as expected, MSD saturates to$\langle r^2 \rangle = 2R^2$, in the long-time limit that is for $t \gg \tau$. Secondly, in the short-time limit($t \ll \tau$), however, we find that MSD does not depend on the radius of the sphere and we recover, in this limit, the celebrated Einstein's diffusion law on the two-dimensional Euclidean space. This result can be explained by the fact that in short-time limit the particle has not enough time to feel the influence of curvature of the sphere.

4. Conclusion

In this paper, we have discussed the diffusion of the confined particles on the spherical surface of liquid-droplets in Pickering emulsions, using the Green's function techniques. In particular, we have found that, for short-times, when they move, these particles do not feel the influence of the geometrical shape.

References

[1] Pickering S U 1907 *J. Chem. Soc.* Trans. **91** 2001–21
[2] Ramsden W 1903 *Proc. R. Soc.* Vol 72 (London) 156–64
[3] Binks B P 2002 *Curr. Opin. Colloid Interface Sci.* **7(1)** 21–41
[4] Chevalier Y and Bolzinger M A 2013 *Colloids Surf.* A **439**23–34
[5] Lin Y, Skaff H, Emrick T, Dinsmore A D and Russell T P 2003*Science***299(5604)**226–9
[6] TarimalaS, Ranabothu S R, Vernetti J P and Dai L L 2004 *Langmuir***20*(13)**5171–73
[7] Balakrishnan G, Nicolai T, Benyahia L and Durand D 2012 *Langmuir* ***28*(14)**921–26
[8] Chirikjian G S and Kyatkin A B 2000 *Engineering applications of noncommutative harmonic analysis: with emphasis on rotation and motion groups*(CRC press)p 23
[9] Buelow T 2001 *Tech. Rep.* (CIS) 172
[10] Gradshteyn I S, Ryzik I M 1980 *Table of Integrals, Series and Products* (New York Academic Press) pp 804–33

Confining a fluid membrane vesicle of toroidal topology in an adhesive hard sphere

Lila Bouzar[1,2], **Ferhat Menas**[2,3] **and Martin Michael Müller**[4,5]

[1] Département de Physique Théorique, Faculté de Physique, Université des Sciences et de la Technologie Houari Boumedienne, BP 32 Al-Alia, Algiers 16111, Algeria
[2] Laboratoire de Physique et Chimie Quantique, Université Mouloud Mammeri, BP 17, Tizi-ouzou 15000, Algeria
[3] Département de Physique, École Nationale Préparatoire aux Études d'Ingéniorat, BP 05 Rouiba, Algiers 16012, Algeria
[4] Equipe Biophysstat, ICPMB-FR CNRS 2843, Université de Lorraine, 1 boulevard Arago, Metz 57070, France
[5] Institut Charles Sadron, CNRS-UdS, BP 84047, Strasbourg 67034, France

E-mail: `Martin-Michael.Mueller@univ-lorraine.fr`

Abstract. We discuss how the equilibrium shapes of a confined toroidal fluid membrane vesicle change when an adhesion between membrane and confining sphere is taken into account. The case without adhesion was studied in Ref. [1]. Different types of solution were found and assembled in a phase diagram as a function of area and reduced volume of the membrane. Depending on the degree of confinement the vesicle is either free, in contact along a circle (*contact-circle solutions*) or on a surface (*contact-area solutions*). All solutions without adhesion are up-down symmetric. When the container is adhesive, the phase diagram is altered and new kinds of solution without up-down symmetry are found. For increasing values of adhesion the region of contact-circle solutions shrinks until it vanishes completely from the phase diagram.

1. Introduction

Mitochondria are essential biological constituents which produce the chemical energy carriers of the cell. Their basic structure consists of two closed lipid membranes, one inside the other [2]. To model this system we consider a fluid membrane vesicle confined in a hard sphere. The case of a membrane of spherical topology has been studied by Kahraman et al. [3, 4]. These studies have shown that—under specific conditions—the topology of the membrane can change to a toroidal one. In nature, topology changes are frequent as well, particularly in mitochondria [5]. To shed some light on these processes, we consider confined fluid membranes of toroidal topology and their behaviour in equilibrium as a function of area, reduced volume and adhesion coefficient between the membrane and its confinement.

2. Model

Within the classical curvature model the simplest version of the bending energy of a fluid membrane is given by [6]

$$E_b = \kappa \int dA\, 2H^2 \,, \tag{1}$$

where κ is the bending rigidity, H the mean curvature, and dA the surface element. We fix the total area and volume of the membrane using the Lagrange multipliers σ, which corresponds to a surface tension, and P, which is the pressure difference between the inside and the outside of the vesicle.

If we suppose that the membrane is axisymmetric, we can use the angle arc-length parametrisation: the cross section of the membrane is given in terms of the tangent angle ψ as a function of arc-length s. The whole surface can be obtained *via* a rotation about the azimuthal angle ϕ. One portion of the membrane is in contact with the confining sphere. The scaled energy of the free part of the membrane writes[1]

$$E = \int ds\, \mathcal{L} \,, \tag{2}$$

where $\mathcal{L}$ is the scaled Lagrangian which equals

$$\mathcal{L} = \rho \left(\dot{\psi} + \frac{\sin\psi}{\rho} \right)^2 + 2\sigma\rho + P\rho^2 \sin\psi + P_\rho \left(\dot{\rho} - \cos\psi\right) + P_z \left(\dot{z} - \sin\psi\right) \,. \tag{3}$$

All variables are scaled with respect to the corresponding quantities of the spherical container. The conjugate momenta P_i follow from $P_i = \frac{\partial \mathcal{L}}{\partial \dot{q}_i}$. The symbols ρ and z denote the radius and height of the membrane in cylindrical coordinates.

With the help of Hamilton's formalism the shape equations can be written as a system of coupled non-linear differential equations

$$\dot{\psi} = \frac{P_\psi}{2\rho} - \frac{\sin\psi}{\rho} \,, \tag{4}$$

$$\dot{P}_\psi = \left(\frac{P_\psi}{\rho} + P\rho^2 - P_z \right) \cos\psi + P_\rho \sin\psi \,, \tag{5}$$

$$\dot{P}_\rho = \frac{P_\psi}{\rho} \left(\frac{P_\psi}{4\rho} - \frac{\sin\psi}{\rho} \right) + 2\sigma + 2P\rho\sin\psi \,, \tag{6}$$

$$\dot{P}_z = 0 \,, \tag{7}$$

to which the geometrical relations

$$\dot{\rho} = \cos\psi \,, \tag{8}$$

$$\dot{z} = \sin\psi \tag{9}$$

have to be added. This system can be solved with a standard shooting method using a forth-order Runge-Kutta algorithm. Equilibrium solutions in partial contact with the container (contact-area solutions) have to fulfill the boundary conditions at the contact lines between the free and the confined part of the membrane [7][2]

$$\psi_0 = \psi_c \,, \tag{10}$$

$$K_{\perp 0} = 1 + \sqrt{\frac{2\omega}{\kappa}} \,. \tag{11}$$

Eq. (10) states that the tangent to the membrane at the contact line, ψ_0, has to equal the container tangent ψ_c. Eq. (11) imposes a relation between the curvature $K_{\perp 0}$ of the membrane

[1] The details of the method can be found in appendix A of Ref. [1].
[2] Similar conditions hold for the free and contact-circle solutions (see again Ref. [1]).

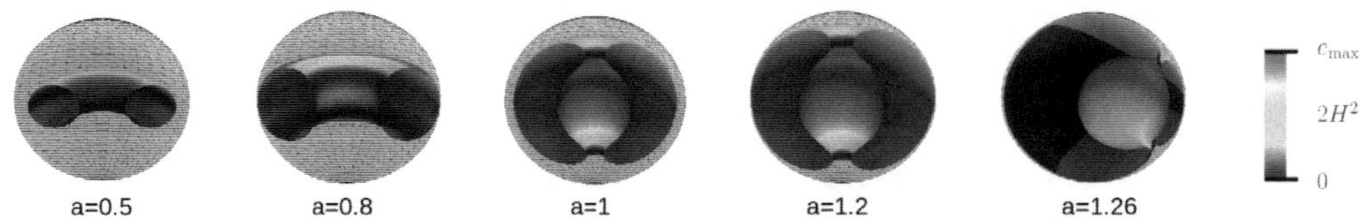

Figure 1. Equilibrium profiles of a membrane without adhesion ($\omega = 0$) for fixed reduced volume $v_r = 0.6$ and increasing area. Colors indicate the density of the bending energy.

that is perpendicular to the contact line, the curvature of the confining sphere, and the adhesion coefficient ω.

The axisymmetry of these solutions can be broken using conformal transformations as in Ref. [8]. These transformations conserve the curvature energy but change the area and the volume of the membrane. Since the membrane is confined in a hard sphere, the transformed solution must fullfil the boundary conditions (10) and (11). In Ref. [1] we have shown that—for the case without adhesion—this restriction reduces the possible transformations to a subgroup defined by the constant parameter $\vec{\Lambda}$

$$\vec{r}' = \frac{\frac{\vec{r}}{r^2} + \vec{\Lambda}}{\left(\frac{\vec{r}}{r^2} + \vec{\Lambda}\right)^2} \left(1 - \Lambda^2\right) + \vec{\Lambda}\,, \tag{12}$$

where $\vec{r}'$ is the transformed shape of the original surface $\vec{r}$. To break axisymmetry $\vec{\Lambda}$ has to have a component perpendicular to the symmetry axis. $\vec{\Lambda} = 0$ yields the identity transformation. When $\Lambda = |\vec{\Lambda}|$ tends to 1, the surface becomes a sphere with a hole of infinitesimal diameter at the edge.

The *total* force on these transformed shapes due to the confining sphere does not necessarily vanish. With the help of finite element simulations we were, however, able to confirm that the shapes approximate the non-axisymmetric solutions remarkably well [1].[3]

3. Results

The complete phase diagram for the case without adhesion can be found in Ref. [1]. Figure 1 shows, for example, shapes that a toroidal vesicle of fixed reduced volume adopts for increasing area. When the area of the membrane equals the container area ($a = 1$), the vesicle is sickle-shaped and does not touch the confinement. In this phase without contact, we exactly retrieve the behaviour of a free toroidal vesicle [8]. When we increase the area, the membrane touches the sphere along a circular contact line (contact-circle solutions). The confinement induces a discontinuity in the derivative of the curvature perpendicular to the contact line which is proportional to the radial contact force. An increase in area will increase this force whereas the curvature at the contact line decreases until it reaches the container curvature. Above this limit the contact line extends to a zone of contact ($a > 1.23$ for $v_r = 0.6$).[4] Subsequently, the vesicle breaks axisymmetry (see the profile with $a = 1.26$ in figure 1 for an example). For very large values of the area shapes with self-contact appear.

To determine the axisymmetric equilibrium shapes of confined *adhesive* membrane vesicles one has to solve the same shape equations as in the case without adhesion. The only difference

[3] The simulation program also allowed us to distinguish between energy minima and maxima.

[4] A similar transition from free to contact-circle to contact-area solutions is found at smaller areas for circular tori.

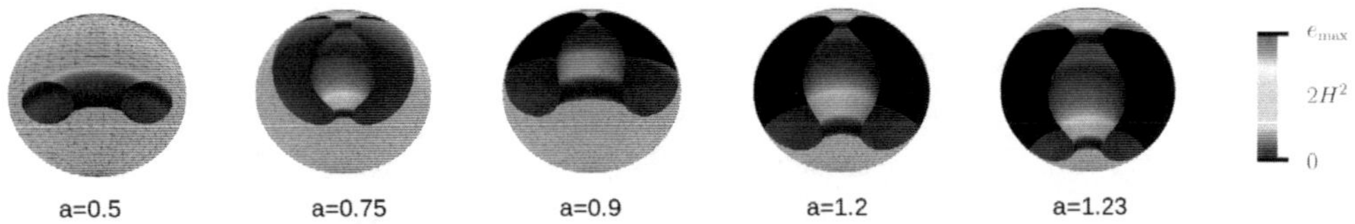

Figure 2. Equilibrium profiles of a membrane with adhesion ($\omega = 1$) for fixed reduced volume $v_r = 0.6$ and increasing area. Colors indicate the density of the bending energy.

lies in one of the boundary conditions, Eq. (11), which implies that the curvature at the contact line is discontinous due to the adhesion. With adhesion one finds regions of solutions that are not up-down symmetric with respect to the midplane. For example, for the reduced volume $v_r = 0.6$ and small area a, the transition from free shapes to confined ones can either be accomplished via symmetric circular tori (similar to the case with no adhesion, see the first two profiles of figure 1), or via shapes that are less symmetric (see figure 2). This is due to Eq. (11): when the adhesion coefficient w is large, the curvature of the membrane at the contact line has to be large. The sickle-shaped profiles have a greater perpendicular curvature far from the midplane. With adhesion they will attach to the confining sphere along a circular contact line which does not lie in the midplane (see the profile with $a = 0.75$ in figure 2). When we increase the area, the contact zone increases because of the negative adhesion energy (see the profiles with $a = 0.9$ and $a = 1.2$ in figure 2) until the symmetry with respect to the midplane is recovered.

A closer look at the transition from contact-circle to contact-area solutions reveals that an increasing adhesion induces a shift of this transition towards smaller areas (see curves C_1^ω in figure 3 for the circular tori). Fixing the reduced volume to $v_r = 0.6$ again, one finds contact-area solutions for $a > 0.85$ when $\omega = 0$. For $\omega = 2$ the transition point shifts to $a = 0.73$. When the adhesion is large one does not find any contact-circle solutions but a discontinuous transition from free to contact-area solutions ($\omega > 6.6$ for $v_r = 0.6$).

4. Conclusion

In this work we have studied toroidal vesicles in an adhesive spherical container using the classical curvature model. We only had a first glance at the rich solution space by focusing on shapes with axisymmetry. A more detailed study would not only have to include non-axisymmetric shapes but also consider questions of stability. Finite element simulations—similar to those made for the case without adhesion—could help to accomplish this task.

References

[1] Bouzar L, Menas F and Müller M M 2015 *Phys. Rev.* E **92** 032721.
[2] Alberts B *et al* 2002 *Molecular Biology of the Cell* (New York, NY: Garland Publishing).
[3] Kahraman O, Stoop N and Müller M M 2012 *New Jour. Phys.* **14** 095021.
[4] Kahraman O, Stoop N and Müller M M 2012 *Europhys. Lett.* **97** 68008.
[5] John G B et al. 2005 *Mol. Biol. Cell.* **16** 1543.
[6] Helfrich W 1973 *Z. Naturforschung* C **28** 693.
[7] Deserno M, Müller M M and Guven J 2007 *Phys. Rev.* E **76** 011605.
[8] Seifert U 1991 *Phys. Rev. Lett.* **66** 2404.

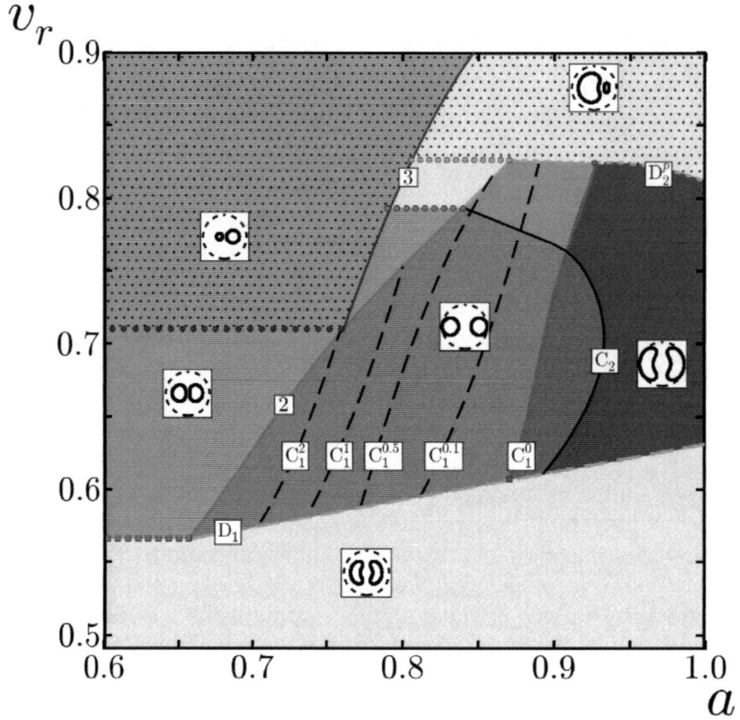

Figure 3. Partial phase diagram showing the axisymmetric extrema together with the conformal transformations of the free vesicles for the case without adhesion (adapted from figure 4 of Ref. [1]). Regions of circular tori are colored in orange-red, regions of sickle-shaped tori are in green. Non-axisymmetric regions are hatched. The shades represent the different degrees of confinemen: light shading corresponds to free solutions, medium shading corresponds to contact-circle solutions, and dark shades to contact-area solutions. The curve C_1^0 separates the regions of contact-circle and contact-area solutions when there is no adhesion between membrane and container. The dashed curves C_1^ω indicate how this boundary shifts to the left when adhesion is switched on. All other line styles correspond to those of Ref. [1].

IOP Conf. Series: Materials Science and Engineering **186** (2017) 012022 doi:10.1088/1757-899X/186/1/012022

Characterization in vitro studies and antibacterial properties on a sol-gel derived silver incorporated bioglass

S Bouhazma[1], S Chajri[1], M Khaldi[1], M Sadiki[2], H Barkai[2], S Elabed[2], S Ibnsouda Koraichi[2], B El Bali[1] and M Lachkar[1]

[1] Engineering Laboratory of Organometallic and Molecular Materials (CNRST, URAC19) Faculty of Sciences, University Sidi Mohamed Ben Abdellah, Po. Box 1796 (Atlas), 30000 Fez,
Morocco.
[2]Laboratory of Microbial Biotechnology, Faculty of Sciences and Technology, University Sidi Mohamed Ben Abdellah, 30000 Fez, Morocco.

Email: sarra.bouhazma@usmba.ac.ma

Abstract. The SiO_2-CaO, SiO2-CaO-P_2O_5 and SiO_2-CaO-P_2O_5-Ag_2O glass systems were synthesized by the sol-gel technique and characterized with different techniques such as X-ray diffraction (XRD), Fourier Transform Infrared spectroscopy (FTIR), and Environmental Scanning Electron Microscopy (ESEM). In vitro bioactivity tests were performed in Simulated Body Fluid (SBF). The antibacterial action of 65S5Ag (65%SiO_2 + 24%CaO + 6%P_2O_5 + 5% Ag_2O) is attributed exclusively to the leaching of Ag^+ ions from the glass matrix. The activity of S_iO_2-CaO-P_2O_5-Ag_2O was compared with that of its binary and ternary counterpart glass system. The concentrations of Ag-bioglass, in the range of 0.05 mg/mL of culture medium, were found to inhibit the growth of these bacteria.

1. Introduction

Bioactive glasses are special glass systems which are generally composed of SiO_2, CaO, P_2O_5, and Na_2O. They can be produced by the traditional melting process or by the more versatile sol-gel process [1, 2]. The bioactive behavior of these glasses is defined as the ability to bond to soft and hard tissues by means of complex chemical and biochemical reactions, which produces strong and compliant interface between the glass and the tissue [1]. Due to their high level of tissue integration and regeneration, bioactive glasses have been used clinically in a variety of situations [2, 3]. Numerous studies have been performed on them to try to improve their properties, for instance addition of other elements [4-11], sol-gel preparation [12-15], or porous bioactive glasses [16, 17].
Sol-gel processing has been intensely studied as an alternative method for preparing ceramics and glasses for a wide variety of applications including the field of bioceramics. This is because of the potential advantages of higher purity and homogeneous materials that are produced at lower processing temperatures. Li et al. showed that CaO-P_2O_5-SiO_2 powders produced by this technique are more bioactive than the melt-derived glasses of the same composition [18]. The high bioactivity of the sol-gel derived materials is related to the textural features of the gels, i.e., pore size and pore volume associated with the large surface area, higher rate of dissolution, and the negative surface charge [19].

Among the problems encountered the microorganism adhesions on implant surfaces represent an initial crucial step in infection and lead to the formation of biofilm, in which microorganisms are more resistant antimicrobial agents [20]. To solve the problem of contamination, it has been proposed to use antimicrobial agents such as antibiotics, fluorine and biocides metal ions [21-23]. The antibacterial activity of silver ions and their biological impact have been demonstrated by many works [24-28]. Silver-containing bioactive glasses have been produced in different ways: sol-gel [29, 30], melting, ion-exchange [31], and their antibacterial activity have been investigated by biological tests using different bacterial stocks.

The main aim of the presented work was to evaluate the in vitro bioactivity and antimicrobial action of a novel sol-gel derived SiO_2–CaO–P_2O_5–Ag_2O. The properties of the silver-incorporated bioglass were compared with other glasses: SiO_2–CaO, SiO_2–CaO–P_2O_5 silver-free bioglass system. The *Escherichia coli* (gram negative bacteria) species was chosen for the preliminary investigation because it was found in very high concentration at biomaterials-related infection sites [32-38].

2. Materials and method

2.1. Preparation of bioactive glasses

The three sol-gel glasses: SiO_2-CaO, SiO_2–P_2O_5–CaO, and SiO_2–P_2O_5–CaO–Ag_2O were synthesized as follows: initially, tetraethoxysilane (TEOS; Aldrich) was added into 0.1 M nitric acid and the mixture was allowed to react for 60 min for the acid hydrolysis of TEOS. Table 1 lists the nominal composition of the prepared glasses. The following reagents were added in the following sequences, allowing 45 min for each reagent to react completely: triethylphosphate (TEP; Aldrich), calcium nitrate tetrahydrate (Aldrich), and silver nitrate (Aldrich). After the final addition, mixing was continued for 1h to allow for the completion of hydrolysis. The resultant solution was kept in a sealed Teflon container for 3 days at ambient temperature conditions for gelation. The gel formed was placed in a sealed container and heated at 70°C for an additional 3-days period. The water was removed and small holes were created on the lid to allow the leakage of gases and heat treated at 150°C for 2 days to remove all the free water. The dried gel was then calcined for 3h at 700°C to stabilize the glass and to eliminate residual nitrates.

Table 1. Composition of the glasses 65S35C, 65S6P, and 65S5Ag (in mol%).

Glasses	SiO_2	CaO	P_2O_5	Ag_2O
65S35C	65	35	0	0
65S6P	65	29	6	0
65S5Ag	65	24	6	5

2.2. Materials

2.2.1. XRD measurements

X-ray diffraction spectra were obtained using a Discover model equipped with a monochromatized Cu-Kα radiation (λ= 1.5418 Å). Scans were performed with a step size of 0.02° over an angular range 2θ from [20-80°].

2.2.2. FT-IR absorption measurements

Fourier transform infrared (FTIR) spectroscopy was performed by FTIR using an IR Bruker VERTEX 70 spectrometer. The FTIR spectra were recorded from 400 to 4400 cm^{-1}. All measurements were recorded with a resolution of 4 cm^{-1}.

2.2.3. SEM study

The microstructure of the surface of the samples was observed on a scanning electron microscope (ESEM Quanta 200 FEI Company), with energy dispersive X-ray (EDX) analysis.

2.2.4. pH measurements

The pH changes of the SBF during the dissolution of the undoped and silver-doped glasses were measured at every hour and up to 6 h using IQ microprocessor pH-meter (210 pH)-meter (IQ Inc. USA).

2.2.5. Antibacterial activity test

Antimicrobial activities of samples were investigated against yeast O128B12. The bacteriostatic effect of 65S5Ag was determined by performing a viable count after exposure of the bacterial colony to 65S5Ag for 24 h. A single bacterial colony was used to inoculate a 50 mL starter culture. The initial concentration was approximately 2×10^{8} colony forming units/mL (CFU/mL). Assay mixtures containing 65S35C, 65S6P, and 65S5Ag in concentrations ranging from 0.05 to 2 mg/mL were prepared in three replicates. Negative control assay mixtures, containing only the cell inoculum in growth medium, were cultured in triplicate. This concentration range was chosen in consideration with the elution of Ag^{+} from 65S5Ag during the course of the dissolution experiment. Following inoculation, the cultures were incubated for 24 h at 37°C. After the incubation period, a 1 mL sample was taken and serial dilution was plated onto the growth medium employed in the liquid culture. The number of colonies was counted after overnight incubation. Approximate minimal bactericidal concentration (MBC) was estimated as the concentration of 65S5Ag leading to a 99% reduction in viability.

3. Results and discussion

3.1. Crystalline phase in 65S3, 65S6, and 65S5Ag

X-ray diffraction and FTIR evidenced the formation of an apatite-like layer on the glass surface after soaking in SBF. This mechanism could be explained through the following steps: (a) rapid exchange of protons H_3O^+ from the physiological solution with Ca^{2+} and Na^+ ions in bioglass to form the Si-OH groups, (b) loss of soluble silica as $Si(OH)_4$ by breaking of Si-O-Si bridging links and subsequent formation of surface silanol groups in the process, (c) condensation and repolymerization of surface silanols to form SiO_2-rich surface layer, (d) migration of Ca^{2+} and PO_4^{3-} ions through the surface silica-rich layer and formation of a Ca-P rich layer on the surface of the bioglass, (e) incorporation of OH^- and CO_3^{2-} ions from the solution and subsequent crystallization of the Ca-P layer to form HCA [39]. An XRD study of as-received samples confirmed that all of them were amorphous, except 65S5Ag, the crystalline phase in this bioglass was identified to be $CaAg(PO_3)_3$. X-ray diffraction patterns obtained for the binary and ternary glasses after 4 days soaking showed one wide diffraction maxima 32°, corresponding to (211) reflections of an apatite-like phase. For quaternary glass 65S5Ag, after just 4h, the (200), (211) and (310) reflections of the apatite phase appeared owing to the increase in its crystallinity. Data obtained by XRD show that an apatite-like phase is formed on glass surface. The formation of the apatite was also confirmed by FTIR and SEM/EDX examination.

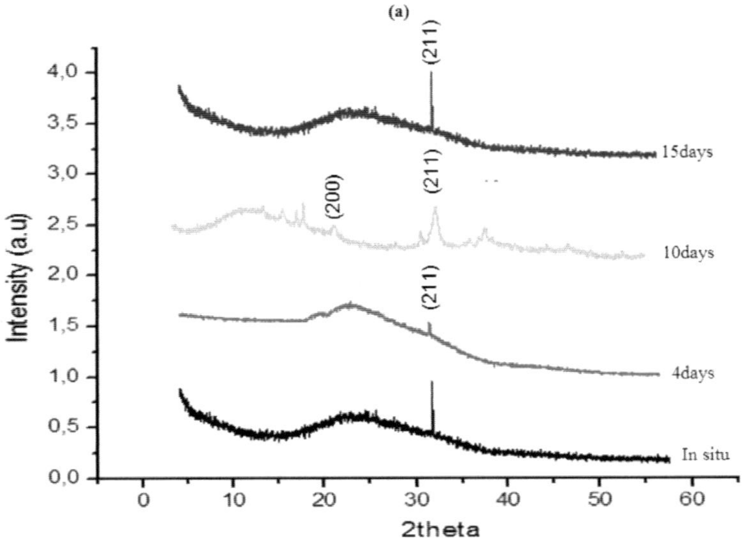
(a)
(211)
(200)
(211)
(211)
15days
10days
4days
In situ
Intensity (a.u)
2theta
4,0
3,5
3,0
2,5
2,0
1,5
1,0
0,5
0,0
0
10
20
30
40
50
60

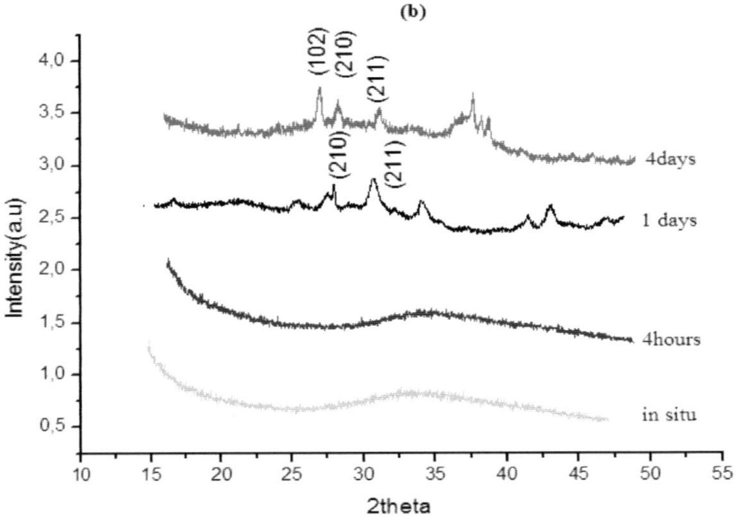
(b)
(102)
(210)
(211)
(210)
(211)
4days
1 days
4hours
in situ
Intensity(a.u)
2theta
4,0
3,5
3,0
2,5
2,0
1,5
1,0
0,5
10
15
20
25
30
35
40
45
50
55

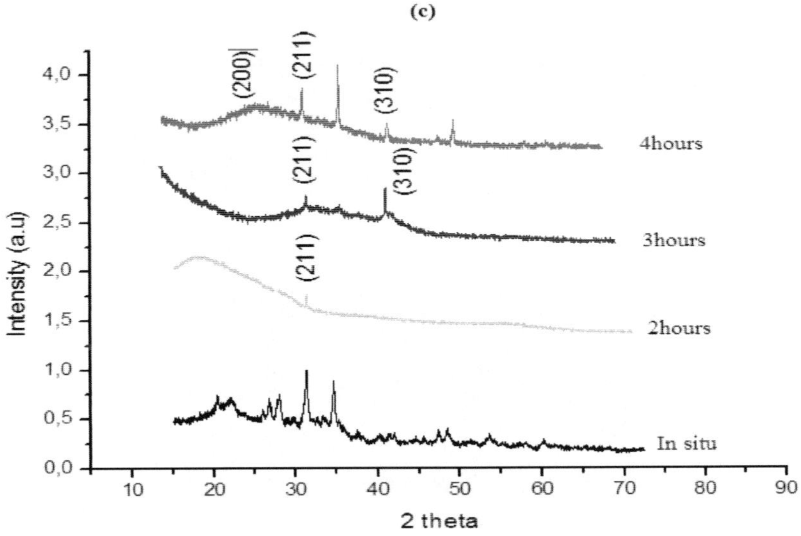

Figure 1. XRD spectra of the sol-gel derived bioglasses sintered at 700°C for 3h before and after immersion in SBF for different times. (a): 65S35C, (b): 65S6P and (c): 65S5Ag.

Figure 2 shows FTIR spectra. All glasses after soaking in SBF solution for different times reveal P-O bending (amorphous) (560 cm^{-1}) bands. Emerging of P-O bending (crystalline) (475cm^{-1}) bands reveals the formation of hydroxycarbonated apatite (HCA) layer. Presence of O-H stretching (1630-3453 cm^{-1}), C-O stretching (800 cm^{-1}) and (1426 cm^{-1}) bands shows the crystalline nature of HCA layer.

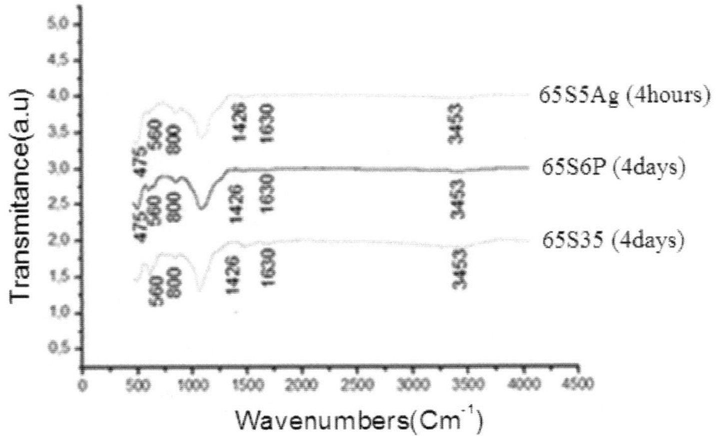

Figure 2. FTIR spectra of the sol-gel derived bioglasses sintered at 700°C for 3h.

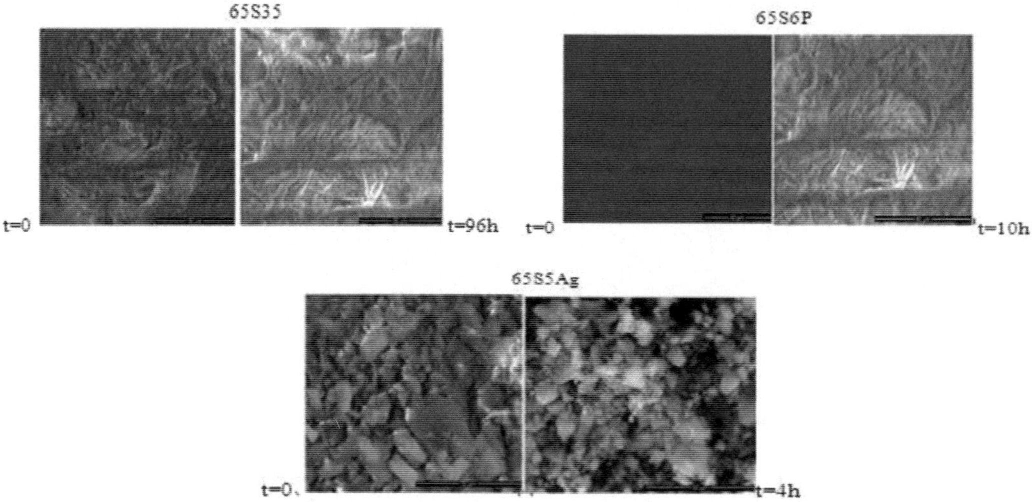

Figure 3. SEM micrographs of the the bioglasses samples before and after soaking into SBF.

3.2. pH measurements and concentration of Ca and P

Figure 4 shows the variation of pH of the SBF with immersion times, this variation is similar for all glasses. The cumulative variation of Ca and P ionic concentrations and variation of pH with the soaking time in SBF for samples 65S35C, 65S6P, and 65S5Ag. As can be observed in all cases, pH variation with time was also similar for the three samples, increasing up to 9 during the first 10 hours and to 10 after 96 hours of immersion and then no significant change took place. However, variation of Ca and P concentration in SBF was different for each composition of glass. For 65S35C, Ca concentration reached a maximum after 4 days and did not change significantly until the end of the test, while for 65S6P and 65S5Ag, Ca and P concentrations, gradually decreased up to 6 hours and 4 hours successively.

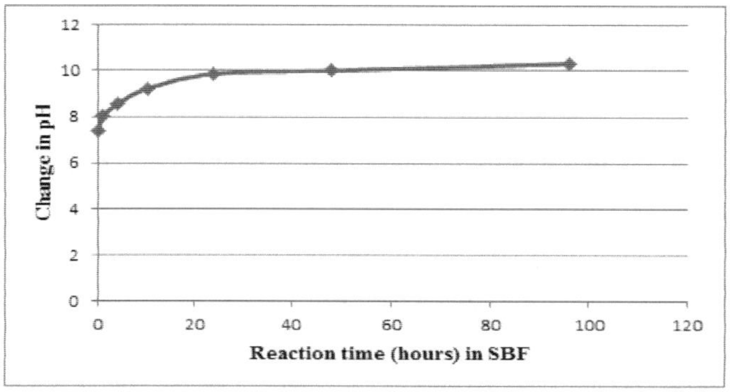

Figure 4. Variation of pH in the SBF with soaking time, for samples 65S35, 65S6P, and 65S6Ag.

(a)

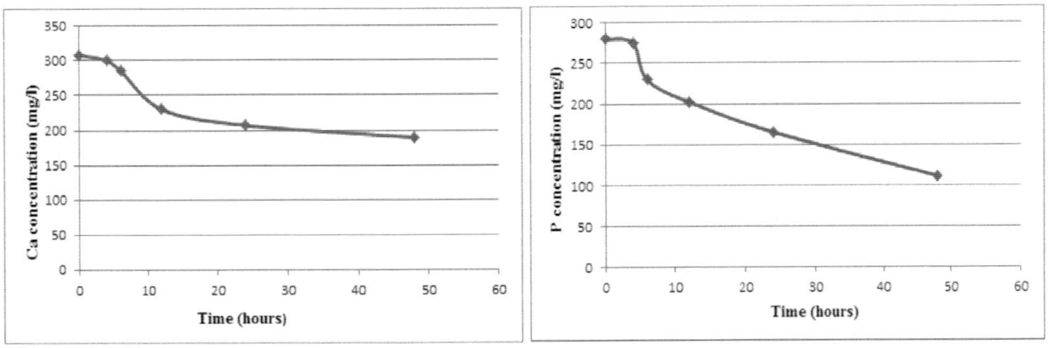

(b)

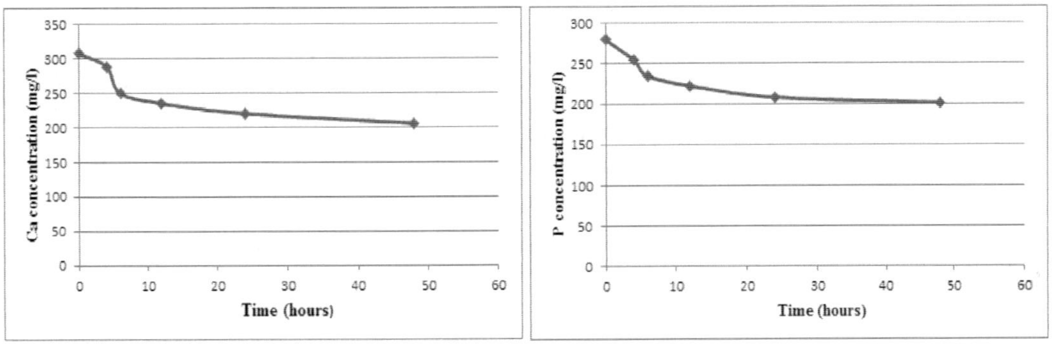

(c)

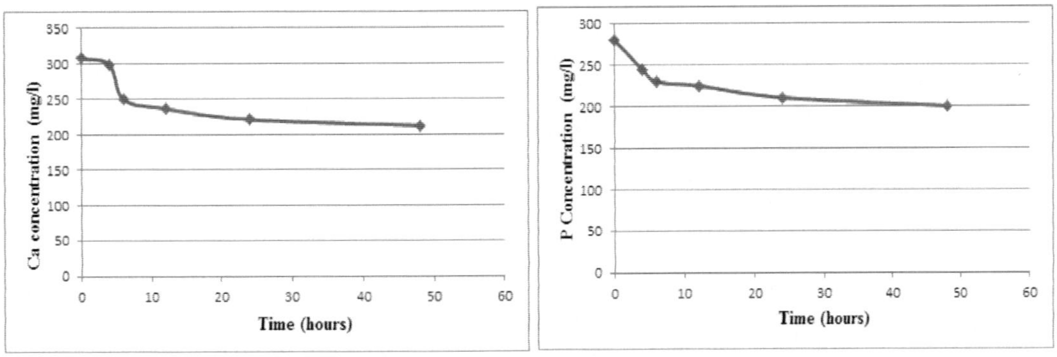

Figure 5. Ca and P concentration with immersion time in SBF for gel glass samples: (a) 65S35, (b) 65S6P, and (c) 65S5Ag

IOP Conf. Series: Materials Science and Engineering **186** (2017) 012022 doi:10.1088/1757-899X/186/1/012022

3.3. Antibacterial activity test

The MBC, MIC was estimated for 65S5Ag as 0.06 mg/mL for *E. coli*. The silver-free bioglasses (65S35C and 65S6P) show neither bacteriostatic nor bactericidal effects on the micro-organism studied. As a matter of fact, the silver free bioglass did not affect the viability of the *E. coli* bacterial strain under investigation. Comparing the results obtained for 65S5Ag and 65S35C, 65S6P, it can be inferred that the silver released from 65S5Ag is instrumental in the antibacterial action of the 65S5Ag. Further, it can be concluded that the other ionic species, such as Ca, Si and P released along with Ag^+ from the glass, do not alter the antibacterial property of the 65S5Ag.

4. Conclusion

Glasses 65S35C, 65S6P, and 65S5Ag were obtained successfully by the sol-gel method. They are bioactive by forming a hydrated carbonate apatite layer on their surface on exposure to SBF. The glasses have been characterized by XRD, FTIR and SEM. An *in vitro* antimicrobial activity test has shown that the introduction of silver to the three-component system, SiO_2-P_2O_5-CaO inhibits the growth of the *E. coli* bacteria. The dissolution study has confirmed that the material is capable of releasing silicate species, which is very important for its mitogenic effect *in vivo*. The rate of dissolution of silver ions is relevant to eventual clinical applications, since the prerequisite of an effective, topical antimicrobial agent is an immediate and concentrated release of that agent.

Acknowledgments

The authors would like to acknowledge the support and technical assistance of Interface Regional University Center (University Sidi Mohammed Ben Abdellah, Fez), and National Center for Scientific and Technical Research (CNRST-Rabat, Morocco).

References

[1] Hench L L, Splinter R J, Allen W C and Greenlee T K 1972 Bonding mechanism at the interface of ceramic prosthetic materials. J. Biomed. Mater. Res. **2** 117-141.

[2] Hench L L Bioactive materials: the potential for tissue regeneration. J. Biomed. Mater. Res. **41** 511-518. 1998.

[3] Xynos I D, Edgar A J, Buttery L D K., Hench L L, and Pollak J M 2000 Ionic products of bioactive glass dissolution increase proliferation of human osteoblasts and induce insulin-like growth factor II mRNA expression and protein synthesis. Biochem. Biophys. Res. Commun. **276** 461-465.

[4] Verne E, Nunzio S Di, Bosetti M, Appendino P, Brovarone C V, Maina G and Cannas M 2005 Surface characterization of silver-doped bioactive glass. J. Biomater. **26** 5111-5119.

[5] Singh R K. and Srinivasan A 2010 Bioactivity of SiO_2-CaO-P_2O_5-Na_2O glasses containing zinc-iron oxide. Appl. Surf. Sci. **256** 1725-1730.

[6] Varanasi V G, Saiz E, Loomer P M, Ancheta B, Uritani N, Ho S P, Tomsia A P, Marshall S J and Marshall G W 2009 Enhanced osteocalcin expression by osteoblast-like cells (MC3T3-E1) exposed to bioactive coating glass (SiO_2-CaO-P_2O_5-MgO-K_2O-Na_2O system) ions. J. Acta Biomater. **5** 3536-3547.

[7] Fredholm Y C, Karpukhina N, Law R V and Hill R G. 2010 Strontium containing bioactive glasses: Glass structure and physical properties. J. Non-Cryst, J. Solids. **356** 2546-2551.

[8] Watts S J, Hill R G, O'Donnell M D and Law R V 2010 Influence of magnesia on the structure and properties of bioactive glasses. J. Non-Cryst. Solids. **356** 517-524.

[9] Gentleman E, Fredholm Y C, Jell G, Lotfibakhshaiesh N, O'Donnell M D, Hill R G and Stevens M M 2010 The effects of strontium-substituted bioactive glasses on osteoblasts and osteoclasts in vitro. J. Biomater. **31** 3949-3956.

[10] O'Donnell M D and Hill R G 2010 Influence of strontium and the importance of glass chemistry and structure when designing bioactive glasses for bone regeneration J. Acta Biomater. **6** 2382-2385.
[11] Aina V, Bertinetti L, Cerrato G, Cerruti M, Lusvardi G, Malavasi G, Morterra C, Tacconi L and Menabue L 2011 On the dissolution/reaction of small-grain Bioglass (R) 45S5 and F-modified bioactive glasses in artificial saliva (AS) J. Appl. Surf. Sci. **257** 4185-4195.
[12] Li Z, Chen X, Lin C and Zhao N 2012 The in vitro bioactive of sol-gel bioactive glass powders with three-dimensional lamellar structure J. Adv. Powder Technol. **23** 13-15.
[13] Silver I A, Deas. J and Erecinska M 2001 Interactions of bioactive glasses with osteoblasts in vitro: effects of 45S5 Bioglass (R), and 58S and 77S bioactive glasses on metabolism, intracellular ion concentrations and cell viability J. Biomater. **22** 175-185.
[14] Chen X, Meng Y, Li Y and Zhao N 2008 Investigation on bio-mineralization of melt and sol-gel derived bioactive glasses J. Appl. Surf. Sci. **255** 562-564.
[15] Arcos D, Vallet-Regi M 2010 Sol-gel silica-based biomaterials and bone tissue regeneration J. Acta Biomater. **6** 2874-2888.
[16] Yun H S, Park J W, Kim S H, Kim Y J and Jang J H 2011 Effect of the pore structure of bioactive glass balls on biocompatibility in vitro and in vivo J. Acta Biomater. 7 2651-2660.
[17] Arcos D, Vila M, Lopez-Noriega A, Rossignol F, Champion E, Oliveira F J and Vallet-Regi M 2011 Mesoporous bioactive glasses: Mechanical reinforcement by means of a biomimetic process J. Acta Biomater. **7** 2952-2959.
[18] Li R, Clark A E and Hench L L 1991 An investigation of bioactive glass powders by sol-gel processing J. Appl. Biomater. **2** 231-239.
[19] Pereira M M and Hench L L 1996 Mechanism of hydroxyapatite formation on porous gel-silica substrates J. Sol-Gel Sci. Technol. **7** 59-68.
[20] Harris L and Richards R 2006 Staphylococci and implant surfaces: a review. Injury. **37** 3-14.
[21] Montali A 2006 Antibacterial coating systems Injury. **37**, S81-S86.
[22] Ge X., Leng Y, Bao C, Xu S L, Wang R and Ren F 2010 Antibacterial coatings of fluoridated hydroxyapatite for percutaneous implants J. Biomed. Mater. Res. **95A** 588-599.
[23] Jeyachandran Y L, Venkatachalam S, Karunagaran B, Narayandass S K, Mangalaraj D, Baoand C Y and Zhang C L 2007 Bacterial adhesion studies on titanium, titanium nitride and modified hydroxyapatite thin films J. Mater. Sci. Eng. **C 27** 35-41.
[24] Djoric S S and Burrel R E 1998 Behavior of silver in physiological solutions J. Electrochem Soc. **145(5)** 1426-1430.
[25] Feng Q L, Wu J, Chen G Q, Cui F Z, Kim T N and Kim J O 2000 A mechanistic study of the antibacterial effect of silver ions on *Escherichia coli and Staphylococcus aureus* J. Biomed. Mater Res. **52** 662-668.
[26] Berger T J, Spadaro J A, Chapin S E and Becker R O 1976 Electrically generated silver ions: quantitative effects on bacterial and mammalian cells Antimicrob Agents Chemother. **9(2)** 357-358.
[27] Williams R J, Doherty P J, Vince D G, Grashoff G J and Williams D F 1989 The biocompatibility of Silver Crit. Rev. Biocompat.; **5** 221-243.
[28] Kraft C N, Hansis M, Arens S, Menger M D and Vollmar B 2000 Striated muscle microvascular response to silver implants: a comparative in vivo study with titanium and stainless steel J. Biomed Mater Res. **49** 192-199.
[29] Kawashita M, Tsuneyama S, Miyaji F, Kokubo T, Kozuka H, Yamamoto K 2000 Antibacterial silver-containing silica glass prepared by sol-gel method J. Biomaterials. **21** 393-398.
[30] Bellantone M, Coleman N J and Hench L L 2001 A novel sol-gel derived glass featuring antibacterial properties J. Bioceram., Key Eng. Mater. **192-195** 597-600.
[31] Hosono H and Abe Y 1994 Silver ion selective porous lithium titanium phosphate glass-ceramics cation exchanged and its application to bacteriostatic materials J. Mater Res. Bull. **29**(11) 1157-562.

[32] Adams A P, Santchi E M and Millencamp M A 1999 Antibacterial properties of silver chloride coated nylon wound dressing J. Vet Surg. **28** 219-225.
[33] George N, Faoagali J, Muller M and Silvazine M 1997 Silver sulfadiazine and chlorexidine activity against 200 clinical isolates Burns. **23** 493-495.
[34] Tiller J C, Liao C J, Lewis K and Klibanov A M 2001 Designing surfaces that kill bacteria on contact Proc. Natl. Acad. Sci. USA. **98** 5981-5985.
[35] Bosetti M, Masse A, Tobin E and Cannas M 2002 Silver coated materials for external fixation devices: in vitro biocompatibility and genotoxicity J. Biomaterials. **23** 887-892.
[36] Chu C S, McManus A T, Pruitt Jr. B A and Mason Jr. A D 1988 Therapeutic effects of silver nylon dressing with weak direct current on Pseudomans aeruginosa infected burn wounds J. Trauma. **28** 1488-1492.
[37] Deitch E A, Marino A A, Malakanok V and Albright J A 1987 Silver nylon cloth: in vivo and in vitro evaluation of antimicrobial activity J. Trauma. **27** 301-304.
[38] Blaker J J, Nazhat S N and Boccaccini A R 2004 Development and characterization of silver-doped bioactive glass-coated sutures for tissue engineering and wound healing applications J. Biomaterials. **25** 1319-29.
[39] Hench L L, Splinter R J, Allen W C, Greenlee T K 1971 Bonding mechanisms at the interface of ceramic prosthetic materials J. Biomed. Mater. Res. **5** 117-141.

IOP Conf. Series: Materials Science and Engineering **186** (2017) 012023 doi:10.1088/1757-899X/186/1/012023

Thermal stability of a modified sol-gel derived hydroxyapatite nanopowders

S Herradi, B El Bali, M Khaldi and M Lachkar

Engineering Laboratory of Organometallic and Molecular Materials (CNRST-URAC19), Faculty of Sciences, University Sidi Mohamed Ben Abdellah, Po. Box 1796 (Atlas), 30000, Fez, Morocco

Email: Smaiel_her@hotmail.fr

Abstract. Hydroxyapatite $Ca_{10}(PO_4)_6(OH)_2$ (HA) powder was successfully synthesized by a modified sol-gel method using a solution of calcium nitrate in ethanol, along with a solution of diammonium hydrogen phosphate in water and NH_4OH as starting materials. The Ca/P molar ratio was maintained at 1.67. The powder was subjected to furnace and microwave heating to compare the decomposition of HA and study the crystallite sizes. It was found that microwave heated powders were pure HAP up to 230°C with absence of secondary phases. However, XRD patterns show that furnace heated powders convert completely to β-TCP when treated at 750°C and 1000°C. This result was confirmed by the absence of hydroxyl bands in the FT-IR spectra for these temperatures.

1. Introduction

Hydroxyapatite (HA) [$Ca_{10}(PO_4)_6(OH)_2$] is a biocompatible polycrystalline ceramic material that has chemical composition and crystal structure similar to the human bone and teeth [1]. In addition, HA exhibits biocompatibility for bone bonding, osteoconductivity, and activity [2, 3-5], due to its ability to form strong chemical bonds [2] with host material/tissue/bones, and thus has been recognized as a potential graft material [2]. Hydroxyapatite $Ca_{10}(PO_4)_6$ OH_2 has a network structure formed by the chains of $Ca(I)O_9$ and $Ca(II)O_6OH$ polyhedral interconnected by discrete PO_4-tetrahedra. It is observed that the cation or anion substitution at Ca, P or OH sites, or variation in the stoichiometry, leads to significant differences in surface and bulk properties of HA [6]. Calcium substitution allows the addition of metallic sites to the material and the anion substitutions such as PO_4^{3-} groups by HPO_4^{2-} and CO_3^{2-} and OH^- by F^-, modify the surface properties and the thermal stability of the material [7].
Various chemical synthesis methods including hydrothermal [8], sol-gel [9], sonochemical [10], precipitation [11], microemulsion [12], microwave [13], mechanochemical [14] and surfactant-assisted approach [15] are available for the preparation of HA nanostructures. The sol-gel method is preferred due to its low synthesis temperature, high product purity, homogenous molecular mixing and the ability to generate nano sized particles compared to other alternatives [16]. Sol-gel method for the preparation of HA can usually form fine-grain microstructure containing a mixture of nano- to submicron particles with crystalline structure. It has been reported that these crystals are very efficient to improve the contact and stability at the artificial/natural bone interface observed in vitro and in vivo environments [17].

The present study reports the synthesis of HA done at room temperature, by a sol-gel method. Furthermore, this work reports thermal stability studies of HA particles in the given temperature range, so that one can choose between microwave and furnace heating for different purposes.

2. Materials and methods

2.1. Synthesis of HA nanopowders
Calcium nitrate (Ca(NO3)2·4H2O,LOBA Chimie) di-ammonium hydrogen phosphate (NH4H2PO4, FLOKA),), ethanol and ammonia solution (to adjust the pH) were the reagents used for the synthesis. Briefly, 1.67 M calcium nitrate solution in ethanol (pH=10) was added dropwise to 1M NH4H2PO4 (pH= 10) (stirring at room temperature, when the white precipitate solution was obtained, it was aged for 92 h at room temperature. The gel obtained after aging process was filtered and washed repeatedly with de-ionized water and acetone. The filtered cake was dried at 80°C for 12h in an oven, and then was crushed using mortar and pestle. The synthesized powders are then subjected to heat treatment in an electric furnace and a domestic microwave oven for 4 h and 20 min and at different temperatures viz. 200, 250, 300, 500, 750, 1000°C and 180, 220, 230°C respectively to study their effects on the decomposition of HA into other phases. Microwave heating of HA done at lower temperatures, ensures phase purity and is much suited for biomedical applications. Electric furnace treatment is time and energy consuming than the microwave. The procedure used to generate HA is depicted in the flow chart given in the Figure 1.

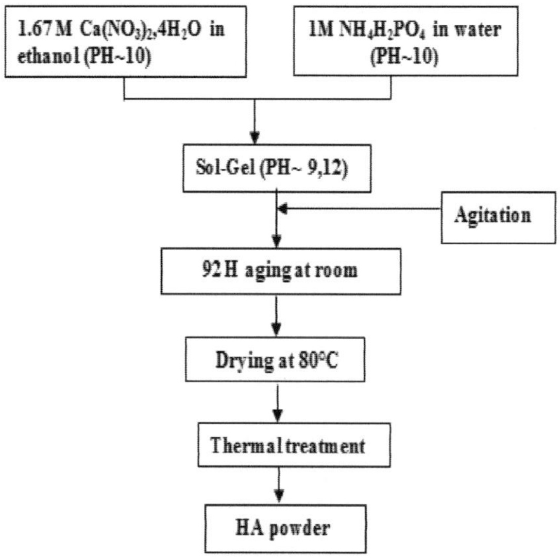

Figure 1. The flow chart of HA preparation.

2.2. *Characterization of HA nanopowders*
To investigate the phase composition and crystallinity of the sintered nano-HA, the X-ray diffraction (XRD) analysis was performed at a Discover model equipped with a monochromatized Cu-Kα radiation (λ= 1.5418 Å) in the 2θ range [10-80°] and a step size of 0.02° and a step duration of 2 s. Fourier transform infrared spectroscopy (FTIR: Bruker VERTEX 70 spectrometer) was used to analyze the functional group of the obtained powders. The spectra were recorded from 400 to 4000 cm^{-1} at 4 cm^{-1} resolution.

3. Results and discussion

The XRD patterns of the furnace and microwave heated powders are shown (in addition to the as-dried powder) in Figures 2 and 3 respectively. For the as-dried powder no characteristic peaks has been detected which means that the sample is amorphous while the furnace heated powders are of pure HA composition up to 500°C [2, 18-22]. However, it converts to β-TCP when treated at 750°C and 1000°C [22-24].

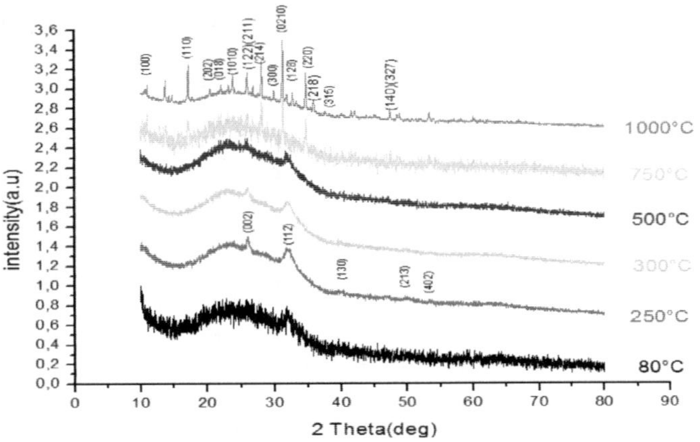

Figure 2. XRD pattern of as-dried and furnace heated powders.

Conversely, for microwave heated samples (Figure 3), it was conspicuously observed that, at all temperatures, there is no decomposition of HA into other secondary phases, and the sharp diffraction peaks indicate the high crystallinity of the heated samples. These results are confirmed by the FT-IR spectra of furnace and microwave heated samples as it is depicted at Figures 4 and 5 respectively. In fact, for the two samples the characteristic absorption bands at 3219 and 1649 cm^{-1} are attributed to adsorbed atmospheric water. The weak peak located at 3570 cm^{-1} corresponds to the vibrations of OH^- ions in the HAP lattice [25]. The two weaker peaks, located at 771 cm^{-1} and 879 cm^{-1}, are associated with the carbonate group and clearly indicate its presence in the samples. The band at 1019-1023 cm^{-1} is assigned to triply degenerated (v_3) asymmetric stretching mode of P–O bond in hydroxyapatite structure. The characteristic doublet IR peaks around 1430 and 1471 cm^{-1} can be attributed to the vibrational frequencies of carbonate ions substituted at the phosphate site in apatite, so called B-type apatite [26, 27]. The band located at 559 cm^{-1} [28, 29] belong to triply degenerate v_4, O-P-O bending mode in the phosphate group.

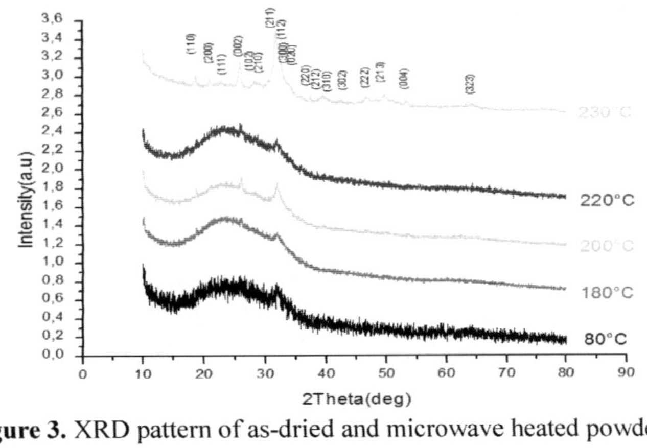

Figure 3. XRD pattern of as-dried and microwave heated powders.

Interestingly, for furnace heated samples, as the temperature increases, the OH^- peaks, at about 3418 cm^{-1}, disappears at 200°C, while the intensities of peaks at 3219 and 1649 cm^{-1} diminish indicating the removal of some amount of hydroxyl components such as crystalline water. We can also see that the peak of carbonate situated at 873°C disappear along with a diminution in intensity of the peaks located at 771, 1430 and 1471 cm^{-1} witch also indicate the removal of carbonate from the crystalline structure, this feature remains the same up to 500 °C. After heating at 750 °C, the OH^- absorption bands disappeared; Moreover, the spectrum of the sample treated at 750°C contains features from β-TCP, i.e. band shoulders around 1019, 1112 cm^{-1} [23], and also we note the presence of three bands situated at 490, 550 and 594 cm^{-1} witch belongs to triply degenerate ν_4 O-P-O bending mode in the phosphate group. As shown in Figure 4, the characteristic peaks of β-TCP can be detected more obviously for the sample calcined at 1000°C, along with the absence of all peaks belonged to carbonate which means that we obtain a free carbonate structure at this temperature.
Contrarily, for microwave heated samples, it is conspicuously observed that there isn't removal of hydroxyl and carbonate components since all peaks relevant to them have the same feature for all microwave treated samples. Moreover, approximately no shift is observed for all peaks, which means that microwave heating is an outstanding technique since it converts the as-prepared apatite structure to hydroxyapatite rather than damaging it. Besides, microwave heating give us the hydroxyapatite structure in much shorter time (20 min) at a smaller temperature (180°C) compared to furnace heating (4h). To our knowledge, this is the first work that study phase stability of HA obtained by sol-gel process and heated by furnace and microwave oven. Furthermore, we report a total conversion of HA to β-TCP at 750°C.

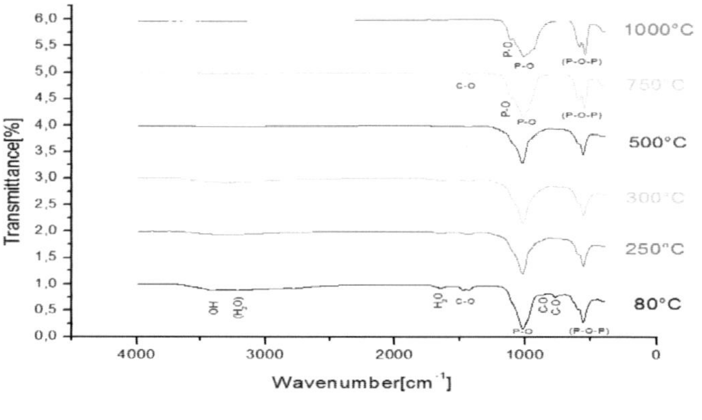

Figure 4. FTIR spectra of as-dried and furnace heated powders.

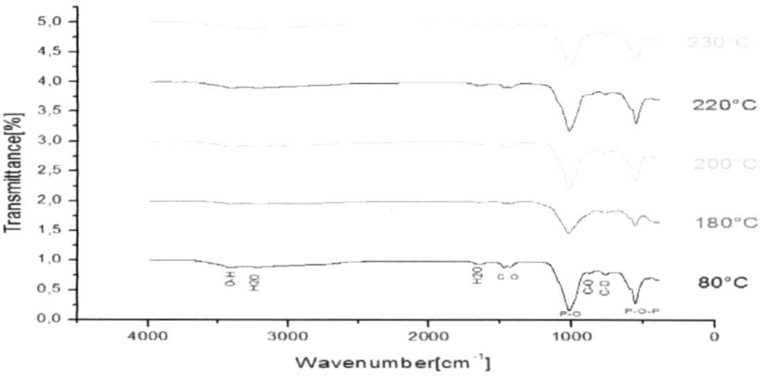

Figure 5. FTIR spectra of as-dried and microwave heated powders.

Another interesting point worthy to mention is that the amorphous structure obtained after drying process is of an apatitic structure (but not hydroxyapatite) since its XRD pattern show that it is amorphous whilst infra-red spectrum shows clearly the band of structural hydroxyl groups at 3570 cm^{-1} along with the other bands assigned to hydroxyapatite. Consequently, phase assertion come only after scrutinizing XRD patterns.

4. Conclusion

Thermal stability of synthesized HA was studied by furnace and microwave heating and it was found that microwave heated HA was pure with absence of secondary phases in all temperature, whilst, it was conspicuously observed that furnace heated HA convert to β-TCP when treated at 750°C and 1000°C. It is concluded that the microwave technique furnishes the product in much shorter time with high purity.

Acknowledgements
The authors would like to acknowledge the support and technical assistance of Interface Regional University Center (University Sidi Mohammed Ben Abdellah, Fez), and National Center for Scientific and Technical Research (CNRST-Rabat).

References

[1] Suchanek W,Yashima M,Kakihana M and Yoshimura M 1996 Processing and mechanical properties of hydroxyapatite reinforced with hydroxyapatite whiskers *Biomaterials*17(17) 1715-23.
[2] Bakan F,Laçin O and Sarac H 2013 A novel low temperature sol-gel synthesis process for thermally stable nanocrystalline hydroxyapatite Powder *Technol.* 233 295-302.
[3] Bezzi G, Elotti G, Landi E, La Torretta T.M.G, Sopyan I and Tampieri A 2003 a novel sol-gel technique for hydroxyapatite preparation *Mater. Chem. Phys.*78 816-24.
[4] Feng W, Sen L M, Peng L Y and Xin Q Y2005 A simple sol-gel technique for preparing hydroxyapatite nanopowders *Mater. Lett.* 59 916-19.
[5] Lei L, Li L, Zhang L, Chen D and Tian W 2009 Structure and performance of nano-hydroxyapatite filled biodegradable poly((1,2-propanediol-sebacate)-citrate)elastomers *Polym. Degrad. Stab.* 4 1494-1502.
[6] Cheikhi N, Kacimi M, Rouimi M, Ziyad M,Liotta L F,Pantaleo G and Deganello G 2005Direct synthesis of methyl isobutyl ketone in gas-phase reaction over palladium-loaded hydroxyapatite *J. Catal.* 232(2) 257-67.
[7] Shimoda S,Aoba T, Moreno E C and Miake Y 1990 Effect of solution composition on morphological and structural features of carbonated calcium apatites *J. Dent. Res.*69(11) 1731-40.
[8] Vázquez-Hernándeza F,Mendoza-Barrera C, Altuzar M, Meléndez-Lira M A,Santana-Aranda M, Olvera M andde la Olvera L 2010 Synthesis and characterization of hydroxyapatite nano particles and their application in protein adsorption *Mater. Sci. Eng. B.* 174 290-95.
[9] Fathi M H and Hanifi A 2007 Evaluation and characterization of nanostructure hydroxyapatite powder prepared by simple sol-gel method *Mater. Lett.*61(18) 3978-83.
[10] Giardina M A and Fanovich M A 2010 Synthesis of nanocrystalline hydroxyapatite from $Ca(OH)_2$ and H_3PO_4 assisted by ultrasonic irradiation *Ceram. Int.* 36 1961-69.
[11] Li Z,Wang P and Wu Z 2005 Preparation of nano sized hydroxyl apatite particles at low temperatures *J. Mater. Sci.* 40 6589-91.
[12] Bose S and Saha S K 2003 Synthesis and characterization of hydroxyapatite nano powders by emulsion technique *Chem. Mater.* 15 4464-649.
[13] Mishra V K,Srivastava S K, Asthana B P andKumar D J 2012 Structural and spectroscopic studies of hydroxyapatite nanorods formed via microwave-assisted synthesis route *J. Am. Ceram. Soc.* 95 2709-15.

[14] Nasiri-Tabrizi B, Honarmandi P, Ebrahimi-Kahrizsangi R and Honarmandi P 2009 Synthesis of nanosize single-crystalhydroxyapatite via mechanochemical *method Mater. Lett.* 63 543-46.

[15] Coelho J M,Moreira J A,Almeida A and Monteiro F J 2010 Synthesis and characterization of HAP nanorods from a cationic surfactant template method*J. Mater. Sci. - Mater. Med.*21 2543-9.

[16] Eshtiagh-Hosseini H, Housaindokht M R and Chahkandi M 2007 Effects of parameters of sol-gel process on the phase evolution of sol-gel-derived hydroxyapatite *Mater. Chem. Phys.* 106 310.

[17] Li P and de Groot K 1994 Better bioactive ceramics through sol-gel process *J. Sol-Gel Sci. Technol.* 2 797.

[18] Jadalannagari S,More S,Kowshik M andRamanan S R 2011 Low temperature synthesis of hydroxyapatite nano-rods by a modified sol-gel technique *Mater. Sci. Eng., C*31(7) 1534-8.

[19] Santhosh SandPrabu S B 2013 Thermal stability of nano hydroxyapatite synthesized from sea shells through wet chemical synthesis *Mater. Lett.* 97 121-4.

[20] Kuriakose T A,Kalkura S N,Palanichamy M,Arivuoli D,Dierks K,Bocelli G andBetzel C 2004 Synthesis of stoichiometric nano crystalline hydroxyapatite by ethanol-based sol-gel technique at low temperature *J. Cryst. Growth* 263(1) 517-23.

[21] Anee T K,Ashok M,Palanichamy M andKalkura S 2003 NA novel technique to synthesize hydroxyapatite at low temperature *Mater. Chem. Phys.* 80(3) 725-30.

[22] Rangavittal N, LandaCánovas A R,González Calbet J M andVallet Regí M 2000 Structural study and stability of hydroxyapatite and β-tricalcium phosphate: Two important bioceramics*J. Biomed. Mater. Res.*51(4) 660-8.

[23] Sanosh K P,Chu M C, Balakrishnan A,Kim T N andCho S J 2010 Sol-gel synthesis of pure nano sized β-tricalcium phosphate crystalline powders *Curr. Appl. Phys.*10(1) 68-71.

[24] Pan H B and Darvell B W 2009 Solubility of TTCP and β-TCP by solid titration *Arch. Oral. Biol.*54 671-7.

[25] Poinern G J E,Brundavanam R L X T,Djordjevic S,Prokic M and Fawcett D 2011 Thermal and ultrasonic influence in the formation of nanometer scale hydroxyapatite bioceramic *Int. J. Nanomedicine* 6 2083.

[26] Manjubala I and Sivakumar M 2001 In-situ synthesis of biphasic calcium phosphate Ceramics using microwave irradiation *Mater. Chem. Phys.*71 272-8.

[27] Ito A, Nakamura S,Aoki H,Akao M,Teraoka K,Tsutsumi S andTateishi T 1996 Hydrothermal growth of carbonate-containing hydroxyapatite single crystals *J. Cryst. Growth* 163(3) 311-7.

[28] Fowler B O 1974 Infrared studies of apatites. Vibrational assignments for calcium, strontium, and barium hydroxyapatites utilizing isotopic substitution *Inorg. Chem.*13 194.

[29] Klee W E and Engel G 1970 Infrared spectra of the phosphate ions in various apatites*J. Inorg. Nucl. Chem.* 32 1837.

IOP Conf. Series: Materials Science and Engineering **186** (2017) 012024 doi:10.1088/1757-899X/186/1/012024

Synthesis, characterization and luminescent properties of $Sr_{1-x}Pb_xWO_4$ solid solution (x=0, 0.5 and 1)

A Hallaoui[1,2], A Taoufyq[1,2], B Bakiz[2], F Guinneton[1], J-C Valmalette[1], S Villain[1], M Arab[1], A Benlhachemi[2], L Bazzi[2] and J-RGavarri[1]

[1] IM2NP, UMR CNRS 7334, University of Toulon, BP 20132, 83957 La Garde-France
[2] LME, Faculty of Sciences, Ibn ZohrUniversity, 80000 Agadir, Morocco

Email: abdelalihallaoui@gmail.com

Abstract. In this work, a study of the role of chemical substitution in the properties of the solid solution $Sr_{1-x}Pb_xWO_4$ (x = 0, 0.5 and 1) is presented. Polycrystalline samples were synthesized by solid state reaction at 1100°C. Using Rietveld method, the structural parameters of all samples were refined. All structures are of scheelite type Scanning electron microscopy showed that a high level of crystallization characterized the samples, with modifications in sizes and shapes depending on composition x. The Infrared and Raman spectroscopies were performed to characterize the evolution of vibrational modes with composition x. Finally, a systematic study of luminescence under X-ray and UV excitations was performed. The energies and intensities of luminescence depended on composition x and on the nature of excitations.

1. Introduction

The first paragraph after a heading is not indented The scheelite tungstate crystals AWO_4 are characterized by tetragonal structures [1,2] (space group $I4_1/a$), with tetrahedral oxyanions WO_4^{2-} and M^{2+}cations forming MO_8groups [3]. They were extensively investigated for their potential applications as laser host materials [4,5], scintillators [6,7], oxide ion conductors [8], materials for microwave applications [9] and magnetic materials [10]. The luminescent properties under different excitations of single-crystals as well as polycrystalline $SrWO_4$ and $PbWO_4$ compounds have been widely studied by several authors [11–15]. Generally, luminescence spectra of scheelite tungstate were decomposed into two main blue and green bands, and to a weaker orange–red emission: the emissions were observed with energies close to 3.0–2.8, 2.7–2.4 and 2.3–2.0 eV [16–20].

In this work we study the luminescent properties of lead and strontium tungstates $Sr_{1-x}Pb_xWO_4$ for the three compositions x= 0 ; 0.5 and 1 under X-ray and UV excitations. We already published the synthesis, characterization and luminescence under X-ray excitation of $Sr_{1-x}Pb_xWO_4$ and $Ca_{1-x}Cd_xWO_4$ solid solutions [21-23]. Presently, we compare the luminescence responses under X-ray and UV excitations of these substituted samples.

2. Experimental

2.1. Synthesis

All samples of $Sr_{1-x}Pb_xWO_4$ with x= 0, 0.5 and 1, were prepared by conventional solid state chemical reaction using WO_3 [Alfa Aesar N° 11828, >99.8%], $SrCO_3$ [Sigma-Aldrich N° 1633-05-2, >99.0%] and PbO [Sigma-Aldrich N° 1317-36-8, > 99.0%]. The elaboration conditions (grinding process, temperature and time of thermal treatment) were optimized and the final process was as follows: the reagents in stoichiometric proportions were thoroughly mixed and ground in an agate mortar for 15 min, then thermally treated at 600°C for 3 h, in pure alumina crucible under air. The samples were ground again thoroughly for 2 h, and then, retreated at 1100 °C for 4 h, under air.

2.2. Characterization techniques

2.2.1. X-ray diffraction: Each sample was analysed by X-ray diffraction using an Empyrean Panalytical diffractometer, equipped with a copper X-ray source (wavelength $\lambda=1.54.10^{-10}$ m, tension V= 45 kV, intensity I= 35 mA), and with a Ni filter eliminating the K_β radiation. The structural parameters of the samples were refined using the Fullprof software [24] based on Rietveld procedure.

2.2.2. Scanning electron microscopy: A systematic analysis of grain sizes and morphologies was performed using a Supra 40 V_pColonne Gemini Zeiss scanning electron microscope (SEM), with a maximum voltage of 20 kV.

2.2.3. Raman spectroscopy analysis: Raman spectroscopy was used to characterize the evolution of the various phases, and to correlate vibration modes with these structural evolutions. The equipment used to perform the various vibration spectra was a spectrometer Horiba Jobin-Yvon HR800 LabRam, spatially resolved to 0.5 µm, by means of an optical microscope with a 100 objective. The latter has a dual function: it allows firstly focusing the laser beam on a small area, and, secondly, visualizing the area of the sample. The 514.5 nm line of an Ar-ion laser was used as the excitation source; the photonic power applied to the samples was limited to 5 µW with an acquisition time of 30 s. Each Raman band was characterized by its wavenumber (in cm-1).

2.2.4. Luminescence experiments: For the luminescence under X-ray excitation, the copper X-ray source of the diffractometer Empyrean-Panalytical was used to irradiate the samples and perform luminescence experiments. The nominal emission conditions (voltage VRX / current IRX) were 45 kV /35 mA. The luminescence emissions of samples were recorded using a UV–visible spectrophotometer MicroHR (JobinYvon) equipped with an optical fiber of 400 µm in diameter.

The equipment used to perform the measurements of luminescence under UV was the previously described spectrometer Horiba Jobin-Yvon HR800 LabRam. The entrance slit, positioned behind the filter, is a diaphragm whose diameter can range from 50 to 500 µm. The 364.5 nm line of an Ar-ion laser was used as the excitation source. The power applied to the samples was fixed to 0.005 mW with an acquisition time set to 100 ms.

3. Results and discussion

3.1. X-ray diffraction

Fig. 1 shows the XRD pattern of $Sr_{1-x}Pb_xWO_4$ powders. All diffraction peaks are characteristic of the scheelite tetragonal structure with space group $I4_1/a$. The diffraction peaks are shifted to smaller 2θ angles as x increases.

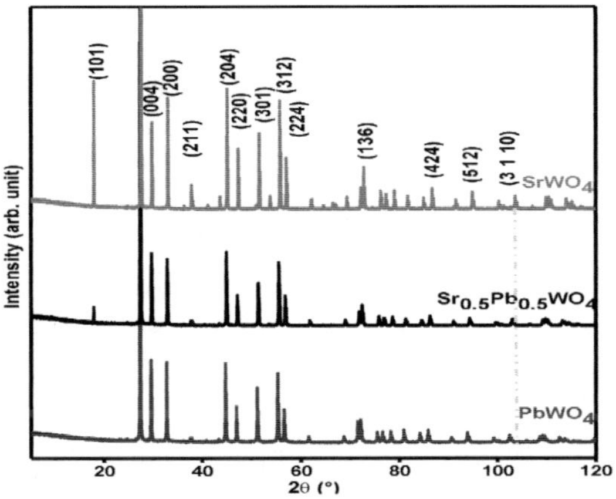

Figure 1.XRD patterns of the tetragonal $Sr_{1-x}Pb_xWO_4$ powders with x = 0, 0.5, 1.

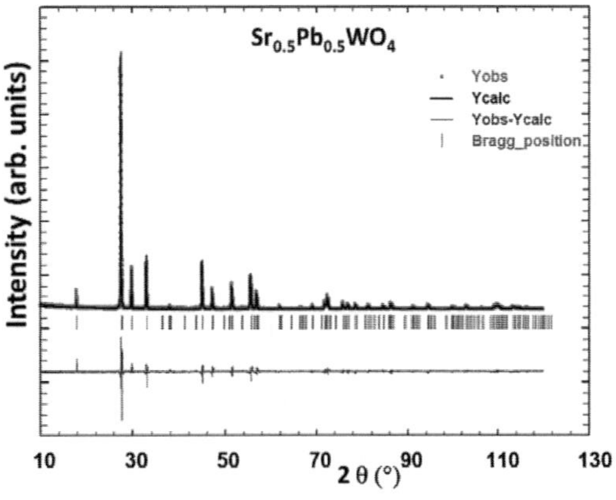

Figure 2. Calculated and observed diffraction profiles from Rietveld analyses for the $Sr_{0.5}Pb_{0.5}WO_4$ phase. Y_{calc}: calculated profile and Y_{obs} : observed profile.

Table 1 reports the refined structural parameters. Fig. 2 shows selected results comparing calculated diffraction profiles to experimental data.

Table 1. Refined structural parameters for $Sr_{1-x}Pb_xWO_4$ compounds with x = 0, 0.5; 1. Cell parameters, Debye-Waller factors B, reliability factors R in %.

x	**0**	**0.5**	**1.0**
a (10^{-10} m)	5.41734(2)	5.43919(3)	5.46220 (3)
b (10^{-10} m)	5.41734(2)	5.43919 (3)	5.46220 (3)
c (10^{-10} m)	11.95599(5)	11.99990(9)	12.04766(8)
Volume Cell (10^{-30} m^3) : V	350.880(2)	355.014 (5)	359.450(4)
B (M) = $(8\pi^2/3).\langle\Delta r^2\rangle$ (Å^2)	0.37(3)	0.77(4)	0.38(2)
B (W) (Å^2)	0.21(3)	0.17(3)	0.20(2)
Reliability factors (*)			
R_B %	5.53	7.64	7.01
R_F (%)	5.34	7.24	8.04
R_{exp} (%)	7.08	8.49	8.45
W (x,y,z)	Fixed values : x = 0.00000 ; y = 0.25000 ; z= 0.12500 (**)		
M (x,y,z)	Fixed values : x = 0.00000 ; y = 0.25000 ; z= 0.62500 (**)		
O (x,y,z)	Fixed values : x = 0.23880 ; y = 0.11410 ; z= 0.04290 (**)		

Notes :
(*)$R_B = 100.\{\sum | I_k^{obs} - I_{ki}^{calc} | / \sum | I_k^{obs} | \}$; $R_F = 100.\{\sum w_i | (I_{iobs})^{1/2} - (I_{ical})^{1/2} | / \sum (I_{iobs})^{1/2} | \}$; $R_{exp} = 100.\{[(N-P) / \sum w_i | y_i^{obs} |^2]^{1/2} \}$ where N, P are the number of observations and parameters respectively.
(**) Sr, W and O coordinates from author [25] have been fixed.

3.2. Raman spectra

Fig. 3 shows the Raman spectra of the three compositions at room temperature. The vibration modes of $SrWO_4$ and $PbWO_4$ were already interpreted by authors *Priya et al.* and *Ling et al.* [26, 27] and

Crane et al. [28]. The decrease with composition x of Raman wavenumbers clearly suggests a continuous modification of chemical bonds W–O due to W–O–Pb interactions: as Pb–O bonds present a more covalent character, the W–O bonds probably present a weaker covalent character, with decreasing rigidity, due to electron displacement in favor of the Pb–O bond. In addition, a mass effect, due to a mass of Pb stronger than the mass of Sr, should play a role in the decrease of wavenumbers.

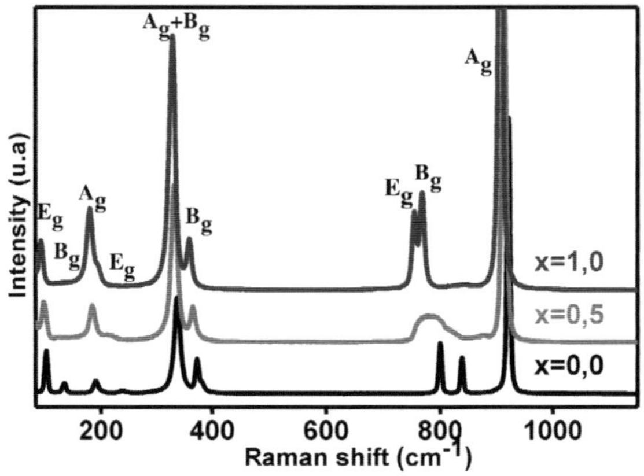

Figure 3. Room temperature Raman spectra of $Sr_{1-x}Pb_xWO_4$, for x = 0, 0.5, 1. (λ (excit.) = 514.5 nm).

3.3. *Scanning electron microscopy*

The SEM analyses of the x= 0.1, 0.5, 1 samples (Figures 4a, 4b and 4c, successively) show a progressive increase of crystallite sizes as x increases. In Figure 4a, we have inserted a magnified image to better illustrate the morphologies. In the case of the x=0.5 sample (Fig. 4b), a majority of large and regular crystallites is observed. For $PbWO_4$ (Fig. 4c) a mix system of crystallites with large and small grains is observed.

Figure 4. Scanning electron microscopy micrographs of the $Sr_{1-x}Pb_xWO_4$ samples (a) x=0 In inset : magnification of image (a) (x5), (b) x=0.5, and (c) x= 1.

3.4. *Luminescence under X-ray excitation*

Figure 5 compares the intensities emitted by the three different materials (x=0, 0.5, 1). We observe a clear difference between the intensity of the substituted compound (x=0.5) and those of x= 0 and 1 samples, with a very strong intensity for the x=0.5 sample. The average energy of the emission band is close to 2.8 eV (blue region).

3.5. *Luminescence under UV excitation*

Figure 6 shows the intensities of luminescence under UV excitation. In this case, no stronger intensity is observed for the substituted sample. This difference between the two emissions under UV and X-ray excitations can be ascribed to the low penetration of UV beam and to the high penetration of X-ray beam in samples. In the case of low energy UV excitation (3.6 eV), the observed luminescence is mainly due to the external areas of crystallites and irregular surfaces of powdered samples, which is not the case for high energy X-ray excitations (from 0 to 45000 eV) that concern larger volumes of material. The average energy of the emission band is close to 2.25 eV (green region).

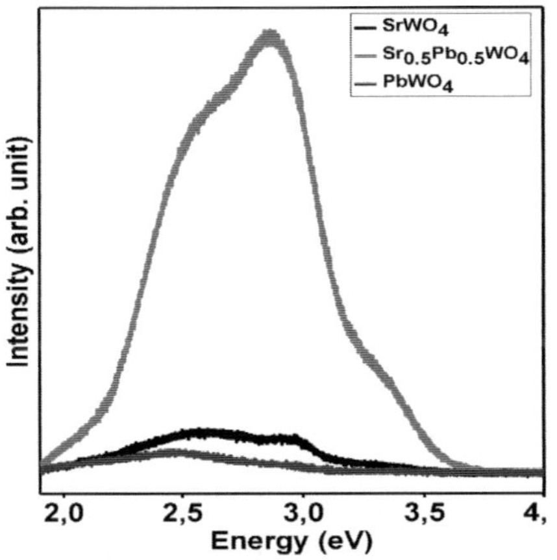

Figure 5. Photoluminescence spectra of $SrWO_4$, $Sr_{0.5}Pb_{0.5}WO_4$ and $PbWO_4$ under X-ray excitation (excitation energies in the continuous range 0 to 45000 eV).

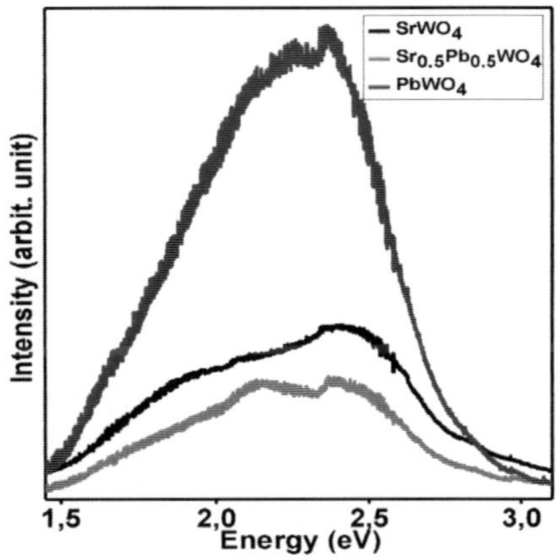

Figure 6. Photoluminescence spectra of $SrWO_4$, $Sr_{0.5}Pb_{0.5}WO_4$ and $PbWO_4$ under UV excitation (λ =364.5 nm or E=3.6 eV.

4. Conclusion

In this work, we have observed a complex influence of the chemical substitution on the luminescence signals under X-Ray excitation and UV excitation, in the series $Sr_{1-x}Pb_xWO_4$. Generally the origin of the emission in tungstate materials is related to the classical $^3T \rightarrow A_1$ transitions in the oxyanions WO_4^{2-}, and cations or anions vacancies in the material. The emission energies under X-ray and UV excitations are in the blue and green regions respectively. The luminescence experiments under X-ray excitation show that the emission of $Sr_{0.5}Pb_{0.5}WO_4$ presents intensity stronger than the intensity observed in $SrWO_4$ and $PbWO_4$. This behavior is fully different from the one observed in experiments under UV excitation: the emission intensity observed in the case of x=0.5 sample is lowered in comparison with the one of $SrWO_4$ and $PbWO_4$.

In these studies, we have shown that the effects of chemical substitution are complex, and that substitution can be used to obtain variable luminescence properties. The extension of these studies of substituted luminescent materials is now in progress: a major objective should be to better understand the different roles of substitution and of crystallization in luminescence efficiency.

Acknowledgments
This work was financially supported by the Regional Council of Provence-Alpes-Côte d'Azur (no. 2012-16322), by the General Council of Var and by Toulon Provence Mediterranean, and was conducted as part of the CNRS-CNRST 2014 project (Chimie 02/14).

References

[1] Liao W, Wang Y-F, Liu Y-M., Li Y-D, Qian Y-T 2000 Chem. Mater. **12**2819.
[2] Zhang Q, Yao W-T, Chen X, Zhu L, Fu Y, Zhang G, Sheng L, Yu S- H 2007Cryst. Growth Des. **7**1423.
[3] Gurman E, DanielsE, KingJ-S 1971 J. Chem. Phys. **55** 1093.
[4] TreadawayM-J, Powel R-C 1975 Phys. Rev. B **11**862.
[5] Chen W, Inagawa Y, Omatsu T, Tateda M, Takeuchi N, Usuki Y2001Opt. Commun. **194** 401.
[6] Lecoq P, Dafinei I, AuffrayENucl. 1995Instrum. Methods **365**291.

[7] TakaiS, SugiuraK, EsakaT1999Mater. Res. Bull.**34**193.
[8] NagirnyiV, FeldbachE, OnssonL-J, KirmM, LushchikA, LushchikC, NagornayaL-L, RyzhikovV-D, SavikhiuF, SvenssonG, TupitsinaI-A1998Radiat. Meas. **29** 247.
[9] Van-UitertL-G, Preziosi S1962 J. Appl. Phys. **33** 2908.
[10] EhrenbergH, WeitzelH, HeidC, FuessH, WltschekG, Kroener T, Van-TolJ, BonnetM1997J. Phys.: Condens. Matter **9** 3189.
[11] Van LooW1975Phys. Stat. Sol. A **27** 565.
[12] Van LooW 1975Phys. Stat. Sol. A **28** 227.
[13] BohacekP, ZazubovichS, SolovievaN, NiklM2007 Opt. Mat. **30** 66.
[14] ChukovaO, NedilkoS2013 Opt. Mat. **35** 1735.
[15] PôrtoS-L, LongoE, PizaniP-S , BoschiT-M, SimoesL-G-P, LimaS-J-G, FerreiraJ-M,SoledadeL-E-B, EspinozaJ-W-M, Cassia-SantosM-R, MaureraM-A-M-A, PaskocimasC-A, SantosI-M-G, SouzaA-G2008 J. Solid State Chem. **181** 1876.
[16] BurachasS, BeloglovskyS, MakovI, SavelievYu, VassilievaN, IppolitovM, MankoV, NikulinS, VassilievA, ApanasenkoA, TamulaitisG2002Funct. Mater. **9** 297.
[17] Kim T, Hole D-E, Townsend P-D, WooJ, WhangC 2005 Phys. Stat. Sol. c**2** 564.
[18] LagutaV-.V, NiklM, ZazubovichS IEEE Trans. Nucl. Sci.2008**55** 1275.
[19] BurachasS, ApanasenkoA, GrinyovB, RyzhikovV, KatrunovK, StarzhinskiyM, IppolitovM, MankoV, TamulaitisG2001Int. J. Inorg. Mater. **3** 1101.
[20] NovosadS-S, KostykL-V, NovosadS-I 2011 J. Appl. Spectrosc. **78**557.
[21] Hallaoui A, Taoufyq A, Arab M, Bakiz B, Benlhachemi A, Bazzi L, Villain S, ValmaletteJ-C, GuinnetonF, GavarriJ-R2015J. of Sol. St. Chem. **227** 186.
[22] Taoufyq A, Guinneton F, Valmalette J-C, Arab M, Benlhachemi A, Bakiz B, Villain S, LyoussiA, NolibeG, GavarriJ-R 2014 J. of Sol. St. Chem. **219**127.
[23] Taoufyq A, MauroyV, GuinnetonF, BakizB, VillainS, HallaouiA, BenlhachemiA, NolibeG, LyoussiA, GavarriJ-R 2015 Mater. Res. Bull.**70** 40.
[24] RoisnelT, Rodríguez-CarvajalJ, DelhezR, MittenmeijerE-J 2000(Eds.)Proceedings of the Seventh European Powder Diffraction Conference, Barcelona 118.
[25] PecharskyV-K, ZavalijP-Y 2005 Springer Science & Business Media 512.
[26] PriyaA, SinhaE, RoutS-K 2013 Solid State Sc. **20** 40.
[27] LingZ-C, XiaH-R, RanD-G, LiuF-Q, SunS-Q, FanJ-D, ZhangH-J, WangJ-Y, YuL-L 2006 Chem. Phys. Lett. **426** 85.
[28] CraneM, FrostR-L, WilliamsP- A, KloproggeJ-T 2002 J. Raman Spectrosc. **33** 62.

Improvement of Direct Torque Control by using a Space Vector Modulation Control of Three-Level Inverter

A Achalhi*[1], M Bezza[2], N Belbounaguia[1] and B Boujoudi[1]

[1] Laboratory of Electronics, Electrotechnics, Automation and Information Processing
[2] Laboratory of Physics of Atmosphere, Materials, and Modelling
Faculty of Sciences and Techniques of Mohammedia.
Hassan II University of Casablanca, Morocco

E-mail: amina.achalhi@gmail.com

Abstract. The performances of Direct Torque Control (DTC) of Induction machine are highly related to the inverter used therewith. The purpose of this paper is to highlight the efficiency of the space vector modulation (SVM) control of three level inverter associated with the direct torque control. The first part of this work is devoted to present the mathematical models of the DTC associated with 2-levels inverter then 3-levels inverter. Simulations on Matlab/Simulink will allow a comparative study to highlight advantages of the use of three levels inverter. The second part is devoted to the improvement of the DTC associated with a 3-levels inverter by application of the space vector modulation strategy (SVM) in order to manage the switching frequency and reduce harmonics. The efficiency of this solution will be attested by simulation on Matlab/Simulink.

1. Introduction

The Direct Torque Control (DTC) of the induction machine is widely adopted in various industrial areas due to its simplicity, quick response, and robustness. However, the major problem of this command is high ripples of torque (and flux) caused by the hysteresis controllers.

Many researchs were achieved in order to improve and reduce torque ripples either by using multilevel inverters, space vector modulation, or using intelligent techniques.

Reference [1] presents a novel Direct Torque Control of Induction Machine algorithm using a three-level Diode Clamped Inverter instead of the two level inverter used in the classical DTC scheme. Simulation results show that the proposed algorithm gives the same dynamical performances as those obtained with 2 level inverter with lower torque ripples.

Reference [2] compares the performance of Direct Torque Control of Induction Machine using 2-level and 3-level inverter, The simulation results show that the torque and flux ripples were reduced using the three level inverter.

In reference [3], authors propose a space vector modulation SVM control of three-level inverter. Simulation and experimental results demonstrate that the proposed strategy permits to keep the harmonics centered on multiples of the switching frequency.

Authors in reference [5] propose a solution for FPGA-based three level SVM inverter. The proposed strategy is verified by simulation in Matlab/Simulink. They also introduce the top-down methodology for FPGA design and give the functional block diagram for FPGA implementation.

This work investigates the efficiency of using multilevel inverter and Space Vector Modulation (SVM) in order to improve performances of DTC of the induction machine.

2. DTC USING 2 LEVEL AND 3 LEVEL INVERTER

2.1. Principal of DTC and mathematical model

The principal of the Direct Torque Control is based on the command of the inverter switches by applying a particular voltage sequence determining its status. The choice of the appropriate voltage vector depends on the machine's electromagnetic state, it is obtained from the hysteresis controllers of flux and torque. It is therefore to maintain the torque and flux inside their hysteresis bands.

The principal scheme of DTC (figure 1) is composed of the induction machine, Estimation block, Hysteresis comparators (flux and torque), Switching table and Inverter.

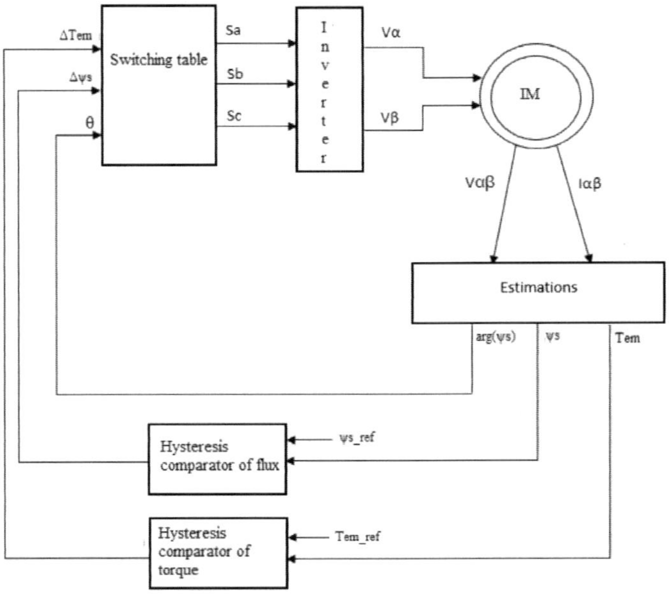

Figure 1. Principal scheme of Direct Torque Control (DTC) of induction machine

The dynamic behavior of an induction machine in the stationary frame is described by the following state equations:

$$\frac{d}{dt}\begin{bmatrix} is\alpha \\ is\beta \\ \psi s\alpha \\ \psi s\beta \end{bmatrix} = \begin{bmatrix} \frac{1}{\sigma Ls}(Rs+\frac{M^2}{LrTr}) & 0 & \frac{M}{\sigma LSLrTr} & \frac{\omega rM}{\sigma LsLr} \\ 0 & -\frac{1}{\sigma Ls}(Rs+\frac{M^2}{LrTr}) & -\frac{\omega rM}{\sigma LsLr} & \frac{M}{\sigma LSLrTr} \\ \frac{M}{Tr} & 0 & -\frac{1}{Tr} & -\omega r \\ 0 & \frac{M}{Tr} & \omega r & -\frac{1}{Tr} \end{bmatrix} * \begin{bmatrix} is\alpha \\ is\beta \\ \psi s\alpha \\ \psi s\beta \end{bmatrix} + \begin{bmatrix} \frac{1}{\sigma Ls} & 0 \\ 0 & \frac{1}{\sigma Ls} \\ 0 & 0 \\ 0 & 0 \end{bmatrix} \begin{bmatrix} Vs\alpha \\ Vs\beta \end{bmatrix} \tag{1}$$

With

$$\sigma = 1 - \frac{M^2}{LsLr} \tag{2}$$

$$Tr = \frac{Lr}{Rr} \tag{3}$$

Mechanical equation:

$$Tem = p*(\psi s\alpha * is\beta - \psi s\beta * is\alpha) \tag{4}$$

$$J\frac{d\Omega}{dt} = Tem - Tr - f*\Omega \tag{5}$$

Estimation of flux and torque:

$$\frac{d\psi s\alpha}{dt} = Vs\alpha - is\alpha \tag{6}$$

$$\frac{d\psi s\beta}{dt} = Vs\beta - is\beta \tag{7}$$

$$\psi s^2 = \psi s\alpha^2 + \psi s\beta^2 \tag{8}$$

$$\arg(\psi s) = a\tan(\psi s\beta / \psi s\alpha) \tag{9}$$

$$Tem = p*(\psi s\alpha * is\beta - \psi s\beta * is\alpha) \tag{10}$$

Switching table:
The switching table allow to choose the appropriate voltage vector in order to maintain the flux and torque in their hysteresis bands. Determining the switch table depends not only on torque and flux hysteresis controllers and sector but also on the level of the inverter used the DTC command.

2.2. Two Level inverter model and the switching table
The operation of the two-level inverter follows a conduction sequence of 180 degrees by the switch of the same arm.
We note Si state of an arm (i= a,b,c).

- Si = 1; case when the switch of the top is closed and the lower one is opened.
- Si = 0; case when the switch of the top is opened and the lower one is closed.

Under this condition, the phase voltage can be written in accordance with the control signal Si:

$$\begin{bmatrix} Van \\ Vbn \\ Vcn \end{bmatrix} = Uc/3 \begin{bmatrix} 2 & -1 & -1 \\ -1 & 2 & -1 \\ -1 & -1 & 2 \end{bmatrix} \begin{bmatrix} Sa \\ Sb \\ Sc \end{bmatrix} \tag{11}$$

The different combinations of the three values (Sa, Sb, Sc) are used to generate eight voltage vectors: V0(000), V1(100),V2(110), V3(010), V4(011), V5(001), V6(101), V7(111).

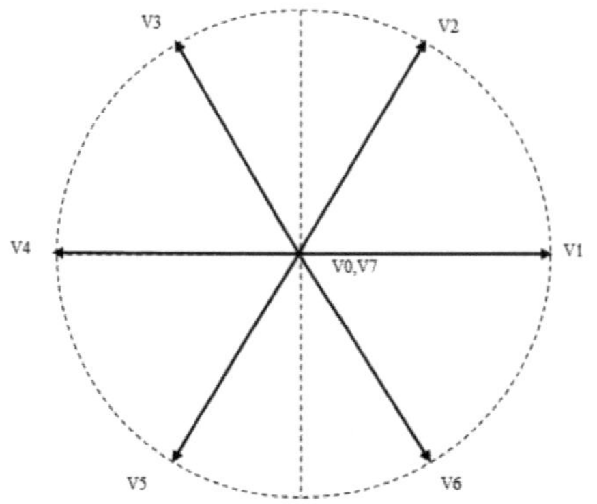

Figure 2. Space vectors representation of the 2 level inverter output voltage

Table 1. Switching table for the 2 level inverter

Δψ	ΔTem	Sectors					
		1	2	3	4	5	6
	1	V2	V3	V4	V5	V6	V1
1	0	V7	V0	V7	V0	V7	V0
	-1	V6	V1	V2	V3	V4	V5
	1	V3	V4	V5	V6	V1	V2
0	0	V0	V7	V0	V7	V0	V7
	-1	V5	V6	V1	V2	V3	V4

2.3. Three-Level inverter and the switching table

The total voltage Udc of the DC input bus of the 3 level inverter is spread over two capacitors whichform a middle point o, to obtain a higher voltage level. The three arms of the inverter are composed of four controlled switches (K1, K2, K3, and K4), these switches are unidirectional in terms of voltage and bidirectional in the current. The three-level inverter has three operating sequences:

- Sequence 1: maximum level generation
  K1 and K2 passing, k3 and k4 blocked:
  Varm = Udc/2
  Vk3 = Vk4 = -E/2
- Sequence 2: medium level generation
  K2 and k3 passing, k1 and k4 blocked:
  Varm = 0
  Vk1 = Vk4= E/2
- Sequence 3: minimum level generation
  K1 and K2 blocked, K3 and K4 passing
  Varm = -E/2
  Vk1 = Vk2 = E/2

Table 2. operating sequences of the three-level inverter

K1	K2	K3	K4	Va0
1	1	0	0	E/2
0	1	1	0	0
0	0	1	1	-E/2

The three-level inverter can be modeled as follows:

$$\begin{bmatrix} Van \\ Vbn \\ Vcn \end{bmatrix} = Udc/3 * \begin{bmatrix} 2 & -1 & -1 \\ -1 & 2 & -1 \\ -1 & -1 & 2 \end{bmatrix} * \left[\begin{bmatrix} F11 \\ F21 \\ F31 \end{bmatrix} - \begin{bmatrix} F10 \\ F20 \\ F30 \end{bmatrix} \right] \tag{12}$$

With
Fk1: state of the top half arm
Fk0: state of the lower half arm
k: arm number
Fk1 = Sk1*Sk2 (states of switch 1 and 2 from arm k)
Fk0 = Sk3*Sk4 (states of switch 3 and 4 from arm k)

The output voltages of the three-level inverter can take three states (-1, 0 and 1) for the voltage levels (-Udc/2, 0 and Udc/2), it means that the voltage vector can take 27 states. The following figure represents the voltage vectors positions in the (α,β) reference.
Theses voltage vectors can be listed in four categories:

- Zero vectors: V0, V7 and V26
- Small vectors: V1, V2, V3, V4, V5, V6, V8, V9, V10, V11, V12, V13.
- Medium vectors: V20, V21, V22, V23, V24, V25.
- Large vectors: V14, V15, V16, V17, V18, V19.

Table 3. Voltage vectors

Vectors	states	Vectors	States	Vectors	States
V0	000	V9	-1-11	V18	00-1
V1	100	V10	1-11	V19	-10-1
V2	110	V11	1-1-1	V20	-110
V3	010	V12	11-1	V21	1-10
V4	011	V13	-11-1	V22	0-11
V5	001	V14	-100	V23	01-1
V6	101	V15	-1-10	V24	10-1
V7	111	V16	0-10	V25	-101
V8	-111	V17	0-1-1	V26	-1-1-1

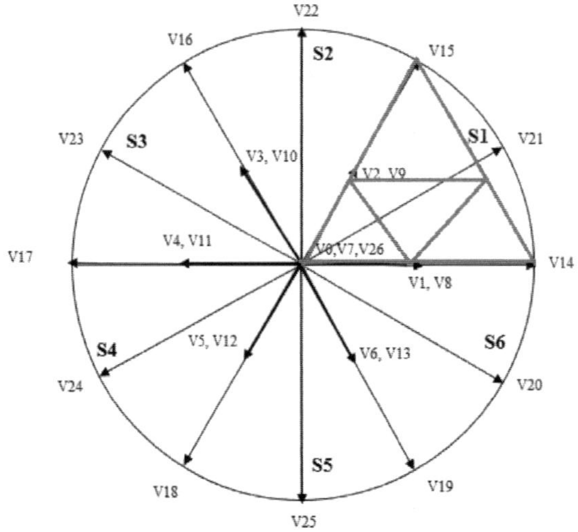

Figure 3. Space vector representation of the 3level inverter output voltage

In this work, the switching table of the 3 level inverter fed DTC- IM drive considers only large, small and zero vectors.

Table 4. switching table for the 3 level inverter

		Sectors					
Δflux	ΔTorque	1	2	3	4	5	6
	-2	V18	V19	V14	V15	V16	V17
	-1	V5	V6	V1	V2	V3	V4
-1	0	V0	V0	V0	V0	V0	V0
	1	V3	V4	V5	V6	V1	V2
	2	V16	V17	V18	V19	V14	V15
	-2	V19	V14	V15	V16	V17	V18
	-1	V6	V1	V2	V3	V4	V5
1	0	V0	V0	V0	V0	V0	V0
	1	V2	V3	V4	V5	V6	V1
	2	V15	V16	V17	V18	V19	V14

3. Space Vector Modulation

The principal of Space Vector Modulation (SVM) is based on the determination of time portions (modulation period) to be allocated to each voltage vector during the sampling period. The SVM is used to determine the ignition and extinction sequences of the inverter switches.

The proposed method is based on the predictive calculation of the stator voltage vector, applied to the induction machine to control the flux and torque, to determine the switching states of the inverter. It, therefore, provides an algorithm based on modulation of space vector for controlling the electromagnetic torque of the induction machine and provides a fixed switching frequency, it improves the dynamic response and static behavior of the DTC by reducing acoustic noises and torque, flux, current and speed ripples during steady state.

In the case of 3 level inverter, the space vectors diagram can be divided into 6 sectors (S1, S6), each sector can be divided into 4 triangles called region (figure 3).

The SVM technique follows three steps:

- Determination of the voltage reference vector and the position.

- Determination of the sector and region.
- Calculation of switching time.
- Generating appropriate gates to control the inverter.

4. Simulation and results

Simulations of DTC were carried out under Matlab/Simulink environment. For a nominal speed of 150 rad/s, we start the simulations with no load then apply a 35N.m load at t =1s.

Figure 4 show the torque response and a zoom of the no load part and the part with a load reference of 35N.m for classical DTC (2 level inverter) (a), DTC associated to 3 level inverter (b) and SVM – DTC associated with 3 level inverter (c).

In order to show the efficiency of the proposed method in reducing ripples, we calculated the harmonics magnitude (in % of the fundamental) for 10 frequency values and plot the results (figure 5).

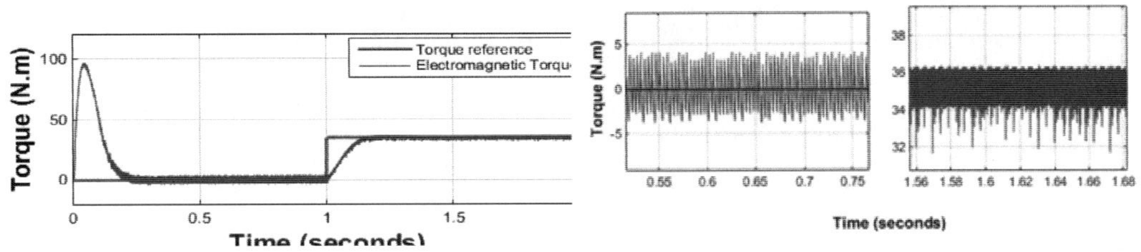

(a)

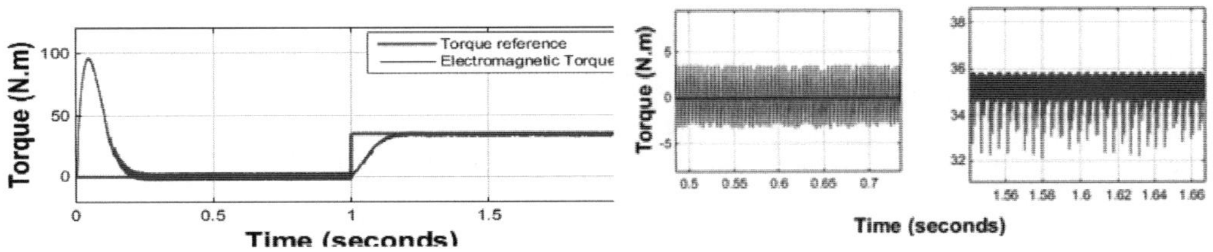

(b)

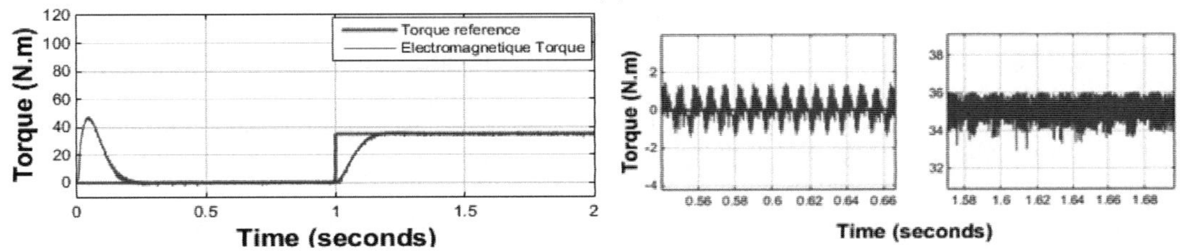

(c)

Figure 4. (a) Torque response in DTC- 2 level inverter, (b) Torque response in DTC- 3 level inverter (c) Torque response in SVM-DTC using 3 level inverter

Results presented in figure 4 show how the use of 3 level inverter (b) improved the classical DTC (a) by reducing torque ripples.

We notice a better improvement of torque response at figure 4 (c) where the SVM was introduced.

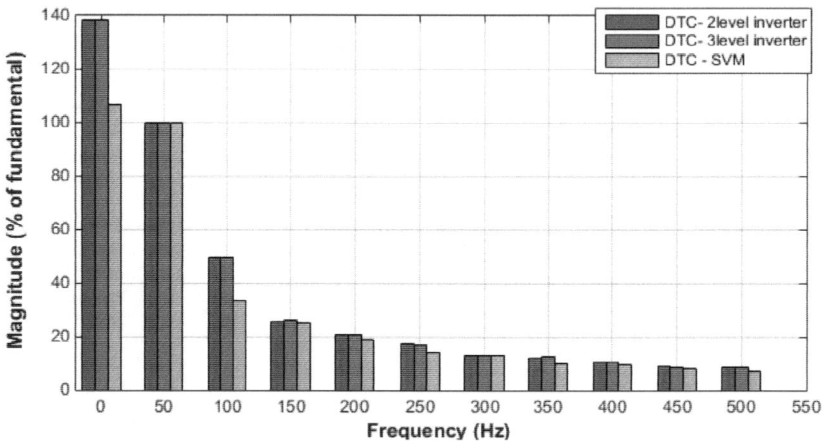

Figure 5. FFT analysis of torque response for the three methods

The FFT analysis in figure 5 also validates the efficiency of using SVM –DTC associated with a3-level inverter to reduce torque ripples. The Total Harmonic Distortion (THD) went from 71.5% in the classical DTC to 70% in DTC associated with 3 level inverter to reach 57.07% in SVM – DTC.

5. Conclusion

In this paper, we presented the model of classical DTC using 2 level NPC inverter, classical DTC using 3 level NPC inverter and SVM- DTC using three-level NPC inverter.

Simulation results under Matlab/Simulink environment show how the SVM-DTC gives better responses by reducing flux and torque ripples.

References

[1] DiptiH.Granatra and Saurabh N.Pandya, *Torque Ripples Minimization in Direct Torque Control based Induction Motor Drive* IEEE Student's Conference on Electrical, Electronics and Computer Science, 2012.

[2] R. Dharmaprakash and Joseph Henry, *Comparison of Direct Torque Control of Induction Motor Using Two-Level and Three-Level Inverter,* Middle-East Journal of Scientifical Research **23**, 2015.

[3] T.Abdelkrim and al. *Etude et Réalisation d'un Onduleur à Trois Niveaux Commandé par MLI Vectorielle*, Revue des Energies Renouvelables Vol. **14** N2, 2011.

[4] Yongchang Zhang and al. *An Improved Direct Torque Control for Three-Level Inverter- Fed Induction Motor Sensorless Drive*, IEEE Transactions on Power Electronics, 2012.

[5] Yun Wan and Jiango Jiang, *The study of FPGA- based Three- level SVM NPC Inverter*, IEEE, IPEMC 2009

[6] D.Lalili and al. *Méthode Simplifiée de la Modulation Vectorielle de l'Onduleur à Cinq Niveau*, 3rd International Conference : Sciences of Electronic, Technologies of Information and Telecommunications, March 27-31, 2005

[7] Schweizer and al. *Comparative Evaluation of Advanced Three-Phase Three-Level Inverter/Converter Topologies Against Two-Level Systems*, IEEE Transations on Industrial Electronics; 2013

[8] V. Ranjih Kumar et al. *Three Level Diode Clamped Inverter Fed Induction Motor with DTC-SVM*. Journal of Engineering Research and Applications, Vol.**3** Issue 5, Spt-Oct 2013.

IOP Conf. Series: Materials Science and Engineering **186** (2017) 012025 doi:10.1088/1757-899X/186/1/012025

[9] C. Hari Krishna et al. *Simplified SVPWM Algorithm Based Diode Clamped 3-Level Inverter Fed DTC-IM Drive*. International Journal of Engineering Science and Technology (IJEST), Vol. **4** No 05, May 2012.

[10] E. Hassankhan and D. A.Khaburi*DTC-SVM Scheme for Induction Motors Fed with a Three-Level Inverter*. International Journal of Mechanical, Aerospace, Industrial, Mechatronic and Manufacturing Engineering Vol. **2** No 8 2008.

[11] D. Lee and al. *Simple Space Vector PWM Scheme for 3 Level NPC Inverters Including the Overmodulation Region*. Journal of Power Electronics, Vol. **11** No 5, Sept 2011.

The optoelectronic properties of a solar energy material: Ag_2HgSnS_4

S Hadjri Mebarki, B Amrani, K Driss Khodja and A Khelil

University d'Oran Ahmed Ben Bella, Laboratoire des Couches minces et Matériaux pour l'Electronique, BP 1524 El Mnaouer Oran , Algeria

Email: abouhalouane@yahoo.fr

Abstract. We used an *ab initio* full potential-linearized augmented plane wave technique within the density functional theory to study the structural and optoelectronic properties of Ag_2HgSnS_4 in a wurtzite-stannite phase. The exchange correlation effects are included through the generalized gradient approximation and modified Becke-Johnson exchange potential. Various physical quantities, such as lattice parameter, bulk modulus, band structure and density of states, are given. Also, we have presented the results of the effective mass for the electrons in the CB and the holes in the BV. We show that the modified Becke-Johnson exchange potential can predict the energy band gap in better agreement with the experiment. In addition the dielectric function and energy-loss function are presented for the energy range of 0-26 eV. The electronic and optical properties indicate that this compound can be successfully used in optoelectronic devices

1. Introduction

Ag_2HgSnX_4, where X=S or Se, and related compounds have attracted increased attention over the past few years due to their variety of applications in optoelectronics and non-linear optics devices, owing to their n-type conductivity, high absorption coefficient, suitable band gap, lower cost and environmental compatibility [1-19].

The quaternary sulfide (Ag_2HgSnS_4) exists in the quasiternary Ag_2S–HgS–SnSe system [20, 21]. For the first time it was synthesized by Haeuseler and Himmrich [22] using the solid state reactions at 970 K. It was ascertained that the sample of Ag_2HgSnS_4 crystallizes into a wurtzite–stannite (space group $Pmn2_1$) structure with the lattice parameters a = 8.203, b = 7.026, and c = 6.710 Å. Its Far-infrared and Raman spectra was studied by the same authors [23]. Parasyuk *et al.* [21] have reported measurements of the absorption edge in Ag_2HgSnS_4 lead to estimates of the bandgap energy Eg from liquid nitrogen temperature to room temperature. Recently, Bozhko *et al.* [24] have used *XPS* method to measure the binding energies of core-level electrons of atoms constituting Ag_2HgSnS_4.

A deeper knowledge of its properties would help improve the understanding of this compound and stimulate the development of new applications.

To the best of our knowledge there are no theoretical calculations exploring the optoelectronic properties of this compound. Consequently, the primary purpose of this work is to provide some additional information to the existing data on the physical properties of Ag_2HgSnS_4 by using the full-potential linearized augmented plane wave method. After introducing the problem, the remaining paper is organized as follows: in Section 2, we describe the calculation procedure. Results are presented in Section 3. Section 4 contains the conclusion.

2. Method of calculations

The first-principles calculations performed in this paper are based on the density functional theory (DFT). The total energies were calculated within the full potential linearized augmented plane wave (FPLAPW) +local orbitals (lo) method, implemented in the WIEN2k code [25]. In this code, the crystal structures were decided under the condition that the total energy is minimized from all atomic configurations. The exchange and correlation effects were treated using the generalized gradient approximation (GGA96) based on Perdew et *al.* [26] in the structural analysis, and modified Becke-Johnson (mBJ) as coupled with (GGA96) approach to treat the optoelectronic properties [27]. The mBJ exchange potential, as an orbital independent and semi-local exchange correlation potential, has been proved to yield quite accurate gaps for numerous semiconductors [28, 29, 30]. The sphere radii used in the calculations for Ag, Hg, Sn and S are 2.46, 2.5, 2.26 and 1.85 u.a., respectively. Within these spheres, the potential is expanded in the form

$$V(r) = \sum_{lm} V_{lm}(r) Y_{lm}(\hat{r}) \quad (1)$$

and outside the sphere, $V(r) = \sum_{K} V_K e^{iKr}$ (2)

where $Y_{lm}(\hat{r})$ is a linear combination of radial functions times spherical harmonics. Within this calculational scheme, there are no shape approximations to the charge density or potential. Special attention has been paid to the convergence of the total energy with respect to the cut off. In consideration of computational cost, we choose the Brillouin-zone sampling mesh parameters for the *k*-point set are 27.

3. Results and discussion

3.1. Structural properties

The structure parameters for wurtzite-stannite phase from experimental observations [21, 22] were optimized and then listed in table 1. The calculated total energies were fitted to the Murnaghan [31] equation of state. Compared with the experimental results, the theoretical lattice parameters *a*, *b*, and *c* are larger with relative errors of 1.58% , 2.38 and 2.32%, respectively. Thus, the theoretical results of our calculations are credible. The bulk modulus (B_0) and its pressure derivative (B_0') are obtained. Unfortunately, no experimental data are yet available to compare with the present calculated results.

Table 1. Calculated lattice parameters *a*, *b*, *c*, bulk modulus *B* and its pressure derivative *B'*, compared to the experiment works of Ag_2HgSnS_4

Approach	a_{eq}(Å)	b_{eq}(Å)	c_{eq}(Å)	B_{eq}(GPa)	B'
This work	8.333	7.193	6.866	46.171	4.596
Experiment	8.207[a]	7.033[a]	6.718[a]		
	8.203[b]	7.026[b]	6.710[b]		

[a] Ref. [21]
[b] Ref. [22]

3.2. Electronic properties

The band structures of Ag_2HgSnS_4, calculated by using the FP-LAPW approach and GGA, are shown in Figure. 1a. This figure shows that the wurtzite-stannite phase is a direct band gap semiconductor.

The conduction band minimum located at the Γ-point and the valence band maximum is located close to the Γ-point resulting in a direct band gap of 0.109 eV. This value is much smaller than the experimental result determined by Parasyuk [21]. It is well known that band structure calculations based on the GGA method underestimate the band gap, due to its incomplete cancellation of the self-interaction. There is an overall topological resemblance of the present band structure and those calculated by Dan and Hing-Hong [32]. In order to improve the band gap prediction of Ag_2HgSnS_4, the modified Becke–Johnson exchange–correlation potential was also used. It is clear from the figure. 1b, that the calculated band gap with mBJ 1.21 eV is consistent with the experimentally reported result of 1.32 eV [21].

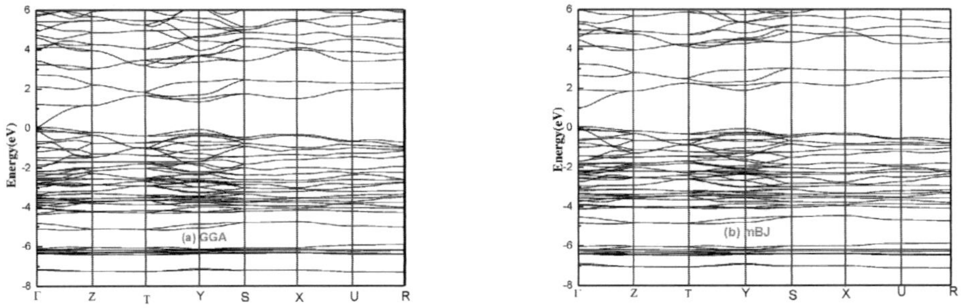

Figure 1. Band structures calculated for Ag_2HgSnS_4. Band structures using (**a**) GGA (**b**) and mBJ. The energy zero is taken at Ef

Figure. 2 shows the partial density of state (PDOS) together with the total DOS of Ag_2HgSnS_4, and we can distinguish two important valence band regions, which we call for simplicity VB1 and VB2 from the top of the valence bands. We find a lower valence bands (VB2 at 7.31-5.94 eV below the Fermi energy E_f) that are formed by Sn -*s* and S -*s* states, and the main contributions is from the S -*s* states. The upper valence bands VB1 are formed by Ag -*d* with some admixture of Hg -*d* and Sn -*s p* states. The conduction band is formed mainly by Sn -*p*. These results are in good agreement with XPS method of Bozhko et *al.* [24].

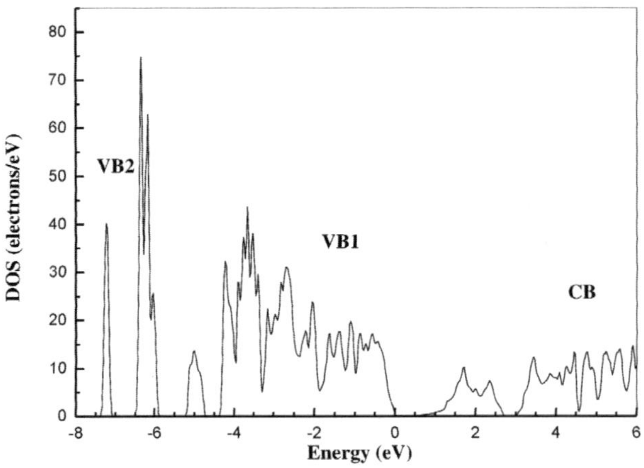

Figure 2. Calculated total density of states of Ag_2HgSnS_4. Fermi level is set at zero energy

Turning to the effective masses, we are interested to study the effective masses of electrons and holes, which are important for the excitonic properties. We have computed the effective masses of electrons and holes both at the conduction-band minima and valence-band maxima. The corresponding values are 0.492 m_0 and 1.151 m_0 for electron and hole, respectively.

3.3. Optical properties

It is known that the optical properties of a medium can be depicted well by frequency dependent complex dielectric function $\varepsilon(\omega)=\varepsilon_1(\omega)+i\varepsilon_2(\omega)$. In order to calculate related dielectric tensor, improved tetrahedron scheme [33] is applied for the BZ integration as provided in WIEN2k package. The imaginary part $\varepsilon_2(\omega)$ was carried out by calculating the electronic structure, using the joint density of states and the optical matrix elements. The real part of dilectric tensor $\varepsilon_1(\omega)$ was obtained from $\varepsilon_2(\omega)$ using Kramers-Krönig relationship. For detail description of this approach used for the calculation of optical properties, work of Ambrosch-Draxl and Sofo will be very helpful [34].

To the best of our knowledge, no theoretical calculations exploring the optical properties have yet been made of Ag_2HgSnS_4. Hence our work is a first attempt in this direction. Using the method above, we have calculated the imaginary $\varepsilon_2(\omega)$ and real $\varepsilon_1(\omega)$ parts for Ag_2HgSnS_4 and the results are shown in figure.3. Our dielectric functions show that the threshold peak appears at 1.21 eV. This energy is related to direct transition between the absolute fourth valence band maximum and the first conduction band minimum Γ_v-Γ_c. All the band transitions which contribute to the structures of $\varepsilon_2(\omega)$ are primarily from Ag $-d$ states to Sn $-p$ states. Additionally, It is clear that there is a significant anisotropy between the three spectra corresponding to different polarizations.

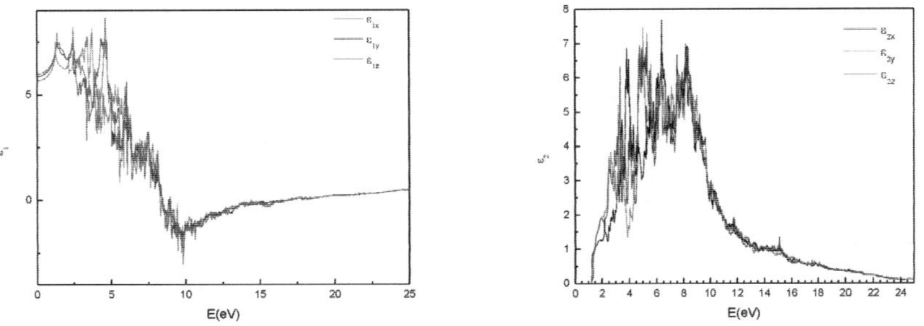

Figure3. The imaginary and real parts of dielectric function of Ag_2HgSnS_4

The static dielectric functions are obtained as the zero frequency limit of the real part of the frequency-dependent dielectric function. The calculated ε_∞ is 5.82. We note that a smaller energy gap yields a larger ε_∞ value. This could be explained on the basis of the Penn model [35]. The Penn model is based on the expression $\varepsilon_\infty \approx 1+(\hbar\omega_p / E_g)^2$. It is clear that ε_∞ is inversely proportional to E_g. Hence, smaller E_g yields larger ε_∞.

The corresponding energy-loss function, $L(\omega)$ is presented in figure 4. The $L(\omega)$ describes the energy loss of fast electrons traversing the material. The sharp maxima in the energy loss function are associated with the existence of plasma oscillations [36]. The curves of L have a maximum near 12.56 eV. This value is similar to that estimated for $Ag_2HgSnSe_4$ [32]

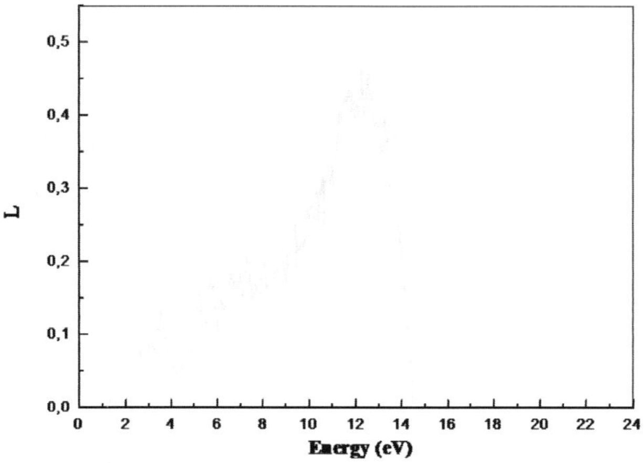

Figure 4. The electron energy loss spectrum of Ag_2HgSnS_4

4. Conclusion

Using first principles calculations, we have obtained the results of Ag_2HgSnS_4 in the wurtzite-stannite phase. Our main results and conclusions can be summarized as follows:

(i) The calculated structural properties agrees nicely with earlier reported work in the literature.

(ii) The band gap calculated show that Ag_2HgSnS_4 is a direct semiconductor with a band gap around 1.21 eV, which is in good accord with experimental value 1.32 eV.

(iii) Density of states analysis showed that the lower part of the valence bands is dominated by S -*s* orbital and upper part by S -*s and Ag -s* orbital in Ag2HgSnS4.

(iv) We have computed the effective masses of the electrons and the holes.

(v) We have calculated the dynamic dielectric function. Further, we have derived the static dielectric constant. All the band transitions which contribute to the structures of $\varepsilon_2(\omega)$ are primarily from Ag $-d$ states to Sn $-p$ states. We have also presented the energy loss function.

The reported calculations provide new structural, electronic and optical results from first principles for Ag_2HgSnS_4. Hence, this study forms part of a large theoretical effort to explore the properties of this compound.

References

[1] Liu M L, Chen I W, Huang F Q, Chen L D 2009 Adv. Mater **21** p 3808.
[2] Lekse J W, Moreau M A, McNerny K L, Yeon J, Halasyamani P S, Aitken J A 2009 Inorg. Chem **48** p 7516.
[3] Uday Bhaskar P, Guresh Babu G, Kishore Kumar Y B, Sundara Raja V 2013 Thin Solid Films **534** p 249
[4] Lekse J W, Leverett B M, Lake C H, Aitken J A 2008 J. Solid State Chem **181** p 3217.
[5] Guo Q, Ford G M, Yang W C, Walker B C, Stach E A, Hillhouse H W, Agrawal R 2010 J. Am. Chem. Soc **132** p 17384.
[6] Ford G M, Guo Q, Agrawal R, Hillhouse H W 2011 Chem. Mater **23** p 2626.

[7] Rincón C, Quintero M, Moreno E, Power C, Quintero E, Henao J A, Macías M A, Delgado G E, Tovar R, Morocoima M 2011 Solid State Commun **151** p 947.
[8] Li Y, Fan W, Sun H, Cheng X, Li P, Zhao X 2011 J. Phys. Condens. Matter **23** p 225401.
[9] Brunetta C D, Minsterman W C III., Lake C H, Aitken J A 2012 J. Solid State Chem **187** p 177.
[10] Shi X I, Huang F Q, Liu M L, Chen L D 2009 Appl. Phys. Lett **94** p 122103.
[11] Steinhagen C, Pantani M G, Akhavan V, Goodfellow B, Koo B, Korgel B A 2009 J. Am. Chem. Soc **131** p 12554.
[12] Tanaka K, Oonuki M, Moritake N, Uchiki H 2009 Sol. Energy Mater. Sol. Cells **93** p 583.
[13] Todorov T K, Reuter K B, Mitzi D B 2010 Adv. Mater **22** p 156.
[14] Pearton S J, Abernathy C R, Norton D P, Hebart A F, Park Y D, Boatner L A, Budai J.D. 2003 Mater. Sci. Eng. R **40** p 137.
[15] Dan L, Xing-Hong Z 2011 Chin. Phys. B **20** p 126102.
[16] Ohmer M C, Pandey R 1998 MRS Bull **23** p 16.
[17] Honig H, Shen H, Yao G, Doverspike K, Kershaw R, Dwight K, Wold A 1998 Mater. Res. Bull **32** p 307.
[18] Levcenco S, Dumcenco D, Huang Y S, Arushanov E, Tezlevan V, Tiong K K, Du C H 2011 J. Alloys Comp **509** p 7105.
[19] Shavel A, Arbiol J, Cabot A 2010 J. Am. Chem. Soc **132** p 4514.
[20] Parasyuk O V 1999 J. Alloys Comp **291** p215.
[21] Parasyuk O V, Chykhrij S I, Bozhko V V, Piskach L V,. Bogdanyuk M S, Olekseyuk I D, Bulatetska L V, Pekhnyo V I 2005 J. Alloys Comp **399** p 32.
[22] Haeuseler H, Himmrich M 1989 Z. Naturforsch **44b** p 1035.
[23] Himmrich M, Haeuseler H 1993 Spectrochim. Acta **47**A p 933.
[24] Bozhko V V, Tretyak A P, Parasyuk O V, Ocheretova V A, Khyzhun O Y 2014 Opt. Mater **36** p 977.
[25] Blaha P, Schwarz K, Madsen G K H, Kvasnicka D, Luitz J 2001, WIEN2k, An Augmented Plane Wave Plus Local Orbitals Program for Calculating Crystal Properties, Vienna University of technology, Vienna, Austria.
[26] Perdew J P, Burke S, Ernzerhof M 1996 Phys. Rev. Lett. **77** p 3865.
[27] Tran F, Blaha P 2009 Phys Rev Lett **102** 226401
[28] Sharma S, Verma A S, Jindal V K 2014 Mater. Res. Bull **53** p 218.
[29] Camargo-Martinez JA, Baquero R 2012 Phys. Rev. B **86** p. 195106.
[30] Khan I, Ahmad I, Rahnamaye Aliabad H A, Maqbool M 2015 Mat. Today Proc **2** p 5122.
[31] Murnaghan F D 1944 Proc. Natl. Acad. Sci. USA **30** p 5390.
[32] Dan L and Xing-Hong Z 2011 Chin. Phys. B **20** p 126102.
[33] Jepsen O and Andersen O K 1971 Solid state Commun **9** p 1763; Lehmann G and Taut M 1972 Phys. Status Solidi b **54** p 496.
[34] Ambrosch-Draxl C and Sofo J O 2006 Comput. Phys. Commun. **175** p 1.
[35] Penn D R 1962 Phys. Rev **128** p 2093.
[36] Philipp H R, and Ehrenreich H 1963 Phys. Rev. **129** p 1550.

IOP Conf. Series: Materials Science and Engineering **186** (2017) 012027 doi:10.1088/1757-899X/186/1/012027

Ultrasonic nondestructive characterization of mortars by the width of the resonances

H Bita, B Faiz, A Moudden ,H Lotfi and El H Ouacha

Laboratory of Metrology and Information Processing, Ibn Zohr University, Faculty of Sciences, Agadir, Morocco

Email: hassan.bita51@gmail.com

Abstract. In this work, we study the width of the resonances of the ultrasound waves reflection coefficient backscattered by a plane structure of the mortar. We establish the relationship between this width with two parameters which are widely used in non-destructive characterization of cementitious materials namely the velocity and attenuation. Monitoring the hydration of three solutions of mortars produced with different sizes of sand grains shows that the experimental results confirmed the theoretical predictions. Linear correlations are established between the width of resonance and the two ultrasonic parameters.

1. Introduction

Techniques using the ultrasonic wave propagation are very suitable tools for non-destructive characterization of cementitious materials. The measured parameters are typically the speed which is related to the dynamic elasticity and attenuation which is related to the microstructure of the material. Several research studies have been conducted to a nondestructive characterization of the cementitious materials [1, 2, 3, 4]. Among these methods, measurements of ultrasonic velocity and attenuation which are proven in evaluating the mechanical and structural properties of these materials [5, 6, 7].
The resonant scattering theory [8, 9,10] provides that the propagation of sound waves in the material generates resonances in its thickness, where each resonance is characterized by its frequency and width. We focus on the resonances of the reflection coefficient of a plane mortar structure that will be put in the vicinity of an isolated resonance in the form of a nuclear function known as the Breit-Wigner. Two parameters appear in the expression of this function, namely the frequency f_0 and its width Γ. During its hydration, the mortar is transformed from the viscous state to the viscoelastic one and then to the solid elastic one. This transformation is accompanied by a change in velocity and ultrasonic attenuation but also the movements of resonance frequencies and variation in their widths. We seek in this work to highlight the behavior of the resonance width with respect to two parameters of the ultrasonic wave in a mortar during its hydration namely the velocity and the attenuation.

2. Materials and methods

2.1. *Experimental device*

The experimental device (figure 1) described in detail in a previous work [11], consists of a parallel sided container enclosing a thick layer of mortar d_m= 15mm, irradiated at normal incidence by an ultrasonic wave emitted by a transducer of central frequency 0,5MHz which acts as emitter and receiver. The recording of the signals reflected by the Plexiglas / mortar / glass structure, is 15min all three days.

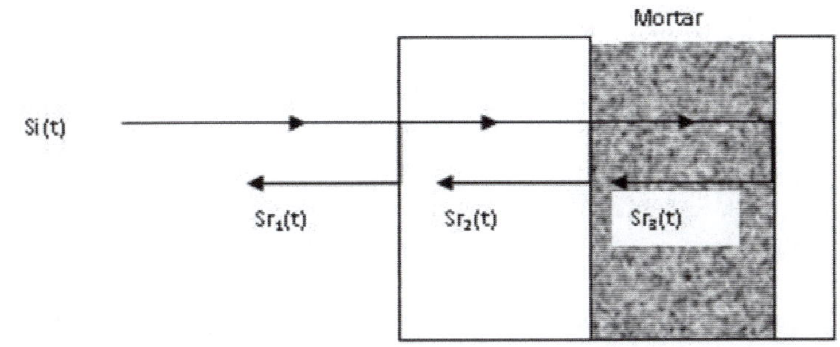

Figure 1. Geometry of the problem and the path of incident and reflected signals

2.2. Techniques of measures

2.2.1. Theory

Assuming that the three layer structures; Plexiglas, mortar and glass are flat, and by placing in the context of the theory of linear systems, the signal reflected by the mortar layer can be written as a sum of the echo reflected by the interface plexiglas / mortar and all echoes that undergoes multiple reflections inside the material. The signal processing in the frequency plan allows to write the reflection coefficient as a series of Debye:

$$R = R_{p/m} + T_{p/m} R_{m/g} T_{m/p} \, exp(-j(2\, k_m d_m) \times \sum_{n=0}^{\infty} (R_{m/g} R_{m/p} exp(-j2k_m d_m))^{n} \tag{1}$$

Experimentally terms of order n $\geq$1 in equation (1) are not visible during all handling. In this case the complex reflection coefficient of the mortar is then:

$$R \cong R_{p/m} + T_{p/m} R_{m/g} T_{m/p} \, exp(-j(2\, k_m d_m) \tag{2}$$

Where $k_m = \frac{\omega}{v_m} - j\, \alpha_m$: The complex wave vector in the mortar.

v_m : The velocity of the wave in the mortar.

α_m : The attenuation of the wave in the mortar.

d_m: The thickness of the mortar layer.

Ri/j ,*Ti/j* are respectively the coefficients of reflection and transmission of the interface *i* / *j*. The letters p, m, and g represent the Plexiglas, mortar and glass environments.

In the vicinity of an isolated resonance, the reflection coefficient can be written as a complex function of Breit-Wigner characterized by its frequency f_0, its width Γ

$$R \cong A - \frac{B}{1-j\,\frac{(f-f_0)}{\frac{\Gamma}{2}}} \tag{3}$$

Where: $A = R_{p/m}$, $B = T_{p/m} R_{m/g} T_{m/p} \, exp(-2\gamma_m d_m)$ *and* $\Gamma \cong \frac{v_m}{2\pi d_m}$

2.2.2. Experimental measurements

The coefficient of experimental reflection of the mortar layer (figure 2) is calculated by using the ratio of Fourier transforms:

$$R = R_{\frac{p}{m}} \times \frac{S_{r2}(f)+ S_{r3}(f)}{S_{r2}(f)} \tag{4}$$

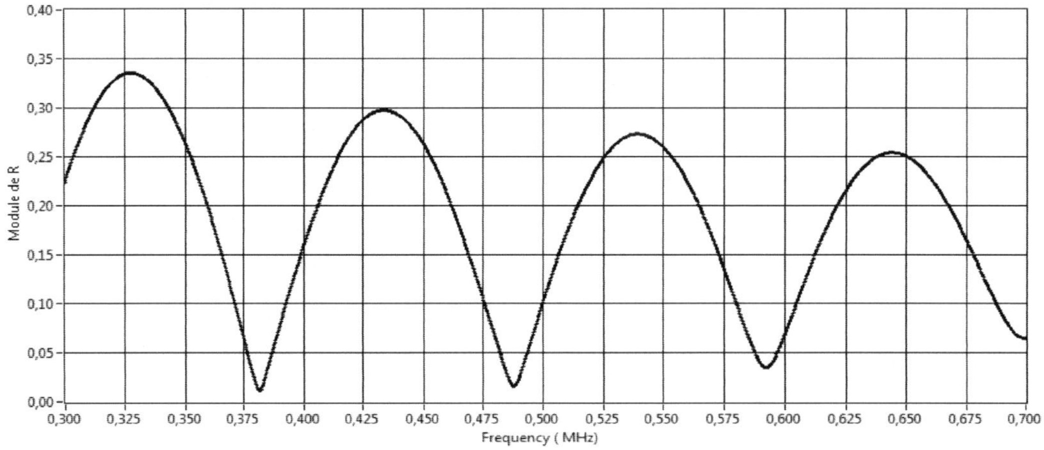

Figure 2. A representation of the reflection coefficient modulus as a function of the frequency

The speed of ultrasounds in 1,5 cm thick mortar is obtained by measuring time delay Δt between the echoes $Sr_2(t)$ and $Sr_3(t)$ and attenuation, by using the reporting method spectra of these two echoes [11].

3. Results and discussion

3.1. Relationship between the width of resonance and ultrasonic velocity

Monitoring the hydration at temperature T = 42°C for three mortars solutions made with grains of sand of different sizes (d =250µm, d = 315µm and d = 500µm), from the young age to curing for 70h , by measuring the ultrasonic velocity and the width of a single resonance mode (n=5) is used to draw the curves shown in Figure 3. The three curves show that the evolution of the resonance width follows a linear law with velocity. The analysis of the parameters of the linear regression shows that the constant of proportionality $c \cong 16{,}67\ m^{-1} \cong \frac{1}{4d_m}$ depends only on the thickness of the mortar. This result is in good agreement with the theoretical interpretation of equation (3). The figure shows also the sensitivity of the resonance width to the variation of the mortar microstructure; the higher the sand grain size, the higher the width of resonance.

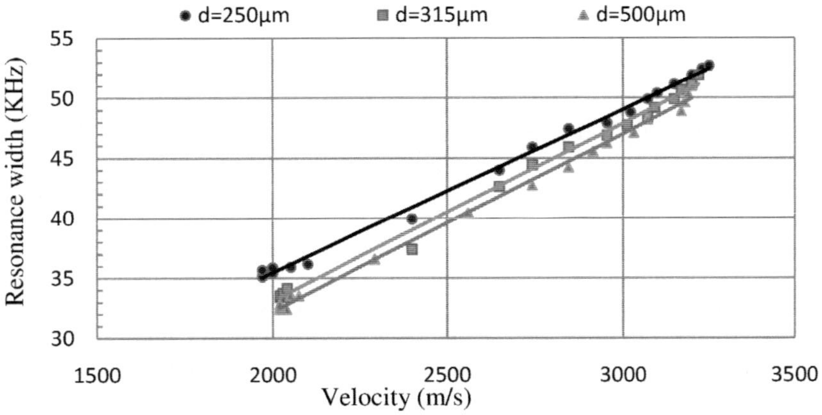

Figure 3. Variations in the width of resonance depending on the velocity in mortars made with the sand grains of different sizes (d = 250µm, d=315µm and d= 500µm)

3.2. Variation of ultrasonic velocity and the width of resonance according to the attenuation

The curves shown in Figures 4 and 5 respectively illustrate the variations of the velocity and the width of resonance according to the attenuation in the three mortars solutions in the curing phase. The two figures show that this variation is linear, with negative slope. The effect of sand grains size on the evolution of the mortar is visible across the width of resonance: it is more important than when the grain diameter is small.

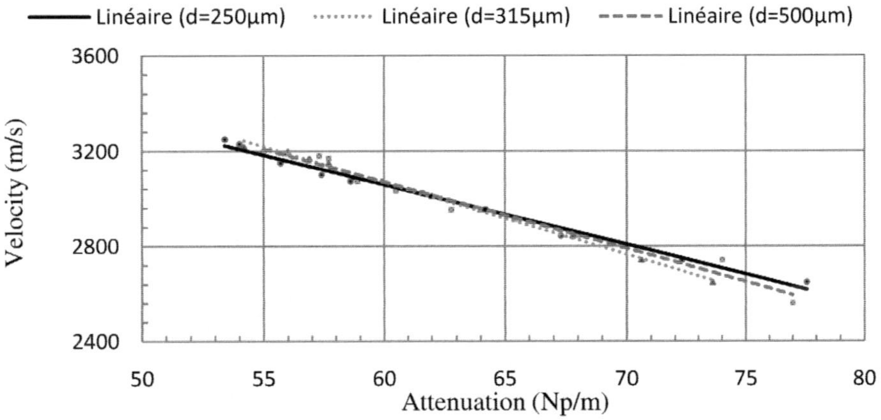

Figure 4. Variation of the velocity according to the attenuation in the mortar curing phase

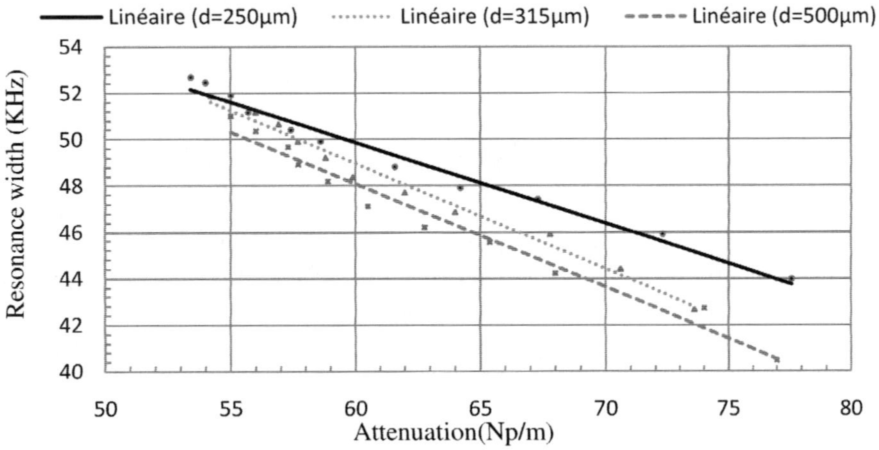

Figure 5. Variation of the resonance width according to the attenuation in the mortar curing phase

The curve of Figure 5 also shows a very important result. In fact, the enlargement of the width of resonance depending on the hydration time is predominantly due to an increase in the ultrasonic velocity. Thus, the effect of attenuation on the evolution of the resonance width is negligible compared to the effect of velocity. Amortization of ultrasound in the mortar is due to the contribution of several factors, the most important is the porosity. Many works on a wide range of porous materials have established empirical laws in linear, exponential, or power between porosity and macroscopic quantities [12,13,14]. Inspired by this work and from our experimental results, we can establish a linear relationship between the compression wave velocity and attenuation firstly $v = v_0(1 - a * \gamma)$ and between the width of resonance and attenuation secondly $\Gamma = \Gamma_0(1 - b * \gamma)$.

With : v_0 Represents the velocity of the compression wave at zero attenuation.

Γ_0 Represents the width resonance at zero attenuation.

γ Represents the attenuation of the wave in the mortar.

The linear regression parameters for three mortar solutions are summarized in Table 1.

Table 1. Parameters of the linear regression

Ultrasonic parameters	V(m/s)			Γ(KHz)		
Parameters of the regression	V_0	a	R^2	Γ_0	b	R^2
d=250μm	4562	5.55E-3	0.99	70.71	5.50E-3	0.98
d=315μm	4884	6.19E-3	0.99	76.20	5.96E-3	0.98
d=500μm	4746	5.88E-3	0.97	74.72	5.94E-3	0.97

Our results show that the variations in the velocity of compression and the width of resonance follow a linear relationship with the attenuation. Similar results were observed in cementitious materials by other authors [15, 16, 17]. In the elastic and isotopes media at low porosity values, the velocity can be approximated by a linear law $v = v_0(1 - bP)$, with P the medium porosity and v_0 the velocity of the wave at zero porosity and b an empirical parameter. In our case the linear regression is made in the mortar curing phase when the hydration reactions are completed, and the linear dependence of the compression velocity and the width of resonance with the attenuation can be explained by improved mechanical and structural properties of the material which acquired in the later stages of hydration a low porosity structure. In addition, the results of correlations can be equated with those of other authors if we assume that changes in the ultrasonic attenuation is linear with porosity in specific conditions which assume that the environments are elastic, isotopes and at low porosity values. We observe that the straight slopes obtained by linear regression are of the same order of magnitude $a \cong b \cong 5{,}83.10^{-3} Np^{-1}$. This allows us to find the empirical relationship established in (paragraph 3.1) between the width of resonance and the velocity $\frac{\Gamma}{v} = \frac{\Gamma_0}{v_0} \cong c$.

3.3. Comparison of experimental results with theoretical predictions

Measuring the width of resonance using the resonances of the theoretical coefficient of reflection (equation (1)) makes it possible to highlight the contribution parameters of the ultrasonic wave and the mortar on the evolution of the width of resonance. The curves shown in Figure 6 show, respectively, that the width of resonance is proportional to the inverse of the thickness of the mortar and the ultrasonic velocity. This is consistent with the formula derived from the Breit-Wigner function and confirms experimentally the variation law established between the width of resonance and the velocity (section 3.1). The curve which shows the effect of the variation of the ultrasonic attenuation in the mortar on the resonance width by assuming that the ultrasonic velocity remains unchangeable, is carried on the Figure 7. We find that this variation follows a logarithmic law and that attenuation change does not affect much the resonance width except for lower values, which explains the very low sensitivity of the resonance width to the attenuation variation of ultrasound in the mortar compared to its variation with velocity.

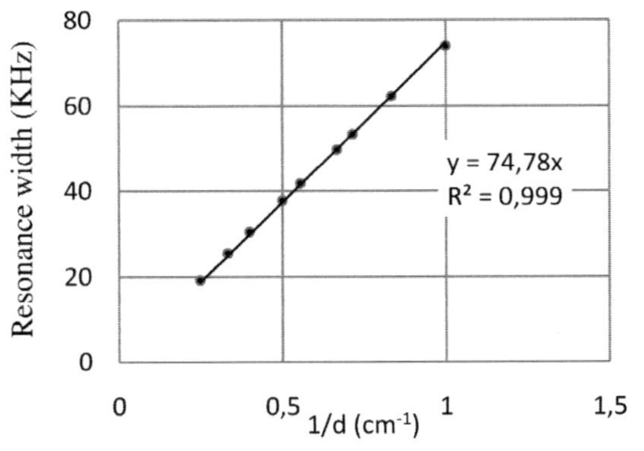

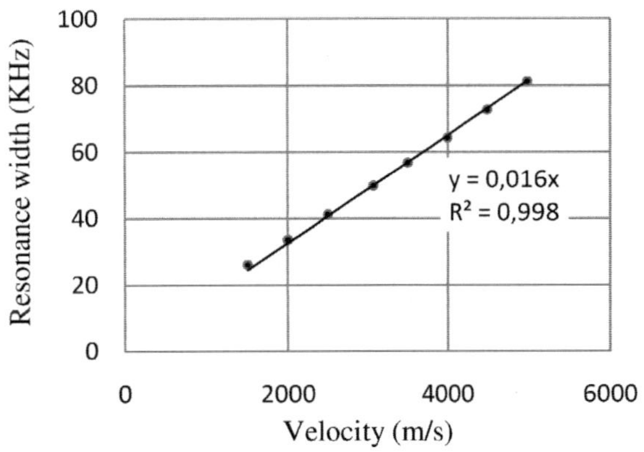

Figure 6. Evolution of the width of resonance as a function of the inverse of the thickness of mortar (velocity = 3072m / s, attenuation = 45.5Np / m) and in function of the velocity (thickness of mortar d = 1.5cm, attenuation = 45.5Np / m)

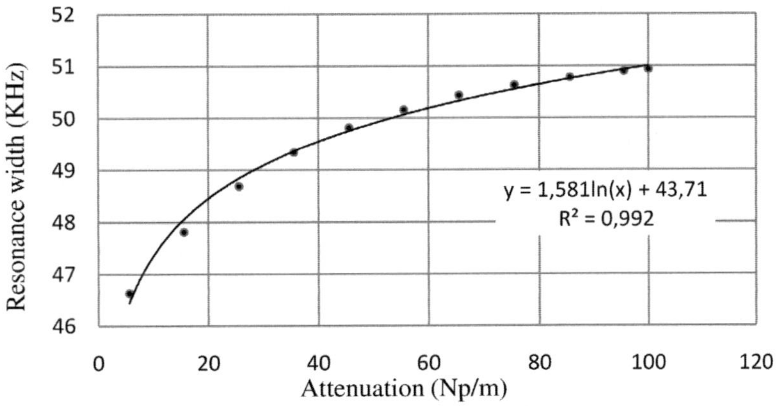

Figure 7. Evolution of the width of resonance as a function of the attenuation (thickness of mortar d =1.5cm, velocity=3072m/s)

4. Conclusion

This study highlights a non-destructive testing technique of cementitious materials based on the measurement of resonance width of the reflection coefficient. The theoretical and experimental study of this parameter shows a linear dependence of the width of resonance with the ultrasonic velocity and with the inverse of the thickness of the mortar in all stages of hydration. A linear correlation with the attenuation is also observed in the curing phase. The sensitivity of the resonance width to variations in the mechanical and structural properties of the mortar shows that the method developed in this work, using the resonance width as characterization parameter, has an alternation with the different methods of existing non-destructive testing.

References

[1] Odile Abraham O. Derobert X., 2003 « *Non-destructive testing of fired tunnel walls: the Mont-Blanc Tunnel case study*», NDT&E International, vol.36, p. 411–418.

[2] Yaman I.O., Hearn N., Aktan H.M., 2002 « *Active and non-active porosity in concrete Part I: Experimental évidence* », Materials and Structures, vol.35, pp.102-109.

[3] Lafhaj Z., Goueygou M., 2009 « *Experimental study on sound and damaged mortar: Variation of ultrasonic parameters with porosity* », Construction and Building Materials, vol. 23, pp. 953 - 958,.

[4] Goueygou M., Lafhaj Z., Soltani F., 2009 «*Assessment of porosity of mortar using ultrasonic Rayleigh waves*», NDT&E International, vol. 42,p p. 353 - 360.

[5] Lafhaj Z., Goueygou M., Djerbi A. and Kaczmarek M.,2006 « *Correlation between porosity, permeability and ultrasonic parameters of mortar with variable water/cement ratio and water content* », Cement and Concrete Research, Vol. 36, pp.625-633.

[6] D.G. Aggelis, D. Polyzos, T.P. Philippidis, 2005« *Wave dispersion and attenuation in fresh mortar: theoretical predictions vs. experimental result*», J. Mech. Phys. Sol. 53 (4) ,pp 857 – 883.

[7] Thomas. Voigt , Zhihui .Sun , Surendra. P. Shah, 2006 "*Comparison of ultrasonic wave reflection method and maturity method in evaluating early-age compressive strength of mortar*", Cement & Concrete Composites, 28 pp 307–316 .

[8] R. Fiorito, W.Madigosky,H. Überall, 1979 "*Resonance theory of acoustic waves interacting with an elastic plate*", J.Acoust.Soc.Am. 66, 1857.

[9] R. Fiorito, W.Madigosky,H. Überall,1985 "*Theory of ultrasonic resonances in a viscoelastic layer*". J. Acoust. Soc. Am. 77, 489 .

[10] S. Derible, P. Rembert, J.L. Izbicki, 1998 "*Experimental Determination of Acoustic Resonance Width via the Argand Diagram* ", Acta Acustica 84 pp. 270–279 .

[11] Hassan Bita, Ali Moudden, Bouazza Faiz, Hicham Lotfi, 2015 «*Non Destructive Characterization of Mortars by the Frequency Offset Method*». Journal of Civil Engineering Research, Vol. 5 No. 6, pp. 136-143.

[12] Dean E.A., 1983« *Elastic Moduli of Porous Sintered Materials as Modeled by aVariable-Aspect-Ratio Self-Consistent Oblate-Spheroidal-InclusionTheory* », Journal of theAmerican Ceramic Society, vol. 66, pp. 847 – 854.

[13] Panakkal J.P., Willems H., Arnold W.,1990« *Nondestructive evaluation of elastic parameters of sintered iron powder compacts* », Journal of Materials Science, vol. 25,pp. 1397-1402.

[14] Kaczmarek M., Goueygou M., 2006 « *Dependence of elastic properties of materials on their porosity: Review of models* », Journal of Porous Media, vol. 9, pp. 335 - 355.

[15] Lafhaj Z., Goueygou M., Djerbi A. and Kaczmarek M.,2006"*Correlation between porosity,permeability and ultrasonic parameters of mortar with variable water/cement ratio and water content*", Cement and Concrete Research, Vol. 36, pp.625-633.

[16] Yaman I.O., Aktan H.M., Hearn N.,2002"*Active and non-active porosity in concrete: Part II. Evaluation of existing models*" RILEM Newspaper of Materials and Structures.Vol 35 (246), pp.110-116.

[17] Marc. Goueygou , Zoubeir .Lafhaj , Fethi .Soltani ,2009 "*Assessment of porosity of mortar using ultrasonic Rayleigh waves*" ,NDT&E International, 42 pp.353–360.

Raman and XRD characterization of Moroccan Marbles

S Khrissi[1], M Haddad[1], L Bejjit[1], S Ait Lyazidi[1], M El Amraoui[1] and C Falguères[2]

[1]Laboraoire de Spectrométrie des matériaux et archéomatériaux (LASMAR), Unité de recherche associée au CNRST (URAC11), Université Moulay Ismail, Faculté des sciences, B. P. 11201, Zitoune, 50000, Meknès, Maroc
[2]Département de préhistoire, Muséum National d'Histoire Naturelle, UMR7194, 75 013 Paris, France

E-mail: mhaddad22@yahoo.fr and khrissi_salam@hotmail.com

Abstract. Various marbles collected from Moroccan careers, and two white reference marbles originating from Carrare-Italy and Drama-Greece, have been studied by different analytical techniques. Chemical composition of rocks has been determined by X-Ray fluorescence. X-Ray diffraction and Raman spectroscopy allowed the identification of both major (calcite and dolomite) and minor (quartz) mineralogical components. Moreover, Raman spectroscopy has been used to identify the calcium carbonate phases and to examine the graphitic ones in the marbles.
Keywords. Moroccan marbles, Characterization, XRF, XRD, Raman Spectroscopy, Calcite, Dolomite.

1. Introduction

Since antiquity, marbles are the most ornamental rocks used as building materials for architecture and sculpture owing to their aesthetical, mechanical and thermal properties [1]. The two major phases constituting the marble are calcite ($CaCO_3$) and dolomite ($CaMg(CO_3)_2$) [2]; the variety of their colors is due to the presence of impurities such as quartz, graphite, mica, clay and iron oxides. The use of marble and ornamental rocks in Morocco date back to the Roman period (Volubilis, Sala, Lixus, Banasa, Thamusida) [3-6]. Currently, limestones and marbles industry is increasingly developped in Morocco [7]. In this sens, we have undertaken their characterization by combining different analytical techniques. Marble samples were collected at different careers in Morocco, and were characterized by mean of X-Ray fluorescence (XRF), X-Ray diffraction (XRD) and Raman spectroscopy. As reference materials, two white marbles, originating from Carrare in Italy and from Drama in Greece, were also characterized.

In this paper we report and discuss the obtained results, and we give a comparison between the studied marbles from Morocco and the two reference white marbles.

2. Materials and methods

2.1. Samples

Several samples of marbles and ornamental stones collected from different regions of Morocco are in course of study by different physicochemical techniques. This article will be limited to the results obtained on five marble samples collected at the careers of Tiskram in the region of Khénifra (KB), the careers of Sahara in the region of Tiflet (TS), the careers of Oued Amlil in the region of Taza (OA). These results will be compared to those corresponding to the reference white marbles from Carrare -Italy (BC) and Drama - Greece (G). In the case of the marbles of Khénifra, the rock is of white color, gray to black and sometimes of morello cherries or nodulated aspect; some areas locally contain fossils. The Marbles collected in Oued Amlil, show a gray to greenish beige color with, locally, heterogeneous levels of gray shades and corrugated aspect. The marble of Tiflet is light gray with a purplish shade.

Figure 1. The samples of studied marbles.

Table 1. Samples codes and careers of their provenance.

Sample code	Color	Provenance
KB	White	Tiskram-Khénifra
TS	Grey	Sahara-Tiflet
OA	Grey	Oued Amlil-Taza
BC	White	Carrare- Italy
G	White	Drama- Greece

2.2. Experimental techniques

X-Ray powder diffraction analysis was carried out on powdered samples at room temperature using the Shimadzu XDR-6100 with Cu Kα radiation at the wavelength 1.5405 Å. Samples were analyzed from 5° to 70°.

Raman spectra were recorded in the range 100 - 2000 cm^{-1} using a Renishaw RM 1000 microraman spectrometer, equipped with a CCD detector and an external Leica DMLM confocal microscope with 10x, 20x, 50x, and 100x objectives. The excitation source is a He-Ne laser operating at 632.8 nm with a resolution better than 2 cm^{-1}. The samples were used directly, without any preparation or prior size fraction.

Elemental analyses were determined using an AXIOS XRF spectrometer with dispersion wavelength. The samples were ground in an agate mortar in pellet form.

3. Results and discussion

3.1. Elemental analyses: XRF

The major and minor element concentrations obtained for the present marbles are summarized in the table 2. In all samples, the most important element content corresponds to CaO with a concentration range of 35.8 - 55.8%. The samples G and OA present the highest MgO concentrations as 28.8% and 7.79% respectively, while the samples BC, KB and TS show contents varying in the range 0.5 - 1.18 %.

Besides SiO_2 (0.2 - 8.48%), Fe_2O_3 (0.11 – 1.92 %) and Al_2O_3 (0.1 – 2.18%), the analyses revealed also trace elements: Na, Sr, Cl, Mn, I, S, P, K and Y.

Table 2. XRF analysis of marble samples

	CaO	Fe_2O_3	SiO_2	MgO	Al_2O_3	Na_2O	SrO	MnO_2	Cl	SO_3	I	P_2O_5	K_2O	Y_2O_3	LOI*
BC	55.8	-	0.20	1.18	0.1	0.07	0.03	-	-	0.05	0.02	-	-	-	42.6
KB	52.00	1.07	8.48	0.58	0.37	0.09	0.24	0.07	0.05	0.03	0.02	0.01	-	0.004	37
TS	52.80	0.11	1.47	0.84	0.78	0.12	0.04	-	0.05	0.15	0.02	0.07	0.07	-	43.5
OA	38.50	1.92	5.37	7.79	2.18	0.20	0.01	-	0.06	0.10	0.02	0.10	0.19	-	43.6
G	35.80	-	0.35	28.80	0.28	-	-	-	-	0.30	-	-	-	-	34.47

**LOI: Loss on ignition*

3.2. Crystallines phases: XRD

The various Bragg reflections are indexed using (JCPDS) files [8]. The predominate lines (2θ = 29.4° and 2θ = 30.93°) are attributed respectively to calcite and dolomite; the marble (BC) is calcitic, the marbles (KB, TS) are calcitic with low contents of quartz, the marble (OA) is calcitic-dolomitic, while the sample (G) from Drama (Greece) is dolomitic. Attributions of calcite and dolomite peaks are in good agreement with those reported by Gunasekaran and al. [9,13].

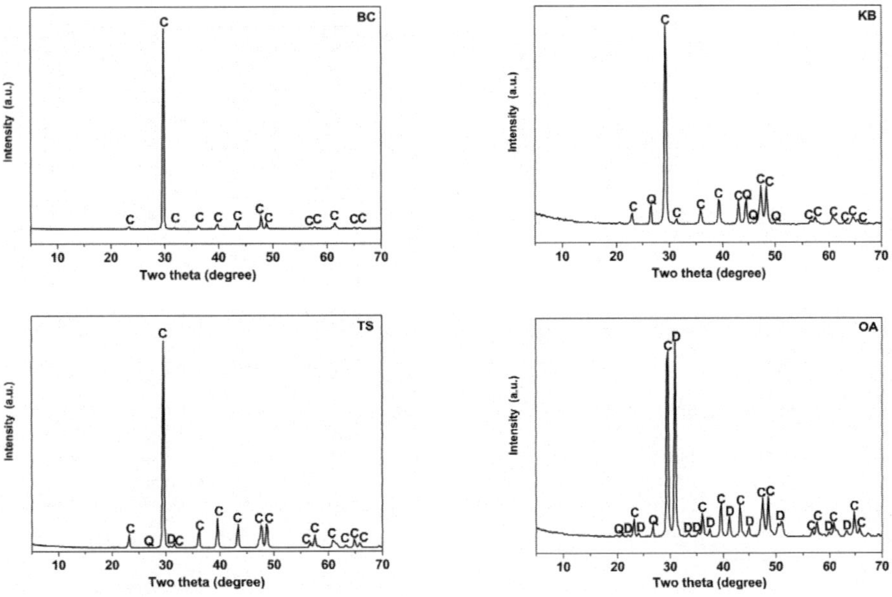

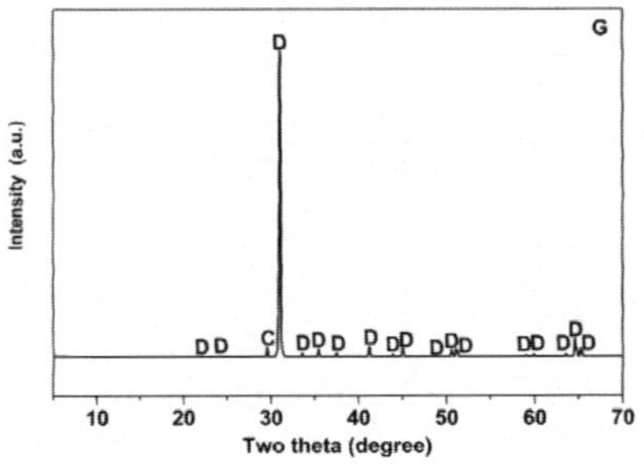

Figure 2. Powder XRD patterns of samples BC, KB, TS, OA and G. C-Calcite, D-Dolomite and Q-Quartz.

3.3. Crystallines phases: Micro Raman

Raman spectra of samples BC, KB, TS and OA (figure 3) exhibit bands located around 153, 280, 710, 1085, 1435 and 1747 cm^{-1} characteristic of calcite [9,10,12,14]. Therefore, these samples can be classified as calcitic limestones. Besides these bands, an additional wide signal at 460 cm^{-1}, signature of quartz [8,15], is observed in the case of the sample OA. The marble TS shows also two supplementary bands at 1335 and 1608 cm^{-1} which can be attributed to carbonaceous matter [16, 17], they are often associated to marbles containing carbonaceous resulting in a more or less grey color [17]. In the case of the sample OA, these bands are probably overlapped by the fluorescence due to the presence of carbonaceous matter.

The sample from Drama - Greece (G) behaves differently with respect to all the local samples. The corresponding Raman spectrum is dominated by dolomitic bands at 180, 305, 728, 1101, 1447 and 1759 cm^{-1} [9,10,12].

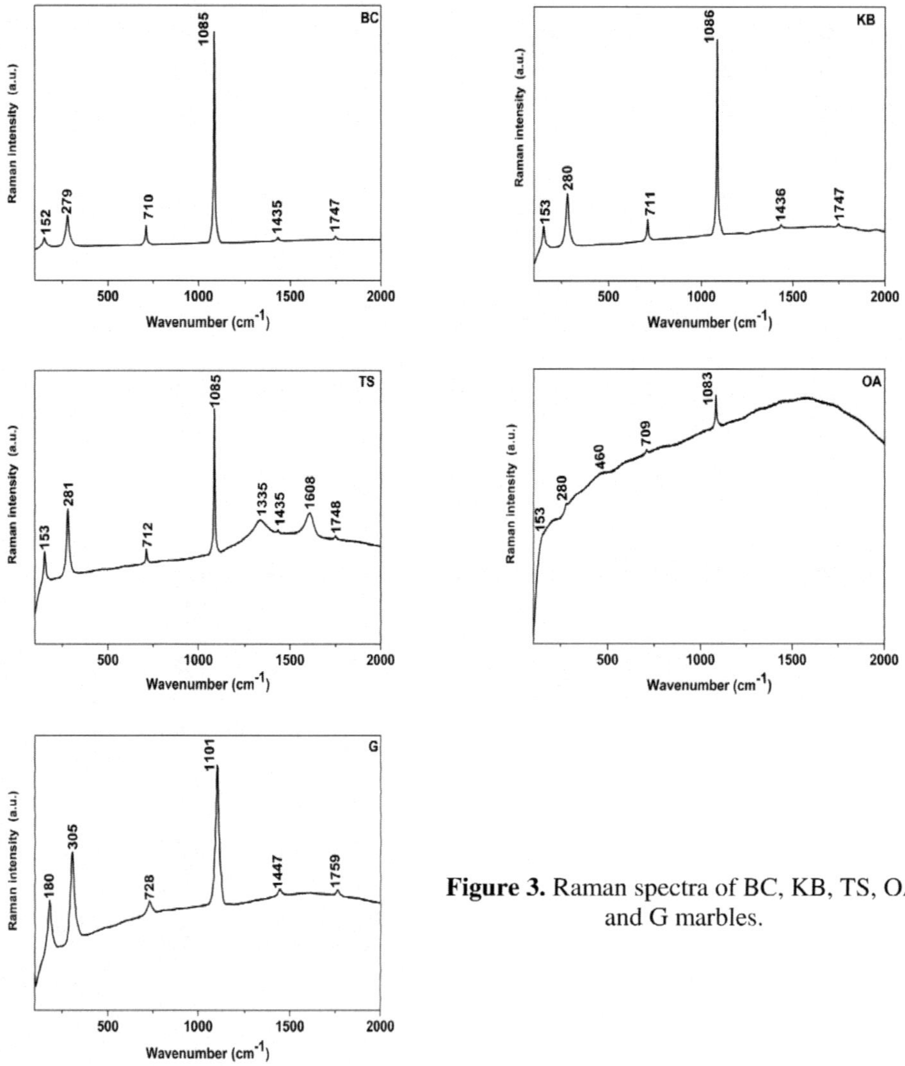

Figure 3. Raman spectra of BC, KB, TS, OA and G marbles.

The strongest band observed on all spectra recorded on the studied Moroccan marbles and the Carrare -Italy reference white marble corresponds to the symmetric stretching υ_1 mode located around 1084 cm^{-1}. This band is related to CO_3 groups vibrating in identical phases [18]. The other weak lines located near 1435 and 710 cm^{-1} can be attributed respectively to the υ_3 (asymmetric stretching) and υ_4 (in-plane-bending) vibrational modes. The observed frequency shifts of about 153 and 280 cm^{-1} originate from the external vibrations of the CO_3 groups involving translatory and rotatory oscillations [18]. A weak band observed at 1748 cm^{-1} may be regarded as the combination band of υ_1 and υ_4 [12,14,19-22].

The results obtained from both XRD and Raman structural analyses of all marbles are combined and summarized on Table 3.

Table 3. Mineralogical composition of samples BC, KB, TS, OA and G.

Region	Sample code	Career	Main Phase	Others
Carrare- Italy	BC	Carrare	Calcite	-
Khénifra-Morocco	KB	Tiskram	Calcite	Quartz
Tiflet- Morocco	TS	Sahara	Calcite	Quartz, Graphite, Dolomite
Taza- Morocco	OA	Oued Amlil	Calcite	Quartz, Graphite, Dolomite
Drama- Greece	G	Drama	Dolomite	Calcite

4. Conclusion

This work is part of an extensive study in progress on the marbles and stones of ornamentation from different careers of Morocco (several tens of samples). The main objective is to characterize these materials and to identify indicators likely to link the sample of marble with his career of provenance. Several techniques in the laboratory have been used (XRF, XRD, Optical absorption, Colorimetry, Raman and EPR). The white marbles references originating from careers well known (Drama and Carrare) have been also analyzed by the same techniques.

This preliminary study shows firstly that there is a clear difference between Moroccan and the reference marbles, and secondly some differences between the marbles of Morocco from different careers. XRD showed that the calcite is the major phase in the all Moroccan marbles and in the marble of Carrare - Italy, while the marble from Drama - Greece is dolomitic. These results were confirmed by micro Raman spectrometry, which also shows the graphitic nature of marbles from certain sites. Quartz with low content has been detected in Moroccan marbles.

References

[1] Al-Naddaf M, Al-Bashaireh K and Al-Waked F 2009 Characterization and provenance of marble chancel screens, northern Jordan *Mediterr. Archaeol. Archaeom.* **10** (2) 75-83.

[2] Vagenas N V and Kontoyannis C G 2003 A methodology for quantitative determination of minor components in minerals based on FT-Raman Spectroscopy the case of calcite in dolomitic marble *Vib. Spectrosc.* **32** 261-64.

[3] El Hassani I E A and El Azhari H 2009 Évaluation des propriétés physico-mécaniques des pierres de construction du Maroc à partir des vitesses des ondes P et de la résistance au choc *Bulletin de l'Institut Scientifique, Rabat, Section Science de la Terre* **31** 41-54.

[4] Ouali H and Ajakane R 2015 Caratérisation et valorisation du calcaire métamorphique de Bou-Acila (Maroc central) *Int. J. Innovation Appl. Stud.* **11**(3) 778-84.

[5] Perrier R 1995 Gisements des roches ornementales du Maroc, *le Mausolée, Mines, Géologie et Énergie, Rabat*, **708** 54-64.

[6] Bouhaouli A, Caraman V and Hilali E 1979 Les marbres du Maroc *Mines, Géologie et Énergie,* **45** 35-102.

[7] Origlia F, Gliozzo E, Gandin A, Meccheri M, Spangenberg J E and Memmi I T 2012 Marbles and carbonate rocks from central Morocco: a petrographic, mineralogical and geochemical study, *Environ. Earth Sci.* **66** (1) 209-22.

[8] JCPDS ASTM card NOS. 5-0586, 36-0426 and 33-1161.

[9] Gunasekaran S and Anbalagan G 2007 Spectroscopic characterization of natural calcite minerals, *Spectrochim. Acta, Part A* **68** 656-64.

[10] Sun J, Wu Z, Cheng H, Zhang Z and Frost R L 2014 A Raman Spectroscopic comparison of calcite and dolomite, *Spectrochim. Acta, Part A* **117** 158-162.

[11] Tomić Z, Makreski P and Gajić B 2010 Identification and spectra-structure determination of soil minerals: Raman stud supported by IR spectroscopy and X-Ray powder diffraction, *J.RamanSpectrosc.***41** 582-86.

[12] Gunasekaran S, Anbalagan G and Pandi S 2006 Raman and Infrared spectra of carbonates of calcite structure *J. Raman Spectrosc.* **37** 892-99.

[13] Gunasekaran S and Anbalagan G 2008 Optical absorption and EPR studies on some natural carbonate minerals, *Spectrochim. Acta, Part A* **69** 383-90.

[14] Behrens G, Kuhn L T, Ubic R and Heuer A H 1995 Raman spectra of vateritic calcium carbonate, *Spectrosc. Lett.* **28** (6) 983-95.

[15] Koralay T and Kılınçarslan S 2016 A multi-analytical approach for determining the origin of the marbles in Temple-A from Laodicea ad Lycum (Denizli-Western Anatolia, Turkey) *J. Cult. Herit.* **17** 42-52.

[16] Sˇtastná A, Prikryl R and Jehlicka J 2009 Methodology of analytical study for provenance determination of calcitic calcite-dolomitic and impure marbles from historical quarries in the Czech Republic *J. Cult. Herit.***10** 82-93.

[17] Jehlička J, Šťastná A and Přikryl R 2009 Raman spectral characterization of dispersed carbonaceous matter in decorative crystalline limestones *Spectrochim. Acta, Part A* **73** 404-09.

[18] Krishnamurti D 1957 *Proc. Ind. Acad. Sci.* 27A 183.

[19] Gunasekaran S and Anbalagan G 2008 Spectroscopic study of phase transitions in natural calcite mineral *Spectrochim. Acta, Part A* **69** 1246-51.

[20] Stopar J D, Lucey P G, Sharma S K, Misra A K, Taylor G J and Hubble H W 2005 Raman effeciencies of natural rocks and minerals: Performance of a remote Raman system for planetary exploration at a distance of 10 meters *Spectrochim. Acta, Part A* **61** 2315-23.

[21] Herman R G, Bogdan C E, Sommer A J and Simpson D R 1987 Discrimination among carbonate minerals by Raman spectroscopy using the laser microprobe *Appl. Spectrosc.* **41** (3) 437-40.

[22] Schaefer C, Bormuch C and Matossi F 1926 Das ultrarote Absorptions spektrum der Carbonate *Z. Phys.* **39** (9) 648-59.

IOP Conf. Series: Materials Science and Engineering **186** (2017) 012029 doi:10.1088/1757-899X/186/1/012029

On-site XRF characterization of archaeological materials in CERA center of Rissani (Morocco)

M El Amraoui[1*], M Haddad[1], LBejjit[1], S Ait Lyazidi[1], and R Lakhal[2]

[1] Laboratoire de Spectrométrie des Matériaux et Archéomatériaux (LASMAR), URAC11, Université Moulay Ismail, Faculté des Sciences BP 11201 - Zitoune, Meknès 50000, Maroc
[2] Ministère de la Culture, Rabat, Maroc

Email: mohamedelamraoui@yahoo.fr

Abstract. Different glazed ceramics and glass artifacts, coming from the excavations of the Sijilmassa site and exhibited in CERA (center for Alaouite studies and research) in Rissani (Morocco), were analysed by portable X-ray fluorescence spectrometry. All glazed ceramics are lead glaze and their ceramic bodies are calcareous clay. Glass objects are different nature: lead-glass, lead-lime-alkaline and potassium-lime glass. Classical elements were used to colour glaze and glass: tin for white, cobalt ore for blue, copper for green, manganese and iron for black and iron for yellow.
Keyword. Glazed ceramic; Glass bead; oil lamp, Sijilmassa site; Portable X-ray Fluorescence.

1. Introduction

On-site characterization of archaeological and cultural heritage materials by portable instrumentation is growing in a remarkable way. This non-invasive approach, avoiding sampling and risk of loss or damage during transportation of artefacts outside sites or museums, contributes also to a great collaboration of curators and managers of cultural heritage. X-ray fluorescence spectrometry (XRF) is among the main used techniques in these investigations, allowing multi-elemental identification and quantitation of the chemical constituents of samples [1]. It is suitable for ceramics, glasses, gems, pigments, frescos, coins, etc [2-9].

The purpose of the present work is the characterization of a set of archaeological ceramics, belonging to the CERA center, by mean of XRF analysis; the goal is the establishment of a scientific documentation enriching historic data on these artefacts.

2. Materials and instrumentation

2.1. Materials

The CERA center, created in 1990 by the Moroccan Ministry of Culture, harbours various and rich archaeological sets of materials: glazed ceramic, potteries, painted plasters, glasses, coins, metal objects and other materials coming from excavation of the Sijilmassa site.

The investigated materials consist of sets of: i) architectural glazed ceramics (Zellige attributed to the 18th -19th centuries period, ii) oil lamps (9th - 11th century), iii) fragments of glazed ceramics bowl with polychrome floral decoration (Fatimid period, 10th century) and iv) glass pearls along with a blue glass fragment. All artefacts are shown on Fig.1 and described on Table 1.

2.2. Instrumentation

The spectrometer consists of a X-ray tube with Mo anode emitting at lines Kα and Kβ of energy 17.44 and 19.60 KeV respectively. The emitted ray characteristics of each chemical element were detected by means of a Si (Li) detector, with a 30 mm^2 active area and 8 μm beryllium window. The energy resolution is 138 eV at 5.9 KeV and a Nucleus PCA card acquisition system is used. The X-ray generator was operated at 50 KV and 200 μA and the acquisition time for each spectrum was 200s. The nominal beam diameter is <150 μm at position of the sample. All measurements were performed in air. After each measurement, the spectra were deconvoluted by MCA software in order to determine the chemical elements. Besides lines (Kα, Kβ, Lα, Lβ, Mα, etc.) characteristics of chemical elements present in the sample, the spectra include Compton and Rayleigh scattering lines of Mo, generator of excitatory X-ray.

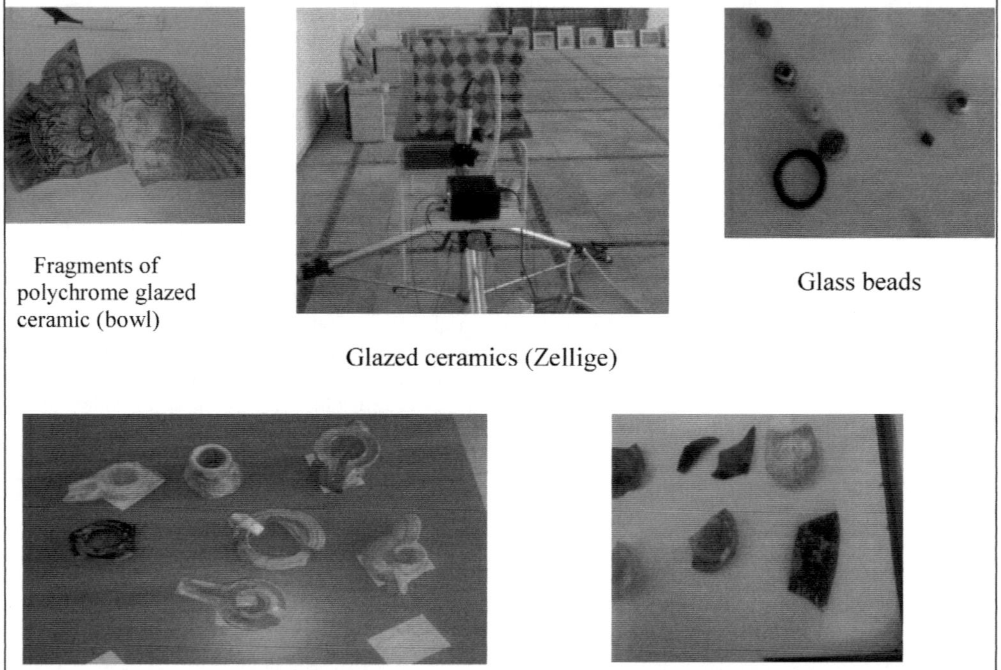

Fragments of polychrome glazed ceramic (bowl)

Glazed ceramics (Zellige)

Glass beads

Oil lamps

glass fragments

Figure 1. Archaeological materials analysed.

3. Results and Discussion

3.1. Glazes and supports *(Zelliges)*

White glaze: The XRF spectrum of white glaze is shown in Figure 2.a. In addition to Kα and Kβ lines of Molybdenum excitation source, Lα, Lβ, Mα, LI, Lη and Lγ lines of lead (Pb) along with Lα and LI lines of tin (Sn) were identified. Consequently, Lead is used as melting agent (flux) and tin is likely used to produce opaque and white glazes. Indeed, using tin as cassiterite (SnO_2) to opacify glazes in medieval and Islamic glazed ceramics has been reported by several authors [10,11]. Iron, calcium; and potassium were also highlighted. Their presence in the glaze can be due to their diffusion from terracotta or as impurities in the sand used as raw material to make the glaze. In addition, glazes can fix the calcium provided by infiltration of water during burial in marl soil. Copper and titanium were also identified in low amounts in the glaze by their lines with very low intensities; they may be present as impurities in the raw materials.

Table 1. The list of examined objects.

Material	*Reference*	*Period*	**Analysed zone**
Oil lamp	CERA93-215		Greeen Glaze
		(9th – 11th) century	Terra cotta
Oil lamp	CERA88-722		Green glaze
		(9th – 11th) century	Terra cotta
		Medraride period	
Oil lamp	CERA02-771	(9th – 11th) century	Black glaze
			Terra cotta
Zelliges	CERA09-610	(18th – 19th)	White glaze
		century	Terra cotta
			Black glaze
			Terra cotta
Bowl	CERA92-186	10th scentury	Yellow glaze
		Fatimid period	Black glaze
			Blue glaze
			Green glaze
Fragments of glases	CERA06-910	11th -14th centuries	Blue
Beads	CERA09-106	11th -14th centuries	Black
		11th -14th centuries	Blue
Bead of necklace	CERA10-906	11th -14th centuries	Blue

Black glaze: Figure 2.b shows that lead was used as flux while manganese and iron were used to produce a black colour of glaze. Manganese and iron produce colour of glaze ranging from brown to black depending on the ratio of their concentrations and the nature of firing atmosphere, oxidizing or reducing [12]. Silicon, calcium, potassium, titanium and copper are also identified.

Terracotta: Calcium and iron are present in high amounts in terracotta of zelliges as it's shown in Fig.2.c. Therefore, a ferruginous calcareous clay is used to produce the terracotta of the zelliges. The use of calcareous clays in the manufacture of glazed ceramics, faience and majolica was a common practice in the medieval Islamic world [10,13,14].

Besides their compatibility with the white colour of the glaze, these clays permit to reduce the amount of tin used for glaze opacifying. Also, their viscosity helps to prevent glaze cracking due to thermal expansion coefficient between glaze and terracotta during firing [14]. Potassium, manganese, sulphur, zinc titanium, chlorine, rubidium and strontium are present but in low concentrations. Strontium is in relatively high content because it can substitute to calcium [15]. Lead is found in terracotta, it originates probably from the dissolution of the glaze over time 'leaching phenomenon'; this is an indicator of ageing [4].

3.2. Fragment of glazed ceramic bowl

The polychrome part of a glazed ceramic bowl with blue, green, yellow and black floral decoration (Fatimid period - 10th century) has been analysed. The XRF spectra recorded on the different coloured areas of the glaze are presented in Figure 3; the presence of lead in high amount reveals its use as flux.

Yellow glaze: High intensity Kα and Kβ lines characteristic of iron are observed (Fig. 3.a). Iron gives colours ranging from brown to yellow, depending on amount, nature of the firing atmosphere (oxidizing or reducing) and associated chemical elements in the glass matrix. Calcium, potassium, titanium, copper and manganese were also identified in low quantities compared to lead and iron. The non-appearance of Tin as cassiterite, commonly used as opaquening agent in glazed ceramics, may likely due to the overlapping of its Lα and Ll lines by Kα and Kβ lines of potassium.

Black décor: Manganese and iron have been identified in the XRF spectrum of black glaze (Fig. 3.b). Association of these two transition elements in lead glazes produces black colour [16]. Calcium, potassium, titanium and copper were also identified but in low content.

Blue décor: Cobalt, nickel, arsenic, manganese, copper, iron, calcium, titanium, potassium and tin are the principal chemical elements of the blue glaze (Fig. 3.c). Cobalt ore rich in arsenic and nickel is used to produce the blue colour, while tin is likely used to opacify the glaze; cobalt correlated with arsenic has been identified in glazed ceramics of fourteenth century [16,17].

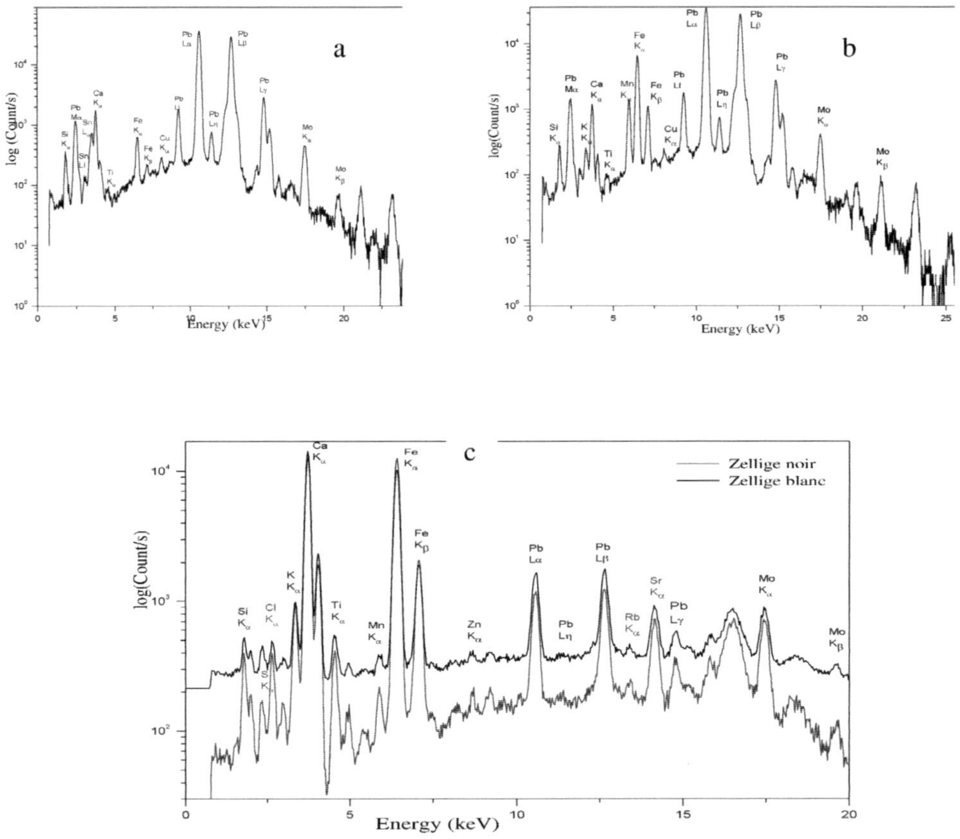

Figure 2. XRF spectra recorded on (a) Zellige's white glaze, (b) Zellige's black glaze and (c) terracotta supports

Green pattern: The spectrum of green decor (Fig 3.d) shows that lead, copper, iron, manganese, nickel, calcium and potassium are the principal chemical elements. Copper and lead, showing high contents, are used respectively for the green colour and as flux of glaze. The green colour can be attributed to the dissolution of copper in the glass matrix [11,18]. In general, the green colour of glazes is often linked to the presence of copper as Cu^{2+}. The presence of iron, manganese, calcium and potassium in the glaze may originate probably from impurities in the raw materials or contamination of terracotta.

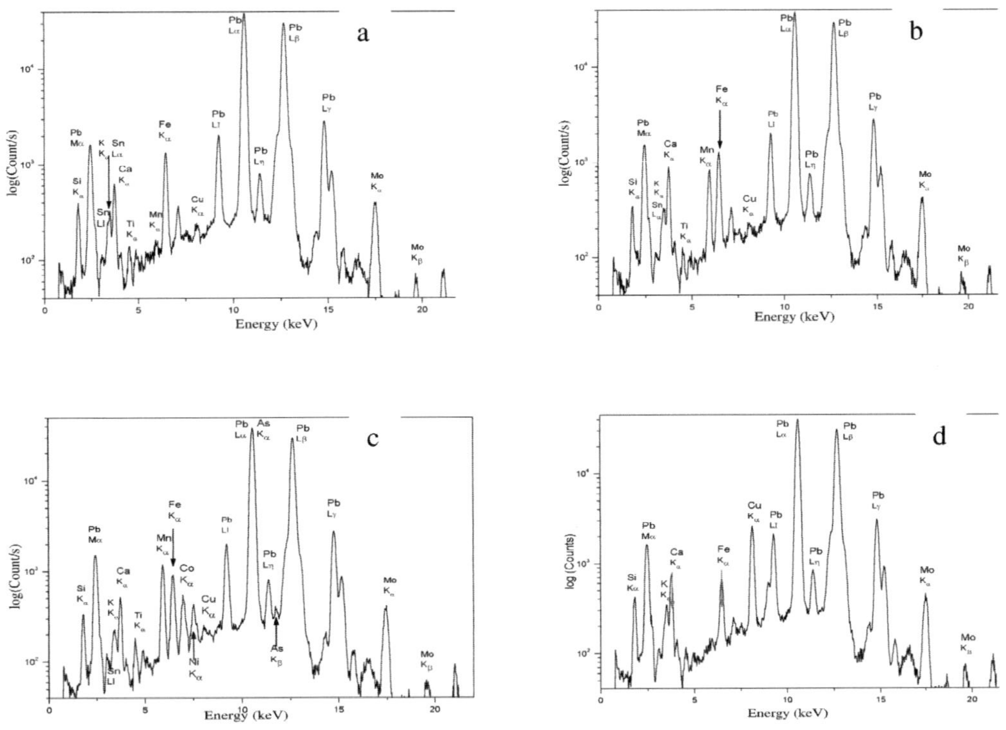

Figure 3. XRF spectra recorded on yellow (a), black (b), blue (c) and green (d) decors of the glazed ceramic bowl fragment.

3.3. Oil lamps

Green glazes: X-ray fluorescence spectra (Fig. 4.a) of green glazes of two oil lamps referenced SM93T2/AD15 and L-3-88/722 show that lead (Pb) is used as melting agent to produce the glaze. The identified Tin (Sn) was probably used as opacifier. Cu is identified in all green glazes studied. Dark green areas of the lamp SM93T2/AD15 present Mn which is likely used to darken the green colour. Zinc, tin, iron, calcium, titanium and potassium were also identified in the glazes. The presence of iron makes the green colour darker, while the presence of calcium improves its mechanical properties and particularly hardness [19]. The zinc gives flexibility to glazed ceramics and prevents cracking due to differences in thermal expansion coefficients of glazes and their supports [20]. This suggests the presence of zinc as an impurity in the raw materials or advanced technological knowledge of craftsmen of the period [21].

Black glaze: Fig. 4.b shows the X-ray fluorescence spectrum measured on the black glaze of the oil lamp referenced BH/K2/771. Characteristic lines of lead show high intensities implying thus that lead was used as flux to produce the glaze. In addition to copper, iron and manganese are present and are likely responsible for the black colour. Copper produce green glaze, while iron associated to manganese produce black colour. The mixture of the two colours gives a more or less deep colour according to Cu/Mn ratio [22]. Calcium, potassium and titanium are also identified; they may originate from clay, during firing process, or they may exist as impurities in the raw materials used to make glaze (sand, pigments, flux...).

Terracotta: X-ray fluorescence spectra of terracotta of oil lamps presented in Fig. 4.c show high amounts of calcium and iron. Bodies are thus made of ferruginous calcareous clay. This type of clay was commonly used to manufacture glazed ceramics with colour of terracotta often cream pink to white, depending on the content of iron [14,23]; the colour of terracotta has a significant effect on the colour of glazes. The spectrum shows the presence of Lead in high content; this occurrence is likely due to phenomena of glaze leaching. Potassium, titanium, manganese and copper are identified in low amounts along with traces of rubidium and strontium.

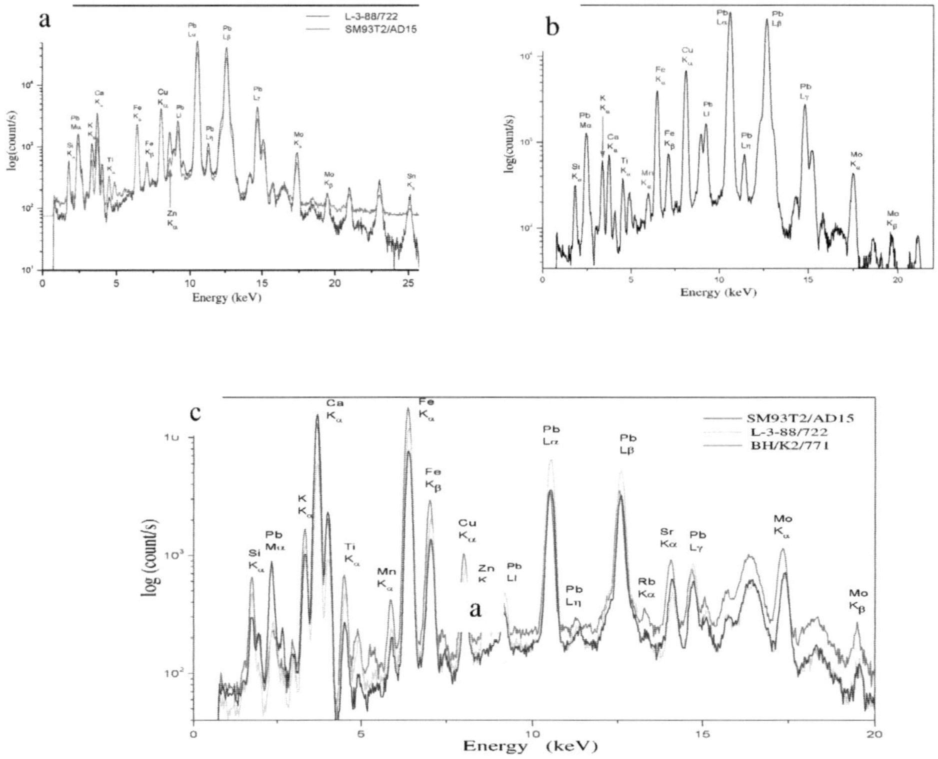

Figure 4. XRF spectra recorded on oil lamps: (a) green glaze, (b) black glaze and (c) terracotta supports.

3.4. Glass fragments and Beads

Blue glass fragment: The XRF spectrum reveals that Si, Cl, K, Ca, Ti, Mn, Fe, Cu, Zn and Pb are the principal chemical elements (Fig. 5.a.). High contents of calcium and iron are also observed while potassium shows q low content. Sodium is not detected because of its low cross section. The melting agent (flux) used to make glass is probably alkaline since the amount of lead is very low. The blue colour of glass can be attributed to copper; this element is reported as responsible of the blue colour in glazes and glasses [24,25]. Zinc contributes to increasing the chemical resistance and hardness, reducing the thermal expansion coefficient and increasing the refractive index [12,19,20]; it gives shine to glass [26]. Chlorine is present and it's probably due to the alteration of glass after its burial in the soil for long periods.

Blue glass bead: The X-ray fluorescence spectroscopy analysis of the blue glass bead shows that it consists mainly of calcium, potassium and iron (Fig.5.b). Characteristic lines of titanium and sulphur have been also identified in the spectrum; sulphur may be due to gypsum which is an indicator of alteration of the glass during burial time in the soil. Lead is also present; it may exist as impurity in the raw materials or may originate from contamination during the manufacture of glass.

Blue pearl necklace: Calcium, iron, lead, silicon and potassium are the main constituents of necklace as revealed by the XRF spectrum (Fig. 5.c). The material seems likely a lead-alkaline-lime glass. Cobalt, manganese, arsenic, nickel and titanium have been also identified. The blue colour of necklace is attributable to cobalt ore associated with arsenic and nickel [24,25].

Black glass bead: The corresponding spectrum is shown in Fig.5.d. The characteristic peaks of lead (Pb) are the most intense; this suggests that lead is used as melting agent in this glass. Silicon, potassium, calcium, manganese, iron, copper and zinc are also identified in the bead. The black colour of the glass bead is likely due to the association of transitional elements mainly manganese and iron.

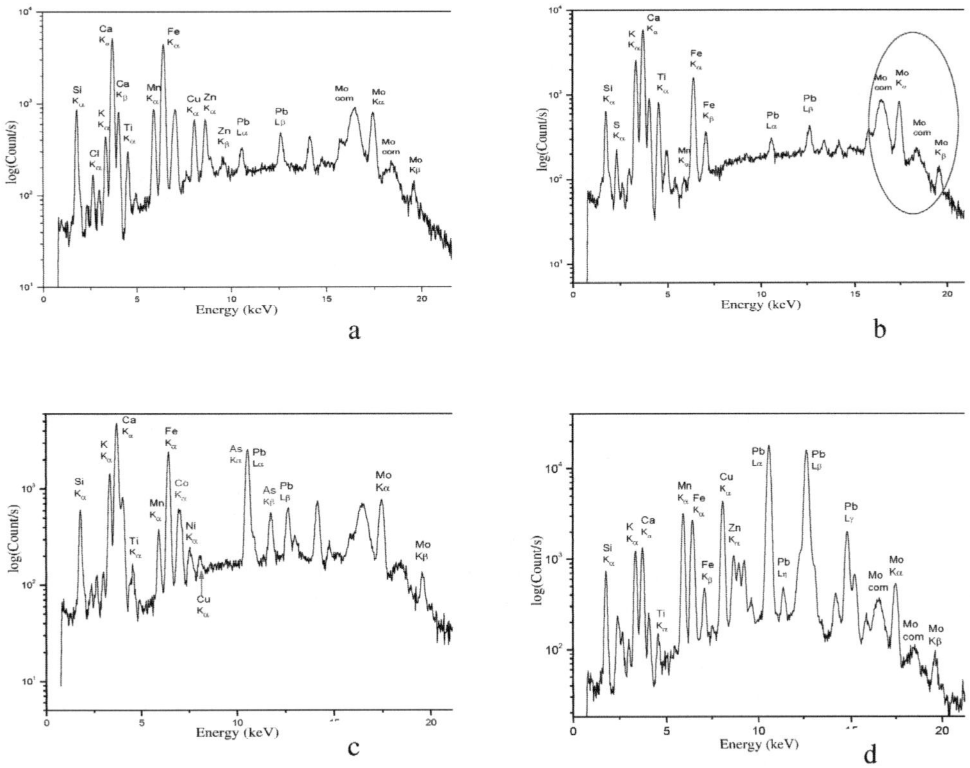

Figure 5. XRF spectra of *glasses:* bleu *fragment (a),* blue bead (b), blue bead necklace (c) and black bead (d).

4. Conclusion

On-site XRF measurements were carried out on a large set of archaeological artefacts exhibited at the CERA centre: architectural glazed ceramics tiles, utilitarian glazed ceramics, oil lamps, glasses and glass beads. These multi-elemental analyses allowed the identification of the chemical constitutions of all materials. The glazed ceramics are lead glaze with bodies consisting of calcareous clays. Different flux, lead, lead-lime-alkaline and lime-potassium are used to produce glasses and glass beads. All elements used to colour glazes and glasses are classic; they are those commonly used in Mediterranean recipes. Further investigations using supplementary analytical technics for a good characterization and dating of these materials are envisaged.

This work is part of the MOLAB program.

Acknowledgement
The authors thank greatly Dr Zucchiatti Alessandro and Dr Paolo PRATI for bringing the portable XRF spectrometer from the INFN of Genes-Italy, and carrying out all on-site measurements.

References
[1] Van Grieken R E and Markowicz A A 2002 Handbook of X-Ray Spectrometry, Second Ed., New York, Basel.
[2] Trojek T, Musílek L and Čechák T 2014 X-ray fluorescence analysis of cultural artefacts Applications to the Czech heritage, *Radiation Physics and Chemistry* **95** 381-84.
[3] Papadopoulou D, Sakalis A, Merousis N and Tsirliganis N.C 2007 Study of decorated archaeologgical ceramics by micro-x-ray fluorescence spectroscopy, *Nucl. Instr. Meth. Phys. Res. A* **580** 743–47.
[4] Trojek T, Hlozek M, Cechak T and Musılek L 2010 X-ray fluorescence analyzers for investigating postmediaeval pottery from Southern Moravia *Appl. radiat. Isotopes* **68** 879-83.
[5] Mirti P, Casoli A and Appolonia A 1993 Scientific analysis of Roman glass from Augusta Praetoria, *Archaeometry* **35** 225-40.
[6] Verità M, Arena M S, Carruba A M and Santopadre P 2008 Roman glass: Art and technology in a 4th century A.D. opus sectile in Ostia (Rome), *J. Cult. Herit.* **9** 16 - 20.
[7] Fontana D M F, Alberghina, Barraco R, Basile S, Tranchina L, Brai M, Gueli A and Troja S. O. 2014 Historical pigments characterization by quantitative X-ray fluorescence, *J. Cult. Herit.* **15** (3) 266-74.
[8] Milazzo M 2004 Radiation applications in art and archaeometry: X-ray fluorescence applications to archaeometry. Possibility of obtaining non-destructive quantitative analyses, *Nucl. Instr. Meth. Phys. Res. B* **213** 683–92.
[9] Cesareo R, Castellano A, Buccolieri G, Quarta S, Marabelli M, Santopadre P, Leole M and Brunetti A 2004 Portable equipment for energy dispersive X-ray fluorescence analysis of Giottos frescoes in the Chapel of the Scrovegni, *Nucl. Instr. Meth. Phys. Res. B* **213** 703–6.
[10] Al-Saad Z 2002 Chemical composition and manufacturing technology of a collection of various types of islamic glazes excavated from Jordan, *J. Archaeol. Sci.* **29** 803–10.
[11] Colomban Ph, Sagon G and Faurel X 2001 Differentiation of antique ceramics from the Raman spectra of their coloured glazes and paintings, *J. Raman Spectrosc.* **32** 351-60.
[12] Sanchez-Ramos S, Bosch Reig F, Gimeno Adelantado JV, Yusa Marco D.J and Domench Carbo A 2002 Study and dating of medieval ceramic tiles by analysis of enamel with atomic absorption spectroscopy, X-ray fluorescence and electron probe microanalysis, *Spectrochim. Acta Part B* **57** 689-700.
[13] Fortina C, Memmi Turbanti I and Grassi F 2008 Glazed ceramic manufacturing in southern Tuscany (Italy): Evidence of the technological continuity throughout the medieval period (10th–14th centuries), *Archaeomtry* **50** (1) 30–47.
[14] Tite M S 2009 The production technology of Italian maiolica – a reassessment, *J. Archaeol. Sci.* **36** 2065-80.
[15] Felici A C, Fronterotta G, Piacentini M, Nicolais C, Sciuti S, Vendittelli M and Vazio C 2004 The wall paintings in the former Refectory of the Trinità dei Monti convent in Rome: relating observations from restoration and archaeometric analyses to Andrea Pozzo's own treatise on the art of mural painting, *J. Cult. Herit.* **5** 17–25.
[16] Zucchiatti A, Azzou A, El Amraoui M, Haddad M, Bejjit L and Aït Lyazidi S 2009 PIXE analysis of Moroccan architectural glazed ceramics (XIV-XIII centuries), *Inter. J. of PIXE* **19** (3-4) 175-87.
[17] El Amraoui M, Azzou A, Zucchiatti A, Haddad M and Bejjit L 2010 Analyse par PIXE de zelliges marocains du XIVème siècle, *Minbar Al Jamiaa*, édition de l'université Moulay Ismaïl, Ed. M Haddad et M Ibnoussina **20** 212-21.
[18] Ben Amara A, Azzou A, Haddad M, Schvoerer M, Ney C, Ait Lyazidi S and Molinie P 2005 Archaeological Zelliges of Meknes (14th-century): Physical measurements of the colour and identification of colouring agents, *J. Phys. IV* **123** 207-11.
[19] Zarzycki J, 1982, Le verre et l'état vitreux, Paris, Masson, 391 p.

IOP Conf. Series: Materials Science and Engineering **186** (2017) 012029 doi:10.1088/1757-899X/186/1/012029

[20] Calvo del Castillo H, Ruvalcaba JL, Calderon T, Salinas Nolasco MF, Mejıa LM and Perdigon K 2008 Characterisation of an enamelled metallic object found in Guerrero Negro (Baja California) by PIXE and RBS techniques, *Nucl. Instr. Meth. Phys. Res. B,* **266** 2311–15.
[21] Van der Werf I, Mangone A, Giannossa LC, Traini A, Laviano R, Coralini A and Sabbatini L 2009 Archaeometric investigation of Roman tesserae from Herculaneum (Italy) by the combined use of complementary micro-destructive analytical techniques, *J. Archaeol Sci.* **36** 2625–34.
[22] El Amraoui M, Azzou A, Haddad M, Bejjit L, Ait Lyazidi S and El Amraoui Y 2007 Zelliges of Dar-El Beïda Palace (18th-century) in Meknes (Morocco): Optical Absorption and Raman Spectrometry study, *Spectrosc. Lett.* **40** 777-83.
[23] Molera J, Pradell T and Vendrell-Saz M 1998 The colours of Ca-rich ceramic pastes: origin and characterization, *Appl. Clay Sci.* **13** 187-202.
[24] Garcıa-Heras M, Rincon JMa, Jimeno A and Villegas M A 2005 Pre-Roman coloured glass beads from the Iberian Peninsula: a chemico-physical characterisation study, *J. Archaeol. Sci.* **32** 727-38.
[25] Costalgiola P, Baldi G, Cipriani C, Pecchioni E and Buccianti A 2000 Mineralogical and chemical characterisation of the Medicean glass mosaic tesserae and mortars of the Grotta del Buontalenti, Giardino di Boboli, Florence, Italy, *J. Cult. Herit.* **1** 287–99.
[26] Eppler R A and Eppler D R 2000 Glazes and Glass coatings, The American Ceramic Society, Westerville, Ohio, 332 p.

IOP Conf. Series: Materials Science and Engineering **186** (2017) 012030 doi:10.1088/1757-899X/186/1/012030

Sliding mode controller for a photovoltaic pumping system

A ElOugli[1], S Miqoi[2], M Boutouba[3] and B Tidhaf[4]

Laboratory of Embedded Electronic Systems and Renewable Energy
University of Mohamed I,Oujda, Morocco

Email: a.elougli@yahoo.fr

Abstract. In this paper, a sliding mode control scheme (SMC) for maximum power point tracking controller for a photovoltaic pumping system, is proposed. The main goal is to maximize the flow rate for a water pump, by forcing the photovoltaic system to operate in its MPP, to obtain the maximum power that a PV system can deliver.And this, through the intermediary of a sliding mode controller to track and control the MPP by overcoming the power oscillation around the operating point, which appears in most implemented MPPT techniques. The sliding mode control approach is recognized as one of the efficient and powerful tools for nonlinear systems under uncertainty conditions.The proposed controller with photovoltaic pumping system is designed and simulated using MATLAB/SIMULINK environment. In addition, to evaluate its performances, a classical MPPT algorithm using perturb and observe (P&O) has been used for the same system to compare to our controller. Simulation results are shown.

1. Introduction

Water pumping system is used in rural areas, where the water sources are spread over many miles of land and power lines are limited and the installation of a new transmission line and a transformer is expensive, so to operate in general renewable energy is recommended. Photovoltaic energy is among the most used renewable energy and has the greatest potential for development.PV generation systems have two major problems; the conversion efficiency of electric power generation is low (in general less than 17%, especially under low irradiation conditions), and the amount of electric power generated by solar arrays changes continuously with weather conditions. Moreover, the solar cell (I-V) characteristic is nonlinear and varies with irradiation and temperature. There is a unique point on the (I-V) or (P-V) curve of the solar array called MPP, at which the entire PV system operates with maximum efficiency and produces its maximum output power. The location of the MPP is not known, but can be located, either through calculation models, or by search algorithms [1, 2]. Therefore MPPT techniques are needed to maintain the PV array's operating point at its MPP.

In this paper we are representing simple photovoltaic water pumping system and we are using a sliding mode control as MPPT to generate the voltage reference to which the PV system should operate. The paper is structured as follows: paragraph 2 is a description of the PV system used, paragraph 3 introduces the concept of sliding mode control and applying it on our system plus a brief representation of the P&O controller and paragraph 4 present the simulation results on Matlab that are discussed, finally we have a conclusion.

2. PV pumping system modelling

Our photovoltaic pumping system consists of a PV panel, a DC-DC Boost converter, a DC motor and a water pump.Figure1 shows the equivalent circuit of the boost system based on PV with water pump. The system's average model is given in the equation (1) [3, 4]:

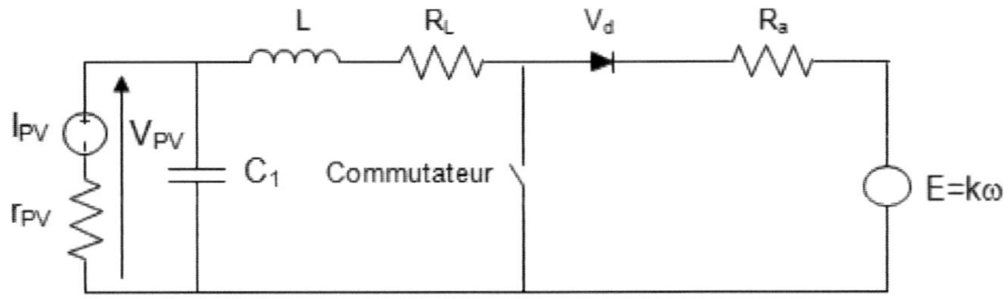

Figure 1. The equivalent circuit of the PV pumping system.

$$\begin{cases} \frac{di_L}{dt} = \frac{V_{pv}}{L} - \frac{(R_a+R_L)}{L} i_L + \frac{(R_a-R_m)i_L+V_d+k\omega}{L} u - \frac{V_d}{L} - \frac{k}{L}\omega \\ \frac{dV_{pv}}{dt} = \frac{I_{pv}}{C_1} - \frac{i_L}{C_1} \\ \frac{d\omega}{dt} = \left(\frac{k}{j} - \frac{k}{j}u\right) i_L - \frac{k_2}{j}\omega - \frac{k_1}{j} \end{cases} \quad (1)$$

where V_{pv} and i_{pv} designthe PV panel voltage and current respectively. L, R_L and i_L are the self-inductance, resistance and current. $R_m$is a resistance characterizing IGBT lost. C_1, V_d and ω are the input capacitance, the diode forward voltage and DC-motor speed. R_a,j,k,k_1 and k_2 are the DC-motor electric resistance, the moment of inertia of the rotor, the electromotive force constant, the disturbance input and the damping coefficient. u is the control input.The values of each of these parameters are listed in the table blow.

Table 1. System parameters.

Model	*Bosch Solar Module c-Si 60*	R_L	0.5 Ω
Power	180 Wp	R_a	3.254 Ω
V_{MPP}	30.30 V	**L**	6.8 10^{-3} H
I_{MPP}	7.76 A	C_1	6.2 10^{-3} F
V_{oc}	36.8 V	**J**	0.004 kg.m^2
I_{sc}	8.45 A	k_1	0.1kg.m^2.rad/sec
k_2	0.0114 kg.m^2/sec^2	**k**	0.333 V.sec/rad
R_m	17.5 mΩ		

3. MPPT

The control circuit takes voltage and current feedback and generates the duty cycle D; this last defines the output voltage of the Boost converter.

3.1. Sliding mode control

Siding Mode Control is a nonlinear control solution and a variable structure control (VSC). It was proposed by Vladim UTKIN in (Utkin, 1977). The advantages of SMC are various and important: high precision, good stability, simplicity, invariance, robustness… [3, 4]. The design of the control can be obtained in two important steps:

•The choice of the sliding surface. Representation of the system state:

$$\frac{d}{dt}\begin{pmatrix} i_L \\ V_{pv} \\ \omega \end{pmatrix} = \begin{pmatrix} -\frac{(R_a+R_L)}{L} & \frac{1}{L} & -\frac{k}{L} \\ -\frac{1}{C_1} & 0 & 0 \\ \frac{k}{j} & 0 & -\frac{k_2}{j} \end{pmatrix}\begin{pmatrix} i_L \\ V_{pv} \\ \omega \end{pmatrix} + \begin{pmatrix} \frac{(R_a-R_m)i_L+V_d+k\omega}{L} \\ 0 \\ -\frac{k}{j}i_L \end{pmatrix}u + \begin{pmatrix} -\frac{V_d}{L} \\ \frac{I_{pv}}{C_1} \\ -\frac{k_1}{j} \end{pmatrix} \quad (2)$$

$$\dot{x} = f(x) + b(x)u + d \quad (3)$$

The vector state is:

$$x = \begin{pmatrix} i_L \\ V_{pv} \\ \omega \end{pmatrix} \quad (4)$$

The MPP is determined when:

$$\frac{\partial P_{pv}}{\partial V_{pv}} = 0 \quad (5)$$

$$\frac{\partial P_{pv}}{\partial V_{pv}} = \frac{\partial I_{pv}^2 R_{pv}}{\partial V_{pv}} = \frac{I_{pv}}{R_{pv}}\left(2R_{pv} + I_{pv}\frac{\partial R_{pv}}{\partial I_{pv}}\right) = 0 \quad (6)$$

Where $R_{pv} = \frac{V_{pv}}{I_{pv}}$is the equivalent load connect to the PV andI_{pv} is the PV current.

$$2R_{pv} + I_{pv}\frac{\partial R_{pv}}{\partial I_{pv}} = 0 \quad (7)$$

So the sliding surface is defined as:

$$S = 2R_{pv} + I_{pv}\frac{\partial R_{pv}}{\partial I_{pv}} \quad (8)$$

•The determination of the control law.As we can see in figure2, the duty cycle output control can be chosen as:

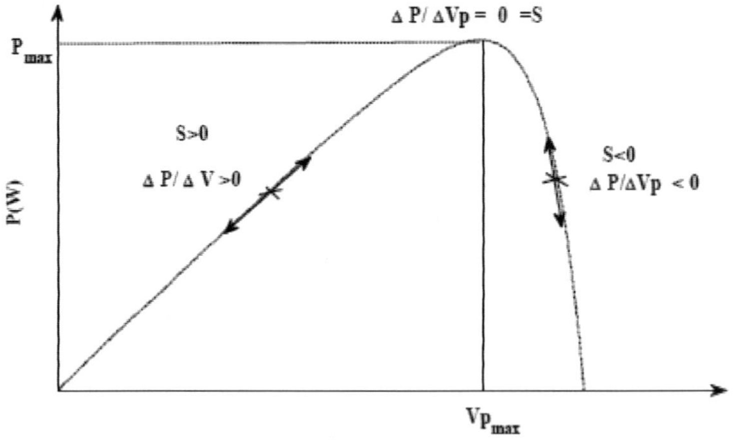

Figure2. Sign of the sliding surface at different positions of the graph P-V.

We have $D = u and D_{eq} = u_{eq}$

$$D = \begin{cases} D - \Delta D for s > 0 \\ D + \Delta D for s < 0 \end{cases} \quad (9)$$

With :

$$\Delta D = u - u_{eq} \quad (10)$$

The equivalent control suggested by Slotine and Li (2005) is determined from the following condition:

$$\dot{s} = \left[\frac{\partial S}{\partial x}\right]^T \dot{x} = \left[\frac{\partial S}{\partial x}\right]^T (f(x) + b(x)u + d) = 0 \quad (11)$$

The equivalent control is:

$$D_{eq} = -\frac{\left[\frac{\partial S}{\partial x}\right]^T (f(x))}{\left[\frac{\partial S}{\partial x}\right]^T b(x)} \quad (12)$$

Since the range of duty cycle must lies in$0 \leq D_{eq} \leq 1$, the real control signal is proposed as:

$$D = \begin{cases} 1 \quad for D_{eq} - \lambda * sign(S) \geq 1 \\ D_{eq} - \lambda * sign(S) for \ 0 < D_{eq} - \lambda * sign(S) < 1 \\ 0 \quad for D_{eq} - \lambda * sign(S) \leq 0 \end{cases} \quad (13)$$

$\lambda$is a positive scaling constant.

3.2. P&O

We have chosen to compare the SMC to the P&O method due to the fact that it is the most used method to control the MPP in the industry, it is known for its simplicity and efficiency. The characteristics of PV panel are measured and then induce a small perturbation on the voltage (or current) to analyse the resulting power variation.Figure3 shows the flowchart of the algorithm for P&O as it should be implemented.

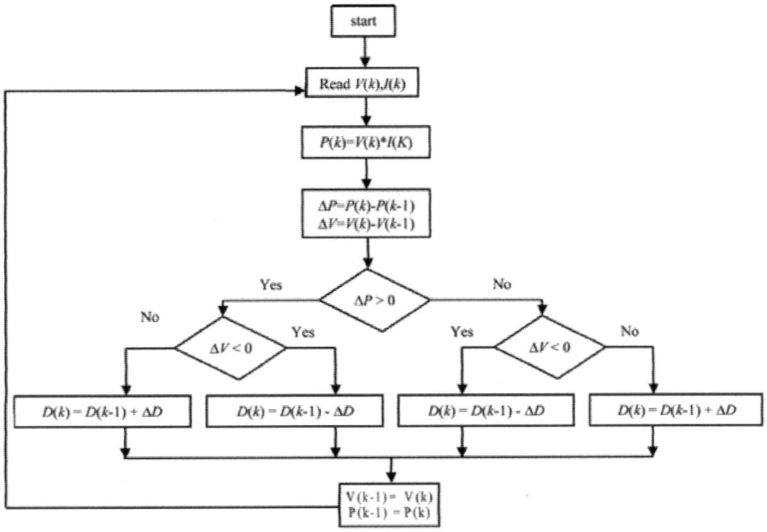

Figure 3. Chart of the algorithm P&O.

4. Results

The results of the MPPT and the speed of the DC motor are taken for a whole day.Figure 4 and figure 5present the climatic conditions on one day (temperature and solar irradiance).

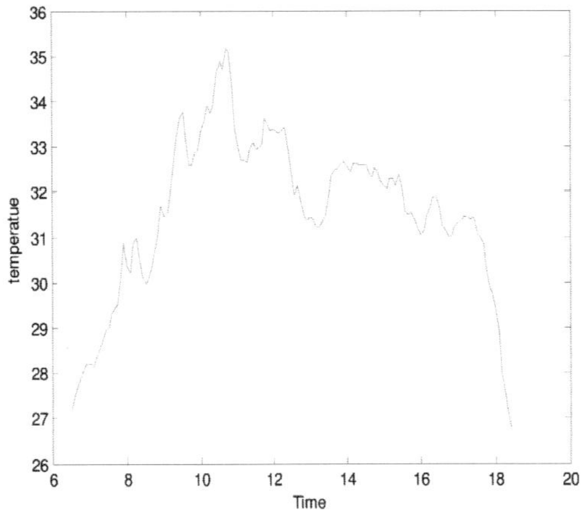

Figure 4. Temperature for the day.

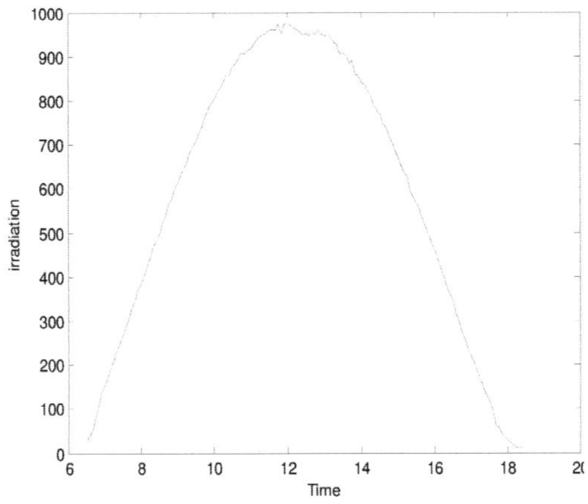

Figure 5. Irradiation for the day.

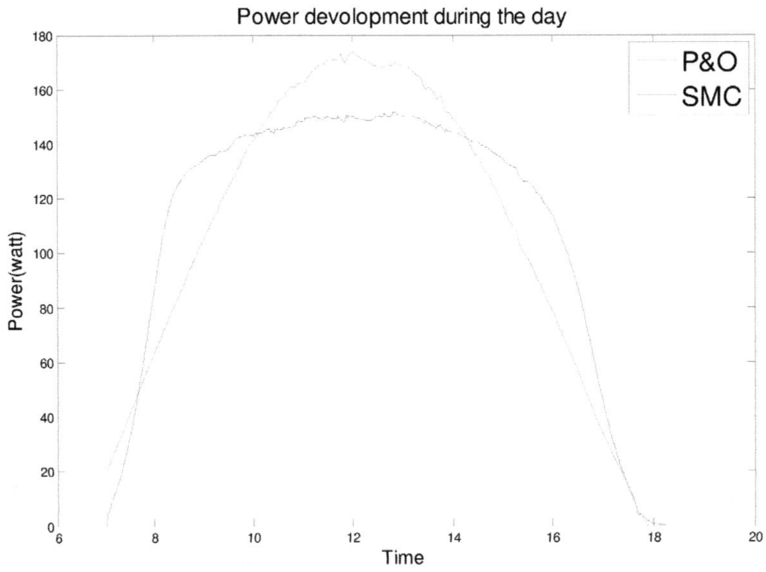

Figure 6. The Power for the day.

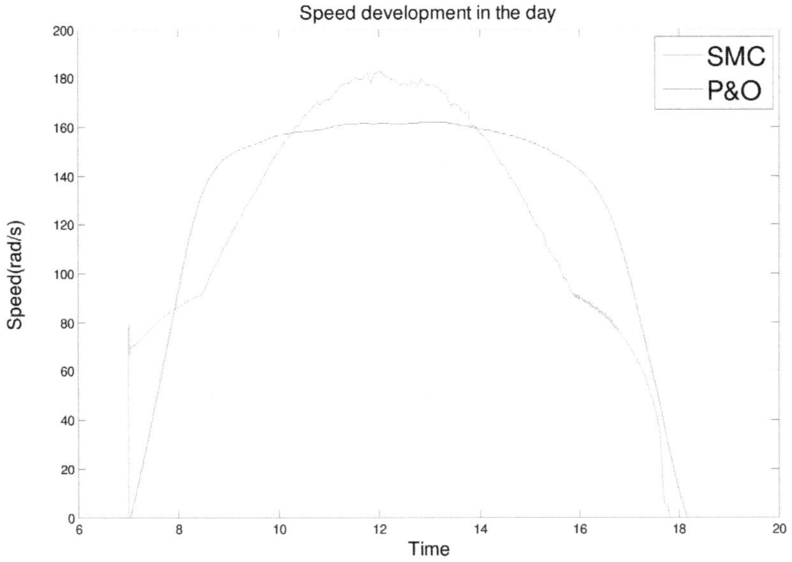

Figure 7. The DC motor speed for the day.

Figure6 and figure 7 represent respectively the development of the Power and the DC motor Speed during day.

5. Conclusion

This paper represents a comparison between two methods of control: P&O and sliding mode control which have been both designed and simulated for the proposed system.It is clear from the results that the sliding mode control gives better performance at tracking the maximum power point, therefore a higher speed for the DC motor and flow rate for the pump.

References

[1] Dolara A, Faranda R and Leva S, Sep. 2009Energy Comparison of Seven MPPT Techniques for PV Systems, *Journal of Electromagnetic Analysis and Applications*, vol.1, no.3, pp.152-162.

[2] Chowdhury S, Chowdhury S P, Taylor G A and Song Y H, July 20-24, 2008Mathematical Modeling and Performance Evaluation of a Stand-Alone Polycrystalline PV Plant with MPPT Facility, *IEEE Power and Energy Society General Meeting - Conversion and Delivery of Electrical Energy in the 21st Century*, Pittsburg, USA.

[3] Nema S, Nema R K and Agnihotri G, 2010Matlab / simulink based study of photovoltaic cells / modules / array and their experimental verification," *International Journal of Energy and Environment*, pp.487- 500, vol.1, Issue 3.

[4] Su J H, Chen J J and Wu D S, Nov. 2002Learning feedback controller design of switching converters via MATLAB/SIMULINK,*IEEE Trans. Educ.*, vol. 45, no. 4, pp. 307-315.

[5] Ghazanfari J and Farsangi M, 2013Maximum power point tracking using sliding mode control for photovoltaic array,*Iranian Journal of Electrical & Electronic Engineering*, vol. 9, no. 3, p. 189.

[6] Chu C C, Chen C L, 2009 Robust maximum power point trackingmethod for photovoltaic cells: A sliding mode control approach,*SolarEnergy*, vol.83, 2009, pp. 1370-1378.

Design and realization of an autonomous solar system

A Gaga[*1], O Diouri[1], N Es-sbai[1] And F Errahimi[1]

[1]Laboratory of Renewable Energies and Intelligent Systems, Faculty of Sciences and Technology, Sidi Mohammed Ben Abdellah University, Fez, Morocco

Email: gaga.ahmad@gmail.com, omar.diouri1@gmail.com

Abstract. The aim of this work is the design and realization of an autonomous solar system, with MPPT control, a regulator charge/discharge of batteries, an H-bridge multi-level inverter with acquisition system and supervising based on a microcontroller. The proposed approach is based on developing a software platform in the LabVIEW environment which gives the system a flexible structure for controlling, monitoring and supervising the whole system in real time while providing power maximization and best quality of energy conversion from DC to AC power. The reliability of the proposed solar system is validated by the simulation results on PowerSim and experimental results achieved with a solar panel, a Lead acid battery, solar regulator and an H-bridge cascaded topology of single-phase inverter.

1. Introduction

An autonomous solar system is a photovoltaic solar power plant which is not connected to the grid. The extraction of the maximum power from solar panels called 'MPPT technique' (Maximum Power Point Tracking) provides an effective method to solve the optimization problem .A multi-junction solar cells are different from silicon PV cells as they are capable of converting solar irradiation into energy at high efficiency [1].PV array represents the essential power conversion unit of a photovoltaic system. The output characteristics of PV array depend on the irradiation, the temperature and output voltage of PV array [2]. Among the MPPT strategies which are the most used: the method of incremental conductance (I.C.), the Perturbation and observation (P&O), and fuzzy logic [3,4].
The performance of a standalone photovoltaic system depends mainly on the technical regulation and control adopted. In fact, there are many methods for estimating charging status 'State Of Charge: SOC'. The simplest method is the use of coulomb metric measurement [5].
The objective of this work is the realization of a complete solar system based on a microcontroller intended for the MPPT control and monitoring the state of charge of the batteries to protect them.
We are implementing the P&O algorithm for MPPT control. For the determination of 'SoC' in the storage system, we are implementing a specific coulomb metric method with an understandable and reliable user display which increases the performance and reliability of the battery.
We are implementing the algorithm control of the five level inverter to achieve an output voltage that resembles a sinusoidal voltage. For the implementation of the whole process, an innovative control application is developed on Labview environment that gives the system a flexible and scalable structure for controlling, monitoring and supervising the complete system in real time.

2. Photovoltaic generator

The principle operation of a photovoltaic generator is based on the photovoltaic effect of semiconductor PN junction. When exposed to sunshine, a DC current is generated; this current varies linearly with the solar radiation [6].

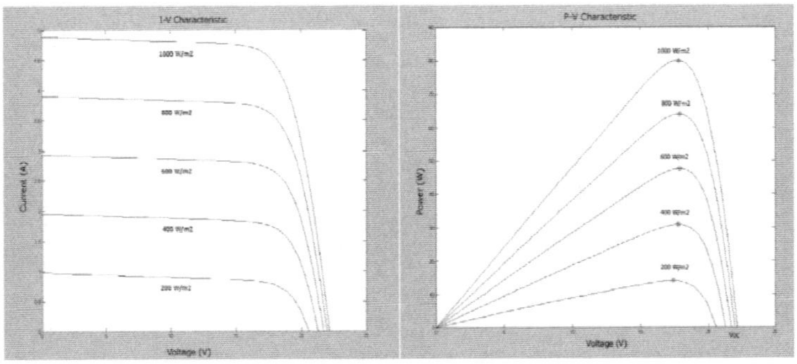

(a) Ipv / Vpv feature *(b) Ppv/Vpv feature*

Figure 1. Characteristic of a solar panel.

The current-voltage characteristic depends on the irradiation and temperature conditions on the surface of the photovoltaic cell. Figure 1 shows the influence of irradiation on the current-voltage characteristics and power-voltage of a photovoltaic module with a constant temperature [7].
The maximum operating point must always be followed to ensure that the PV generator has reached its maximum power output in real-time to maximize permanently the transferred energy.

3. DC / DC Converters

A photovoltaic generator (PV) works in optimum conditions; it must have an adaptable converter. This adaptation is achieved by searching automatically the maximum power point (MPP) of the PV generator [8]. The converter may be a DC / DC "Boost" or "Buck" according to applications. Figure 2 shows the boost converter structure with 'T' is the switching period 'of the duty cycle.

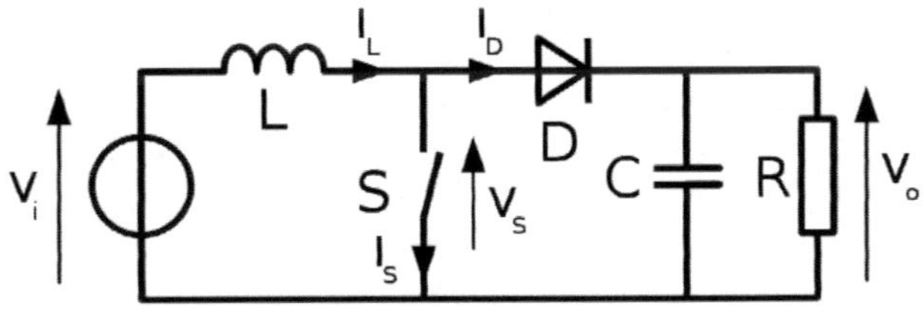

Figure 2. Model of DC / DC boost.

4. MPPT Controller

One of the most commonly used MPPT strategies is the perturb and observation method (P&O), the working principle is simple, we disturb the system power by acting on the control signal "PWM", and observe the direction of the output power variation Figure 3. Indeed, the MPPT controller adjusts the voltage across the photovoltaic source by using a DC-DC boost converter Figure 2, and measures the output power. After that, it varies the voltage across the PV by changing the PWM duty cycle and takes a second power measurement. If the power increases, the maximum power is not yet reached so we must choose the new control signal by changing the PWM duty cycle in some direction. Therefore, we need to adjust the voltage in the positive direction until the power will not increase, that's mean that the variation rate will be null [9]. Figure 4 present in details the MPPT algorithm implemented in this work.

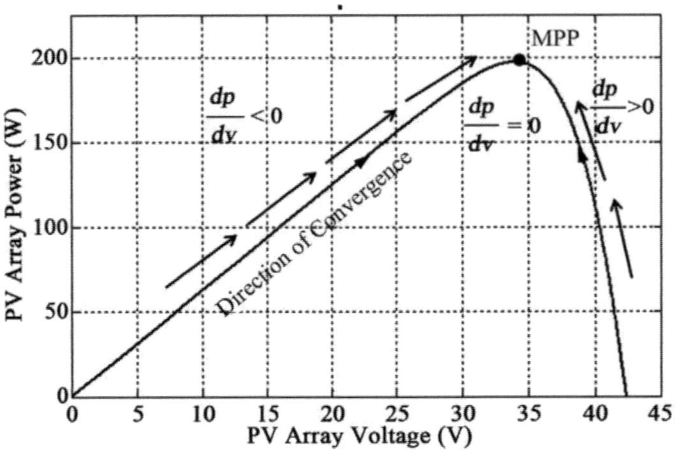

Figure 3. Principle of the P&O algorithm

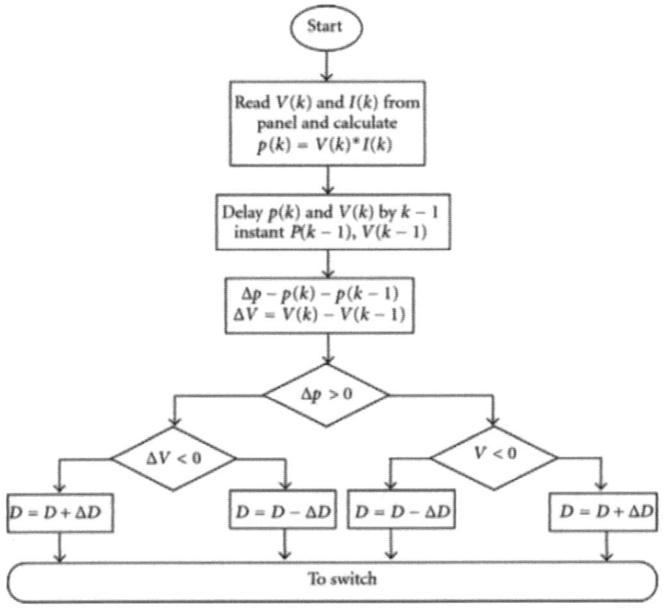

Figure 4. P&O algorithm.

5. Controlling the charge state

The state's control technique battery 'SOC' is the most convenient and most commonly used. It consists of measuring and calculating the arriving and leaving quantities of electricity during charging and discharging in terms of ampere-hours. This technique is called the coulomb metric measurement. The chosen model assumes that the battery charge status 'SOC' is the ratio of the amount of electricity received or returned by the battery and the capacity of the battery (Ah). The calculation of the battery charge status 'SOC' is done using the following recursive relation:

$$\mathrm{SOC}(t+1)n = \begin{cases} \mathrm{SOC}(t) + \eta_b \times \frac{I_b.t}{C} \text{ (Battery charge)} \\ \mathrm{SOC}(t) - \frac{I_b.t}{C} \text{ (discharge cycle)} \end{cases} \quad (1)$$

With :

SOC (t+l) : State of charge at the moment (t+l)
SOC (t) : State of charge at the moment (t)
ηb : medium Faraday efficiency during the charging cycle. The typical value is 0.9.
Ib : Input current or battery output
C :Capacity of the battery

6. Photovoltaic inverter

Figure 5 shows a five levels cascaded H-bridge multilevel inverter. The inverter consists of two H-bridge cells connected in series which are fed by independent voltage sources. The outputs of the H-bridge cells are connected in series such that the synthesized voltage output is the sum of all of the individual cell outputs. The output voltage is given by equation:

$$V = V1 + V2 \tag{2}$$

Where the output voltage of the first cell is labeled V1 and the output voltage of the second cell is denoted by V2. There are five levels of output voltage i.e. V, V/2, 0, -V/2, -V.

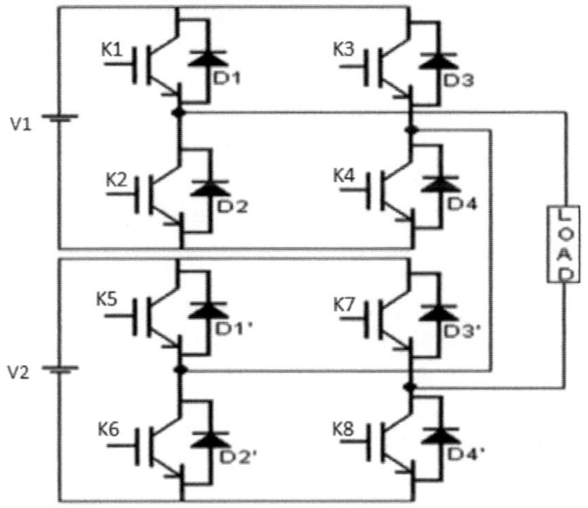

Figure 5. Five level cascaded H-bridge multilevel inverter.

7. Realization and experimental results

The block diagram of the complete solar system is shown in Figure 6, it consists of a PV generator, a battery, a DC-DC converter, an inverter, a MPPT control block and the battery state of charge estimator 'SOC'. A graphical interface is developed in Labview environment to manage the entire system.

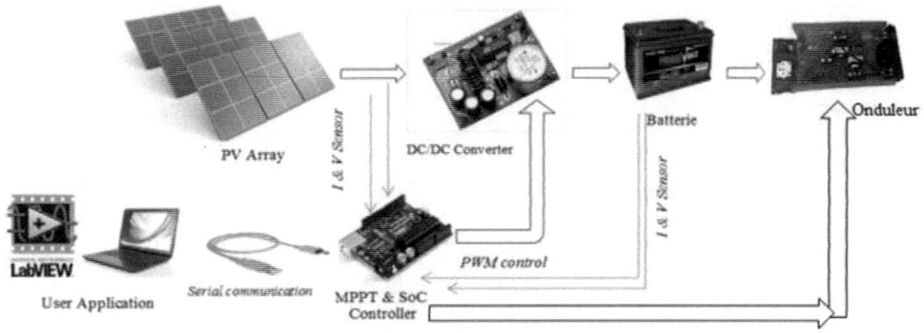

Figure 6. Block diagram of the complete system.

Figure 7. MPPT hardware implementation.

Figure 7 shows the realized experimental system, the characteristics of the used PV module are as follows; maximum power is 30 W, open circuit voltage is 17.6 V, maximum current is 1.7 A. For the battery, nominal voltage is 12 V, capacity is 9 Ah, nominal current is 1 A, the electrolyte density is 1.23 kg/l, end of charge voltage is 14.8 V (20° C) and the end of discharge voltage is 10.5 V (20° C).The MPPT algorithm implementation is divided into two parts; the first one is the firmware which is loaded in the microcontroller, it is principal task establish communication between system and software application developed by LabVIEW which represents the second part of implementation "Graphical User Interface" shown in Figure 9. This software application extends the user possibilities to select other MPPT algorithm than the used one. The hardware system, in Figure 7,controls the photovoltaic power by detecting the panel voltage (V) and current (I) to determine the operating point corresponding to the maximum power.

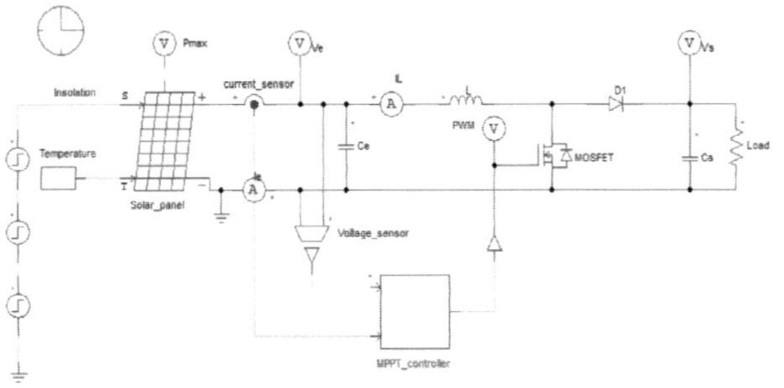

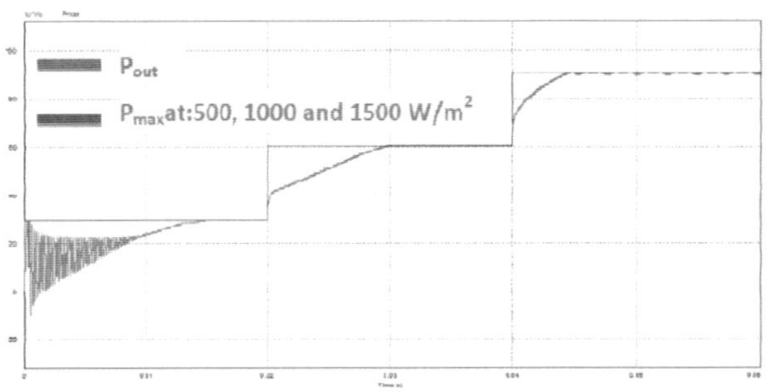

Figure 8. Simulaion of the implementation of MPPT algorithm in PowerSim.

Figure 8 shows the simulation diagram of the complete circuit in the Powersim simulator. The system consists of a PV panel, a DC-DC Boost converter, a resistive load, sensors for voltage and current measurement and the MPPT controller block.
The P&O algorithm shown in Figure 4 is implemented in the MPPT control block. The graph in Figure 8 shows the evolution of the power delivered by the solar panel with different irradiation values(500, 1000 and 1500 W/m2). The maximum power point is reached after a few seconds after starting the MPPT algorithm execution, which proves the performance of the system in terms of dynamics. Once the desired operating point is reached, the DC-DC BOOST converter control is kept constant. Maximum power point values are indicated in blue while the actual values are shown in red.
The acquisition device used in this system is based on an Arduino module that integrates Atmel microcontroller; it measures both current and voltage of the solar panel and generates the control signal of the DC/DC converter which is the PWM signal. For detecting and reading the current supply delivered by the photovoltaic module, a shunt resistor is placed in series with the analogue-digital converter input of the microcontroller card. The shunt resistor of the sensor provides a voltage proportional to the current flowing through it. To measure the PV generator voltage, the node between the two resistors in series must be connected to the analog input of the microcontroller.
Figure 9 reveals the Graphical User Interface (GUI) that we have designed; it allows to the user the full access to all system features. This interface is developed in the Labview environment, it provides several display and options control:

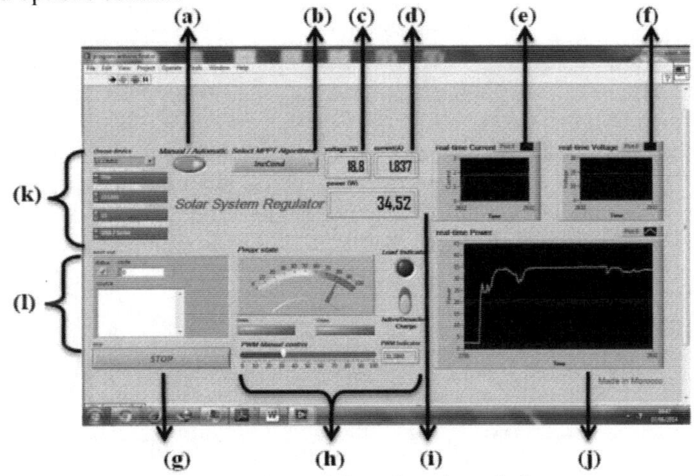

Figure 9. The user interface in Labview

(a) Automatic or manual control
(b) Choosing the MPPT algorithm
(c) (d) & (i) current, digital voltage and power indicators
(e) (f) & (j) current, voltage and power graphical indicators
(k) Initialization of the serial communication
(l) & (g) Error indicator and shutdown button
(h) Indicators and additional orders

All these control options and indicators are the basic interface necessary to control the entire system, additional indicators and buttons control are added to provide extra access to the system. The main state of charge (SOC) controller objective is supervising the real time battery state in a standalone photovoltaic system. The main part of the developed system is a microcontroller which implement the SOC algorithm based on the coulomb metric counting method. To evaluate the proper functioning of

this technique, two tests are made; the first one determines the SOC profile of the battery in charge cycle as shown in Figure 10. The last test determines the battery profile in discharge cycle as shown in Figure 11.
In the charging phase, only the battery is connected, the charge current varies between 0 and 2 A depending on the sunlight. The charge state management program is implemented in the microcontroller and connected to the computer with a Labview application via a USB connection. Several values of the 'SOC' were identified in real time (Figure 10).

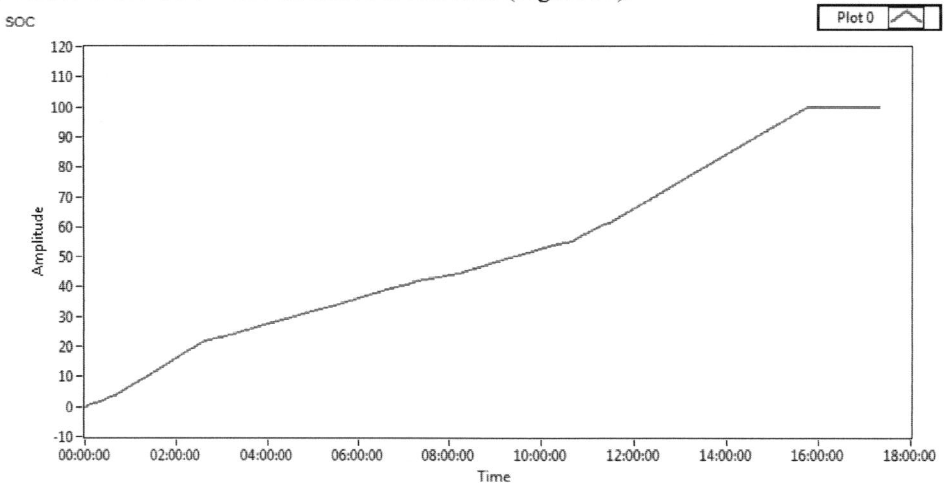

Figure 10. Battery charge cycle

In the discharge phase, a resistive load is connected (10Ω, 25W) at the terminals of the battery that delivers a constant discharge current. Several points of the SOC measures in real time have been identified. Figure 11 shows the extracted values by our device.

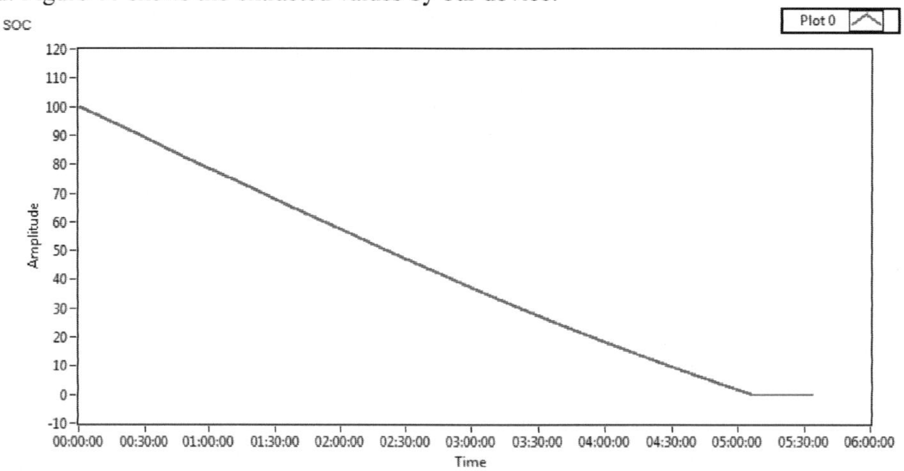

Figure 11. Battery discharge cycle

From these practical results, it appears that the automatic control of the SOC battery charging status during the two processes charging and discharging is primordial to save battery life.
Figure 12 bellow shows the pattern of simulation by Power SIM of a signal-phase five level inverter type H-Bridge and a DC voltage source of 24V. In this topology, two power supplies with a value of 12V each are used. The output obtained has five levels with a maximum voltage of 24V.

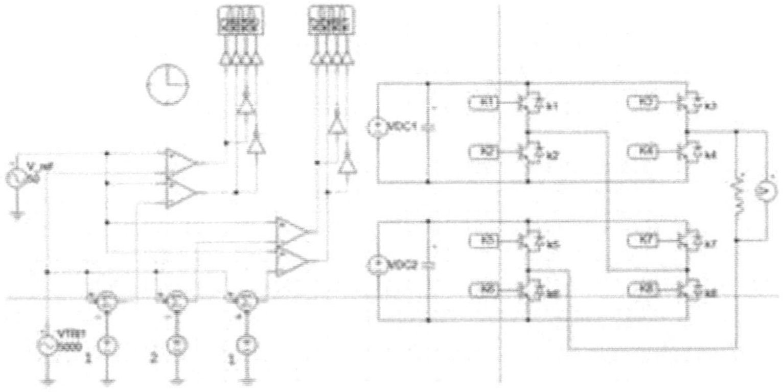

Figure 12. Power circuit and PWM control of five level inverter in PSIM.

Waveform of compound voltage of signal-phase H-Bridge five levels inverter is illustrated in figure 13. The compound voltage shows a maximum value of 24V.

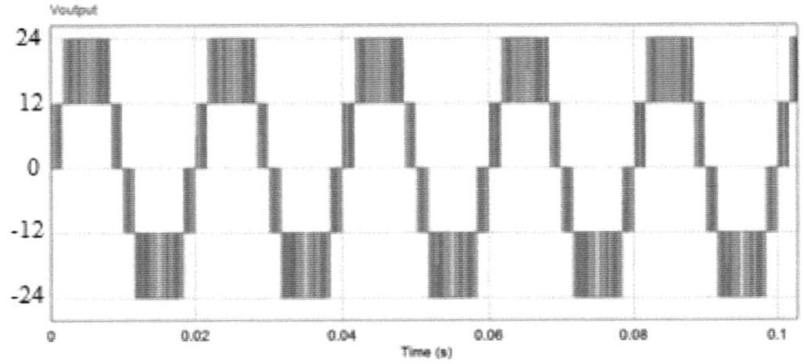

Figure 13. Output voltage of five levels inverter.

8. Conclusion

This paper presents an autonomous and efficient solar system with a solar controller MPPT. The heart of the developed device is a microcontroller that integrates the MPPT control algorithm and the charge state 'SOC' batteries algorithm based on the coulomb-metric method, as well as the control of the various electronic switches of the inverter.

The proposed approach is based on developing a graphical interface in the LabVIEW environment which gives the system a flexible and scalable structure for controlling, monitoring and supervising the entire system in real time. This application is used to implement multiple types of MPPT algorithm without any changes on the equipment part. Thus, the developed system can be a powerful tool in the industrial sector and in the field of research. The reliability of the proposed complete solar system is validated by the results of simulations in the PowerSim environment and experimental results achieved using a solar panel, an average lead-acid battery power and the proposed solar controller.

References

[1] RezkHegazy, El Sayed AHM. Sizing of a standalone concentrated photovoltaic system in Egyptian site. Int J Electr Power Energy Syst 2013;325:330–45.

[2] S. Elmetennani, T.M. Laleg-Kirati, M. Djemai, M. Tadjinec. 'A new MATLAB.Simulink model of triple-junction solar cell and MPPT based on artificial neural networks for photovoltaic energy systems' Ain Shams Engineering Journal, Volume 6, Issue 3, September 2015, Pages 873–881

[3] J.A. Jiang, T.L. Huang, Y.T. Hsiao and C.H. Chen, 'Maximum Power Tracking for Photovoltaic Power Systems', Tamkang Journal of Science and Engineering, Vol. 8, N°2, pp. 147 – 153, 2005.

IOP Conf. Series: Materials Science and Engineering **186** (2017) 012031 doi:10.1088/1757-899X/186/1/012031

[4] J.H.R. Enslin and D.B. Snyman, 'Simplified Feed-Forward Control of the Maximum Power Pint in PV Installations', Proceedings of the IEEE International Conference on Power Electronics Motion Control,Vol. 1, pp. 548 – 553, 1992.
[5] H.J. Bergveld, V. Pop, P.H.L. Notten, Method of estimating the state-of-charge and of the use time left of a rechargeable battery, and apparatus for executing such a method, Patent WO2005085889, filed 23 February (2005)
[6] T.S.USTUN and S. Mekhilef, "Effects of a Static Synchronous Series Compensator (SSSC) Based on Soft Switching 48 Pulse PWM Inverter on the Power Demand from the Grid," Journal of Power Electronics, vol. 10, pp. 85-90, 2010.
[7] H. Knopf, "Analysis, Simulation, And Evaluation of Maximum Power Point Tracking (MPPT) Methods for a solar power vehicle," in Electrical and Computer Engineering, Master of Science in Electrical and Computer Engineering: Portland State University 1999, pp. 177.
[8] Tat Luat Nguyen, Kay-Soon Low, "A Global Maximum Power Point Tracking Scheme Employing DIRECT Search Algorithm for Photovoltaic Systems," IEEE Transactions on Industrial Electronics, vol. 57, no. 10, pp. 3456-3467, Oct. 2010.
[9] D. Sera, T. Kerekes, R. Teodorescu, F. Blaabjerg, "Improved MPPT Algorithms for Rapidly Changing Environmental Conditions," in Proc. 12th International Conference on Power Electronics and Motion Control, 2006, pp. 1614-1619.

Natural convection in a three-dimensional cavity filled with nanofluid and partially heated: Effect of the hating section dimension

M Sannad[1], B Abourida[1], L Belarche[1], H Doghmi[1] and Mouzaouit[1]

[1]LMPEE, National School of Applied Sciences Agadir, BP 1136 Agadir, Morocco.

Email mohamedsannad@gmail.com

Abstract. This study focuses on heat transfer by natural convection in a three dimensional cavity filled with nanoparticles and partially heated from the side with a uniform temperature. The opposite wall of the cavity is maintained in a cold temperature. The effect of nanofluid type on thermal phenomena within the cavity was analyzed for different sizes of the heating section, using the control volume method. The governing parameters are: the Rayleigh number ($10^3 \leq Ra \leq 10^5$), the volume fraction ($0 \leq \Phi \leq 0.1$), the heating section size ($0.5 \leq \varepsilon \leq 1$), and the nanofluid type. The results represent a great interest in terms of the flow and heat transfer through the cavity depending on the chosen parameters sets.

Key words. Natural convection, Nanofluids, Tridimensional cavity.

1. Introduction

The natural convection has been particularly and considerably studied these past years because of its many practical applications in relation with the energy management, air conditioning, cooling of the electronic components etc..... In order to improve the heat transfer by convection in these applications, the researchers started looking deeper into the structure of the matter at the molecular level which has given birth to the development of nanofluids: a colloidal solution composed of particles of nanometric size in suspension in a conventional fluid. The term nanofluid was proposed for the first time in 1995 by Choi [1] and opened the door to a new technology which is characterized by higher thermal properties related. In fact, this new concept has been proposed as a technique to improve the performance of the thermal transmission of conventional fluids. Thus, the interest presented by the nanofluids has given rise to numerous studies [2-4], which have focused on the determination of their thermophysical properties such as the thermal conductivity, specific heat and the dynamic viscosity and their impact on the flow and heat transfer. However, it should be noted that most of the available studies have considered the two-dimensional case when a three-dimensional approach presents more flow visualization opportunities and better approach of the heat transfer.

2. Problem Formulation

A schematic of the physical problem and coordinates are shown in figure 1. A part of the left vertical wall of the cavity is maintained at a hot temperature T_h while the opposite one is maintained at a cold temperature T_c. The other walls are considered adiabatic. The governing parameters are: the volume fraction Φ ($0 \leq \Phi \leq 0.1$), the Rayleigh number Ra ($10^3 \leq Ra \leq 10^5$) and the heating section size ($0.5 \leq \varepsilon \leq 1$).

The considered nanofluids are: water + Al_2O_3; Water + Cu and Water + TiO_2.

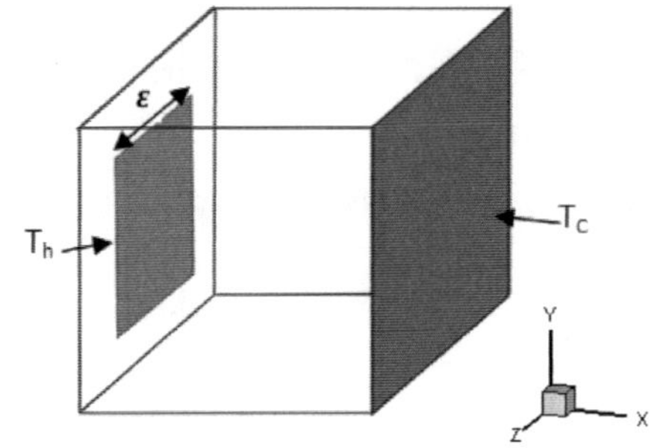

Figure 1. Studied configuration and coordinates.

The governing equations are discretized by finite volume method, adopting the Boussinesq approximation and neglecting the viscous dissipation. The physical parameters are given by the following equations:

- The density:

$$\rho_{nf} = (1-\phi)\rho_f + \phi\rho_{np} \tag{1}$$

- The heat capacitance of the nanofluid:

$$(\rho C_p)_{nf} = (1-\phi)(\rho C_p)_f + \phi(\rho C_p)_{np} \tag{2}$$

- Thermal expansion coefficient:

$$\beta_{nf} = (1-\phi)\beta_f + \phi\beta_{np} \tag{3}$$

- The dynamic viscosity of the nanofluid:

It's given by Brinkman [5] who has extended the Einstein formula to cover a wide range of volumetric concentrations.

$$\mu_{nf} = \frac{\mu_f}{(1-\phi)^{2.5}} \tag{4}$$

- Thermal diffusivity of nanofluids:

$$\alpha_{nf} = \frac{k_{nf}}{(\rho C_p)_{nf}} \tag{5}$$

- The thermal conductivity of nanofluids:

It is given by Maxwell [6] as follows:

$$\frac{k_{nf}}{k_f} = \frac{k_{np} + 2k_f - \left[2(k_f - k_{np})\phi\right]}{k_{np} + 2k_f + \left[\phi(k_f - k_{np})\right]} \tag{6}$$

k_{nf} , k_f and k_{np} are respectively the thermal conductivities of the nanofluid, the basic fluid and solid nanoparticles.

Table 1. Thermophysical properties of water and nanoparticles

	ρ(kg.m^{-3})	β(K^{-1})	k(W.m^{-1}.K^{-1})	C$_p$(J.kg^{-1}.K^{-1})
Pure Water	997.1	21x10^{-5}	0.613	4179
AL_2O_3	3970	0.85x10^{-5}	40	765
Cu	8933	1.67x10^{-5}	400	385
TiO_2	4250	0.90x10^{-5}	8.9538	686.2

Hence, the obtained dimensionless governing equations are:

$$\frac{\partial U}{\partial X}+\frac{\partial U}{\partial Y}+\frac{\partial U}{\partial Z}=0 \tag{7}$$

$$U\frac{\partial U}{\partial X}+V\frac{\partial U}{\partial Y}+W\frac{\partial U}{\partial Z}=-\frac{\partial P}{\partial X}+\frac{\mu_{nf}}{\alpha_f\rho_{nf}}\left(\frac{\partial^2 U}{\partial X^2}+\frac{\partial^2 U}{\partial Y^2}+\frac{\partial^2 U}{\partial Z^2}\right) \tag{8}$$

$$U\frac{\partial V}{\partial X}+V\frac{\partial V}{\partial Y}+W\frac{\partial V}{\partial Z}=-\frac{\partial P}{\partial Y}+R_a*P_r*\theta*\frac{\rho_f\beta_{nf}}{\rho_{nf}\beta_f}+\frac{\mu_{nf}}{\alpha_f\rho_{nf}}\left(\frac{\partial^2 V}{\partial X^2}+\frac{\partial^2 V}{\partial Y^2}+\frac{\partial^2 V}{\partial Z^2}\right) \tag{9}$$

$$U\frac{\partial W}{\partial X}+V\frac{\partial W}{\partial Y}+W\frac{\partial W}{\partial Z}=-\frac{\partial P}{\partial Z}+\frac{\mu_{nf}}{\alpha_f\rho_{nf}}\left(\frac{\partial^2 W}{\partial X^2}+\frac{\partial^2 W}{\partial Y^2}+\frac{\partial^2 W}{\partial Z^2}\right) \tag{10}$$

$$U\frac{\partial \theta}{\partial X}+V\frac{\partial \theta}{\partial Y}+W\frac{\partial \theta}{\partial Z}=\frac{\alpha_{nf}}{\alpha_f}\left(\frac{\partial^2 \theta}{\partial X^2}+\frac{\partial^2 \theta}{\partial Y^2}+\frac{\partial^2 \theta}{\partial Z^2}\right) \tag{11}$$

The adopted thermal boundary conditions are:

$\theta=-0.5$, for $X=1$; on the cold wall

$\theta=0.5$, for $X=0$, on the heating section

$\frac{\partial\theta}{\partial n}=0$ On the adiabatic walls

Where n is the normal to the considered wall.

The adopted hydrodynamic boundary conditions are:

$U=V=W=0$ for all walls.

The local Nusselt number is defined as:

$$Nu_l(y,z)=\frac{k_{nf}}{k_f}\vec{\nabla}\theta.\vec{e}_x \tag{12}$$

The average Nusselt number, Nu$_a$, is defined as the integral of the temperature flux through the vertical cold wall and formulated as:

$$Nu_a=\frac{k_{nf}}{k_f}\int_S \vec{\nabla}\theta.\vec{e}_x dydz \tag{13}$$

3. Numerical method

The Navier-Stokes and energy equations governing the problem are discretized by the finite volume method, using the Boussinesq approximation and neglecting the viscous dissipation. To overcome the difficulty associated with the determination of the pressure, the SIMPLEC algorithm is used to solve the momentum equations coupled with the continuity one. The Alternating Direction Implicit scheme (ADI) is then adopted to solve the algebraic discretized equations. The obtained numerical code was validated by comparing its results with those of Ravnik et al. [2]. The results show that the differences in terms of Nusselt do not exceed 0.5% for $Ra = 10^3$ and 2% to $Ra = 10^5$. Figure 2 shows also a good agreement between the two works in terms of the velocity components V(X) and V(Y) for $Ra = 10^3$ and $Pr = 6.2$.

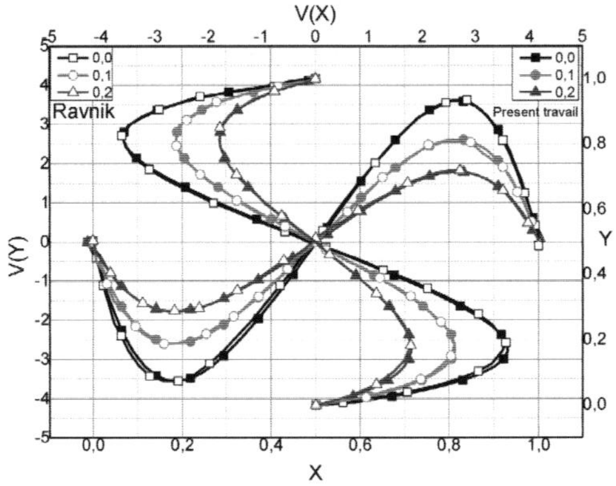

Figure 2. Comparison of the velocity V(X) and V(Y) between our results and those of Ravnik et al.[2]

To check the effect of the grid size, preliminary tests have been conducted for different sets of the governing parameters. The presented results (Table 2) are relative to the nanofluid water-Al_2O_3 (Φ=0.1) and $Ra=10^3$. Hence, the non-uniform staggered grid of 41×41×41 nodes was estimated to be appropriate for the present study since it permits a good compromise between the computational cost (a significant reduction of the execution time) and the accuracy of the obtained results.

Table 2. Variation of the Nusselt number with different grid size

Grid	31x31x31	41x41x41	51x51x51	61x61x61
Nu_a	1.357282	1.353103	1.350895	1.349503

4. Results

A numerical study was conducted to investigate the effects of heating condition on natural convection in an enclosure filled with a nanofluid. The Prandtl number of water which is the base fluid is equal to 6.2, and the nanoparticle volume fraction is varying between 0 and 10%. Three types of nanoparticles are considered in the present work: Al_2O_3 , Cu and TiO_2. Their properties are listed in Table 1. The range of Rayleigh number is $10^3 \leq Ra \leq 10^5$ and the heating section size is $(0.5 \leq \varepsilon \leq 1)$.

In this section, we present the isotherms obtained for different values of the governing parameters and also the speed variation and the resulting Nusselt number.

Figure 3 presents the temperature distribution within the three dimensional cavity for Al_2O_3, $Ra = 10^5$, $\Phi = 0.04$ and ε=0.75. The figure shows that the fluid is moving from the heating wall to the cold one, in such a way that the heat transfer rate is permanently maintained within the cavity.

IOP Conf. Series: Materials Science and Engineering **186** (2017) 012032 doi:10.1088/1757-899X/186/1/012032

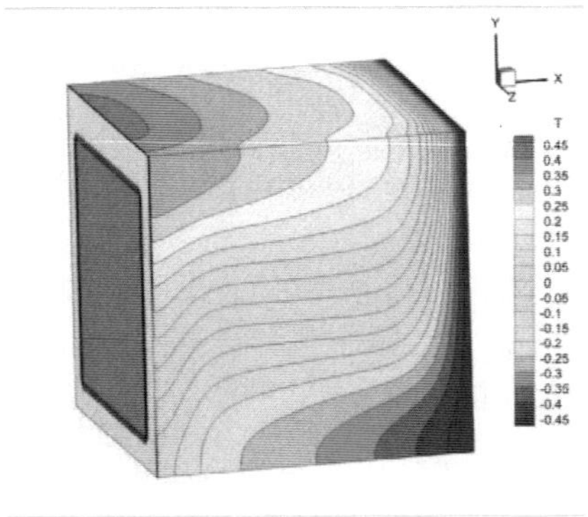

Figure 3. 3D isotherms for Ra=10^5, Φ=0.04 and ε=0.75

Based on different preliminary tests; the plan Z = 0.5 was chosen for the following presentations. In fact, this plan, perpendicular to the vertical wall containing the heating section, is characterized by high activity and presented adequately the fluid motion and the heat transfer within the enclosure.

In order to study the effect of the heating section dimension ε, the hydrodynamic and thermal fields in the cavity are shown in figure 4 for Al_2O_3,Ra = 10^5, Φ=0.04 and three dimensions ε =0.5, ε= 0.75 and ε = 1. Hence, figures 4a and 4b present respectively the streamlines and the isotherms in the plan (Z=0.5) for the three dimensions. For all the considered cases, the flow consists mainly of two counter-rotating cells and secondary flows appearing for ε= 0.75 and in the perpendicular direction. The resulting flow is symmetrical to the mid plan in the case of ε= 1. The isotherms show an increase of the heat transfer with increasing ε.

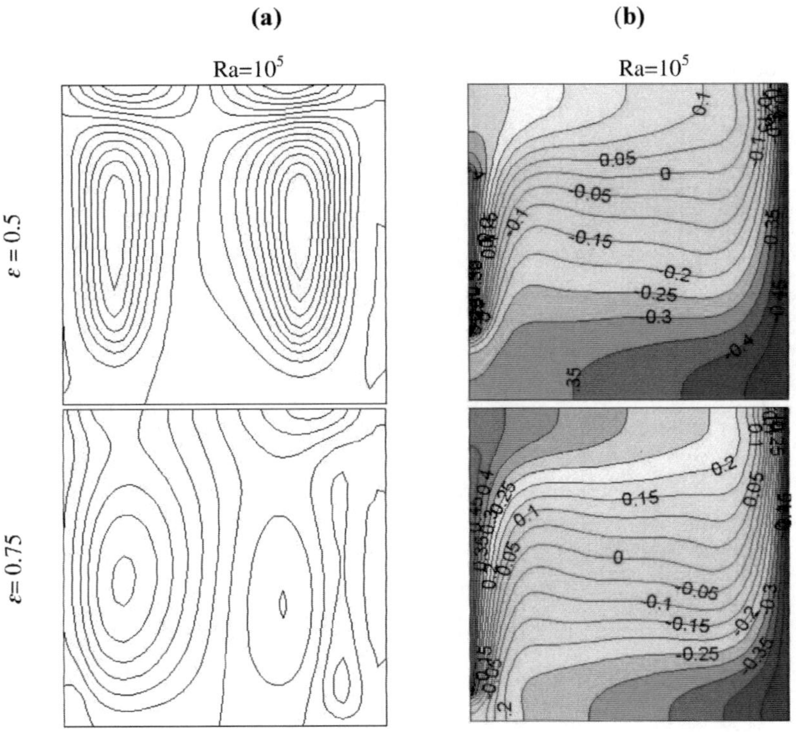

$\varepsilon=1$

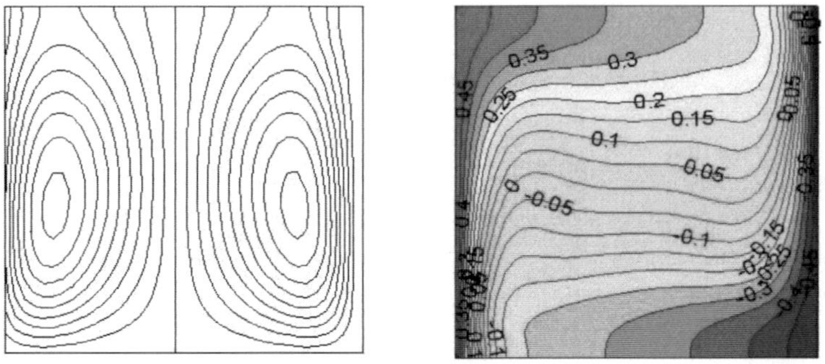

Figure 4. the streamlines (left) and isotherms (right) at Z=0.5 for Al_2O_3 ,Ra = 10^5, Φ=0.04 and different ε.

Figure 5 illustrates the temperature contours in the central plan Z=0,5 for Ra=10^3;10^5, Φ varying between 0 and 0.04 and 0.5≤ ε ≤1. The figures show that for all the considered nanofluids, the temperature distribution has the same behavior in the central part of the enclosure. However, differences are observed near the walls due to the increasing size of the heating section. In fact, increasing ε causes a better heat transfer from the left wall of the cavity to the cold one as shown in the corresponding isotherms. This effect is observed for all the considered nanofluids.

The differences between the pure water and the nanofluid are shown by the variation of the volume fraction between 0 and 0.04.

The figures corresponding to Ra=10^3, show a stratification of the isotherms from the hot wall to the cold one. This is due to the dominance of conduction in this case. The differences due to the increase of Φ are very small in the considered range and disappear totally in the center of the cavity. This trend is encountered for all the considered nanofluids. The same remarks can be said when increasing Ra to 10^5. However, a distortion of the isotherms can be seen in this case, attesting of a better heat transfer by convection through the cavity.

Water +Al_2O_3 for Ra=10^3 | Water +Cu for Ra=10^3 | Water+TiO_2 for Ra=10^3

$\varepsilon=0.5$

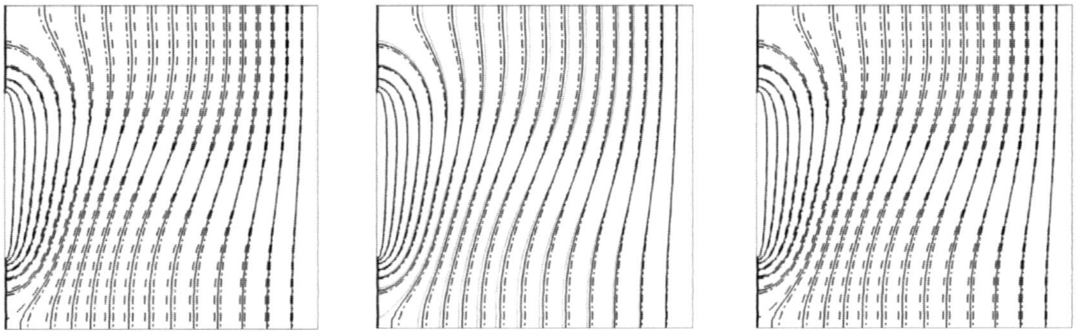

IOP Conf. Series: Materials Science and Engineering **186** (2017) 012032 doi:10.1088/1757-899X/186/1/012032

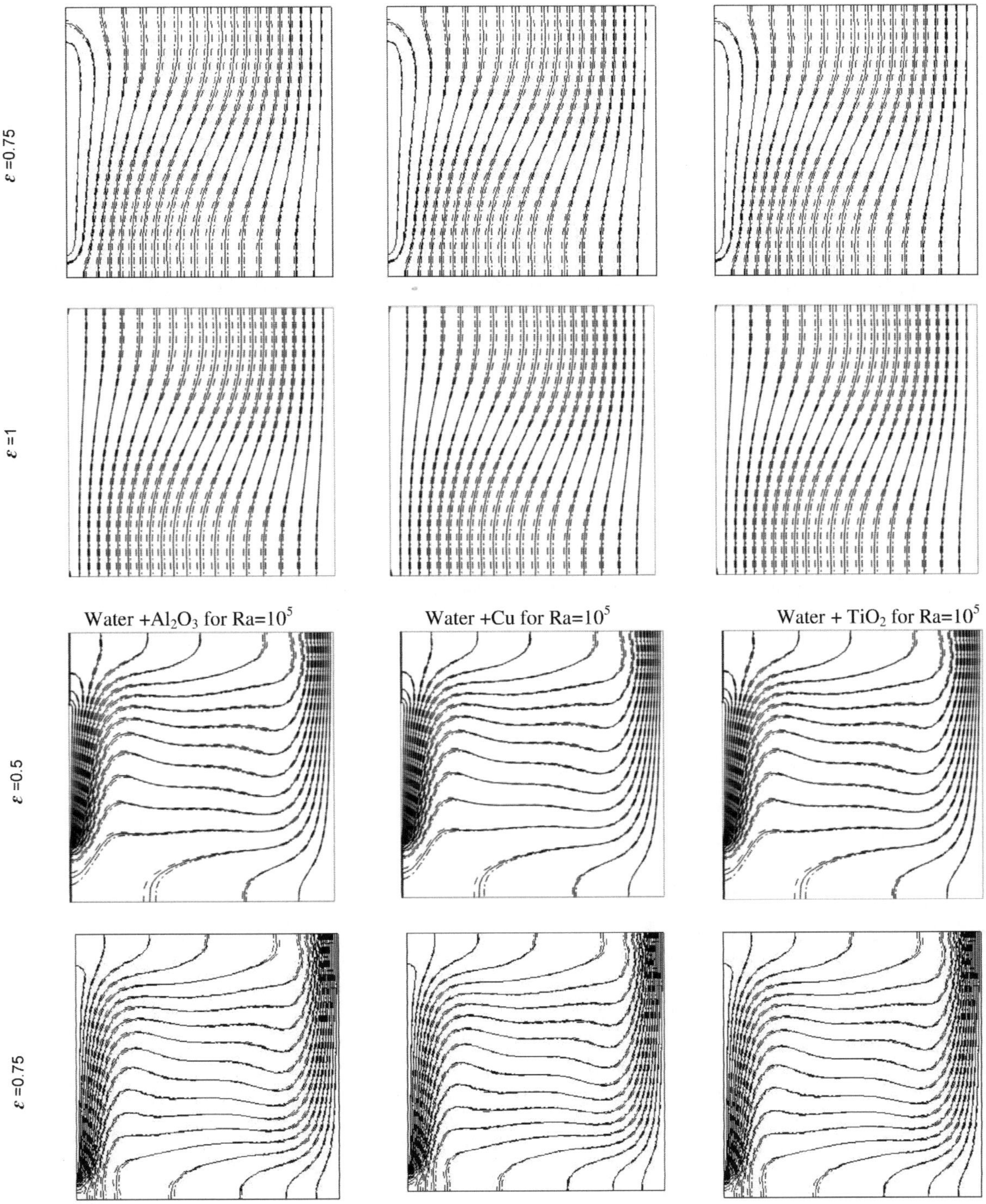

$\varepsilon = 1$

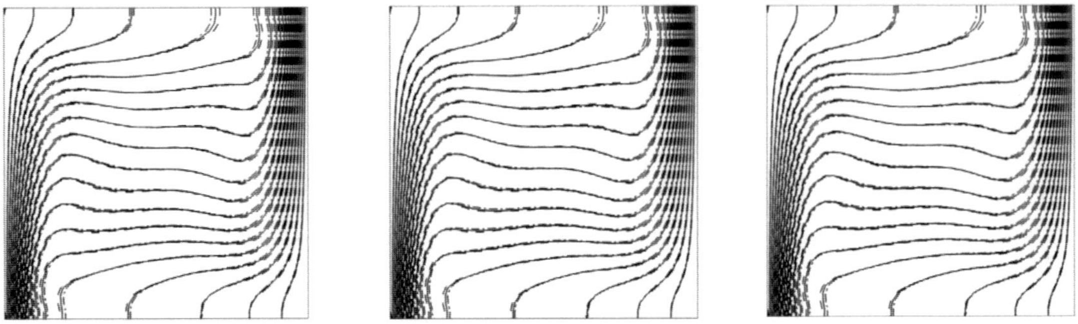

Figure 5. Isotherms for different nanofluids, Φ =4% (- · - · -), Φ =2% (——), and base fluid Φ =0%(– – –) at Ra = 10^3 and Ra = 10^5 and different ε.

As the hot and cold walls of the cavity are positioned in the X direction, the main vortex induced by the natural convection, is evidenced in the Z plan. Hence, the plan Z = 0.5 was chosen to present the profiles of horizontal velocity V(X) and vertical one V(Y). The comparison of the profiles corresponding to the three types of nanofluids is shown in Figure 6, for Ra = 10^3 and 10^5 and the volume fraction varying in the range (0. 0,05. 0,1).

We can notice from the figures that the nanofluid moves slower than pure water. It is also observed that the reduction of velocity of the nanofluid with the volume fraction is more accentuated for small values of Rayleigh number.

The difference of speed between the three types of nanofluids decreases with increasing Rayleigh number. Therefore the nanoparticles speeds approach the speed of pure water with increasing Ra.

When comparing the speed profiles between the different nanofluids only small differences are observed. However, the nanofluid Cu reaches higher speeds than Al_2O_3 and TiO_2.

Ra = 10^3 Ra = 10^5

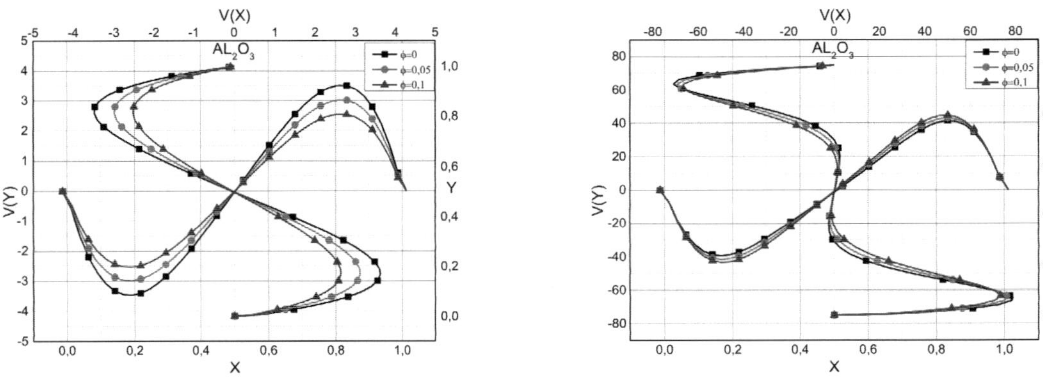

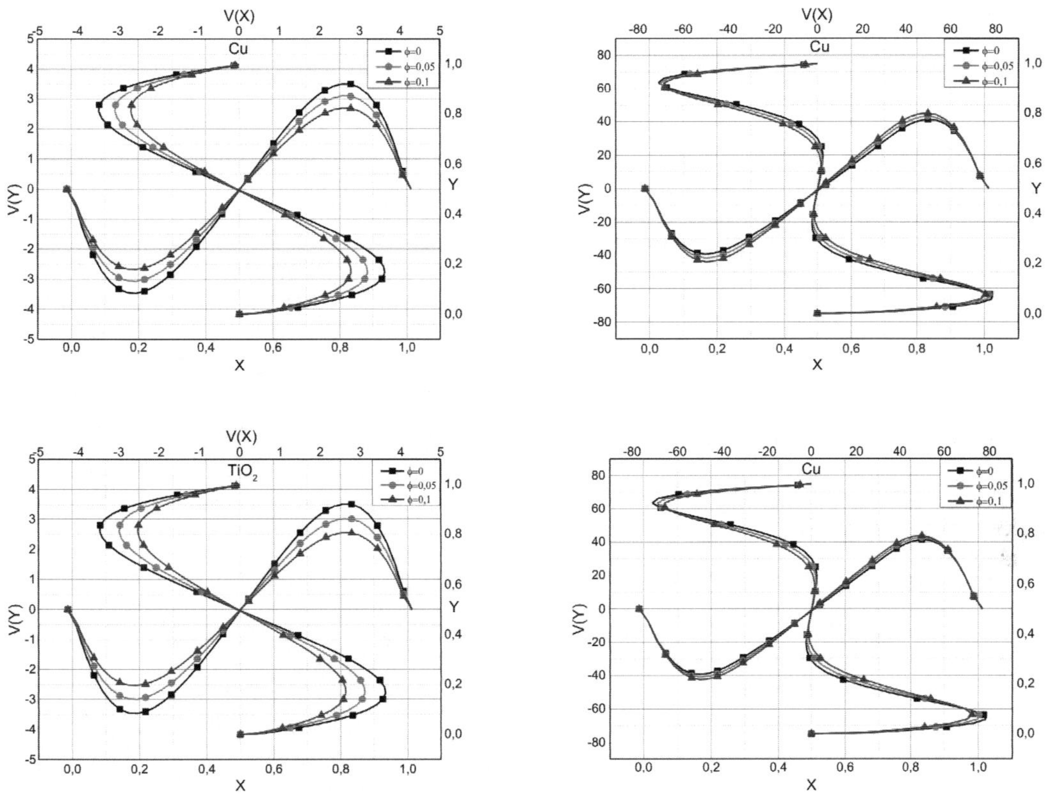

Figure 6. profiles velocity V (X) and V (Y) of the nanofluids in the central plane (Z = 0.5H) for different values of the volume fractions of nanoparticle Φ (Φ=0,0. 0,05 et 0,1), Ra=10^3, Ra=10^5 and ε =0.75.

The Tables 3 and figure 7 show the variation of the average Nusselt number (Nu_a) with the Rayleigh number and the heating section size, for various nanofluid type. For all considered nanofluids, the average Nusselt number increases with the Rayleigh number and the size of the heating portion ε. This increase is due to the improvement of the effective thermal conductivity of the nanofluid, when Ra increases. In addition, the heat transfer rate is maximum for copper (Cu), which has the largest thermal conductivity compared to TiO_2 and Al_2O_3.

Table 3. the variation of the Nusselt number as a function of Ra, ε and volume fraction.

Nu_a	ε =1			ε =0.75			ε =0.5		
Al_2O_3	Φ =0.00	Φ =0.02	Φ =0.04	Φ =0.00	Φ =0.02	Φ =0.04	Φ =0.00	Φ =0.02	Φ =0.04
10^3	1.080170	1.078004	1.177496	1.032905	1.078004	1.126788	0.8485183	0.8878419	0.9300382
10^4	2.121895	2.040174	2.164778	2.020745	2.040174	2.057273	1.544651	1.563698	1.581932
10^5	4.671789	4.422078	4.799010	4.352990	4.422078	4.486072	3.022230	3.078850	3.133210

Nu_a	ε =1			ε =0.75			ε =0.5		
Cu	**Φ =0.00**	**Φ =0.02**	**Φ =0.04**	**Φ =0.00**	**Φ =0.02**	**Φ =0.04**	**Φ =0.00**	**Φ =0.02**	**Φ =0.04**
10^3	1.080170	1.130739	1.053997	1.032905	1.081689	1.134029	0.8485183	0.8906935	0.9356880
10^4	2.121895	2.205745	1.949646	2.020745	2.052631	2.082778	1.544651	1.569689	1.595618
10^5	4.671789	4.769372	4.317255	4.352990	4.446068	4.535076	3.022230	3.093787	3.163945

Nu_a	ε =1			ε =0.75			ε =0.5		
TiO_2	**Φ =0.00**	**Φ =0.02**	**Φ =0.04**	**Φ =0.00**	**Φ =0.02**	**Φ =0.04**	**Φ =0.00**	**Φ =0.02**	**Φ =0.04**
10^3	1.080170	1.118732	1.160530	1.032905	1.070225	1.110517	0.8485183	0.8813071	0.9163874
10^4	2.121895	2.134932	2.142891	2.020745	2.029678	2.036500	1.544651	1.555184	1.565008
10^5	4.671789	4.711801	4.747671	4.352990	4.397934	4.438105	3.022230	3.061568	3.098687

(a)
Ra=10^3

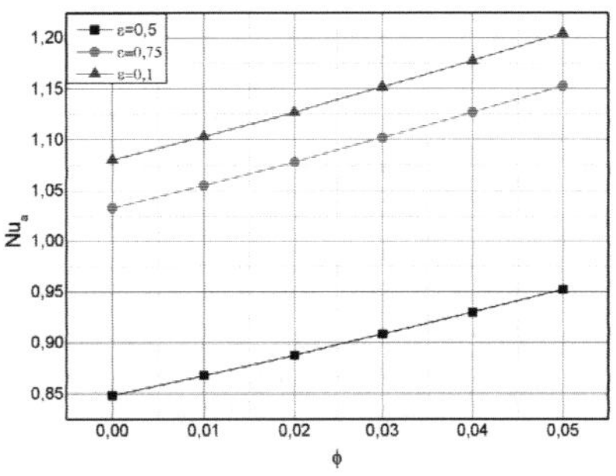

(b)
Ra=10^5

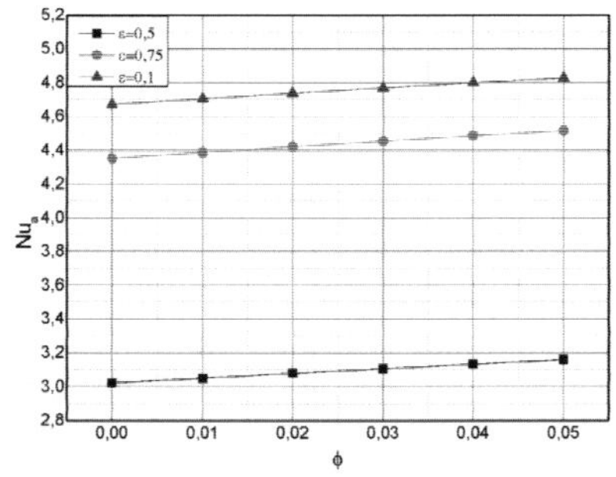

Figure 7. The variation of the average Nusselt number with Φ for Al_2O_3, different heating section sizes and different Ra

5. Conclusion

The study of natural convection in a partially heated three-dimensional cavity, filled with nanofluid was performed numerically. The results show the effect of the governing parameters, namely the Rayleigh number ($10^3 \leq Ra \leq 10^5$), the heating section size ε $0.5 \leq \varepsilon \leq 1$,the volume fraction Φ varying between 0 and 0.1 and nanofluid type Al_2O_3 , Cu and TiO_2..

It should be noted as well as the increase of the heating section size ε and Ra results in increased the amount of heat removed by the same nanofluide. Similarly, increasing the volume fraction causes an intensification of the flow and an increase in heat exchange.

References

[1] Choi SUS. Enhancing thermal conductivity of fluids with nanoparticles. Dev Appl Non Newtonian Flows 1995, vol 66, p.99–106.

[2] RAVNIK, J., ŠKERGET, L., et HRIBERŠEK, M. Analysis of three-dimensional natural convection of nanofluids by BEM. Engineering Analysis with Boundary Elements, 2010, vol. 34, no 12, p. 1018-1030.

[3] KHANAFER, Khalil et VAFAI, Kambiz. A critical synthesis of thermophysical characteristics of nanofluids. International Journal of Heat and Mass Transfer, 2011, vol. 54, no 19, p. 4410-4428.

[4] ÖĞÜT, Elif Büyük. Natural convection of water-based nanofluids in an inclined enclosure with a heat source. International Journal of Thermal Sciences, 2009, vol. 48, no 11, p. 2063-2073.

[5] H. C. Brinkman, The viscosity of concentrated suspensions and solution, Journal of Chemical Physics, 2005, vol.20, 65,p. 863-869.

[6] MAXWELL, James Clerk. A treatise on electricity and magnetism. Clarendon press, 1881.

[7] PATANKAR, Suhas. Numerical heat transfer and fluid flow. CRC press, 1980.

[8] PATANKAR, V. Suhas et D. SPALDING, Brian. A calculation procedure for heat, mass and momentum transfer in three-dimensional parabolic flows.International journal of heat and mass transfer, 1972, vol. 15, no 10, p. 1787-1806.

Comparative study of two CPV optical concentrators, using a Fresnel lens as primary optical element

S El Himer, S El-Yahyaoui, A Mechaqrane and A. Ahaitouf

Laboratory of Renewable Energies and Intelligent Systems, Faculty of Science and Technology, Sidi Mohamed Ben Abdellah University, PO. Box 2202 Fez Morocco

Email: sarah.elhimer@usmba.ac.ma

Abstract. In this work, the performances of two optimized reflective secondary optics elements a CPC (Compound Parabolic Concentrator) and a Cone for use in a CPV concentrator system are studied using ray-tracing simulation for the same primary optical element: a Fresnel lens. These optical elements are compared in terms of concentration, acceptance angle, exit angle and output light distribution. Our results show that the power distribution at the end of the concentrator is more uniform in the case of the cone. The optical efficiency is higher when the secondary element is placed at a distance $\mathbf{f} + \frac{\mathbf{R}}{\mathbf{tan\theta}}$ with f the focal length; R the input radius of the secondary optical element and θ the acceptance angle of the secondary optical element. Also, we found that the length and the input radius of each optical element decrease when the Fresnel lens diameter increases but the input radius of the CPC stills the larger. Finally, our calculation show that the CPC is longer than the cone while the Fresnel lens diameter is less than 200 mm and beyond this value both the cone and the CPC mostly present the same length.

1. Introduction

The main idea that motivates a CPV system is the cost reduction by using optical elements to focus light on solar cells thus reducing the used surface of expensive solar cells materials. The optical concentrators generally use a primary and a secondary element to respectively focus and homogenize the light on the receiver.

In this work, a compound parabolic concentrator (CPC) and a cone are used as secondary optical concentrator elements for CPV and are associated with a Fresnel lens as a primary stage. The achievements concern the performances comparisons in term of concentration, acceptance angle, output light distribution and the exit angle.

A part of this work concerns the sizing of each element starting from a targeted geometrical concentration ratio of 2000x. For the cone design, the number of inside reflections is also considered as a parameter.

Finally, the optimal position of the considered optical secondary concentrators facing the Fresnel lens is studied.

IOP Conf. Series: Materials Science and Engineering **186** (2017) 012033 doi:10.1088/1757-899X/186/1/012033

2. **Optical elements**

The Fresnel lens is characterized by its f_number, FN/# defined by: [1]

$$FN/\# = \frac{f}{d} \quad (1)$$

$$\tan\theta = \frac{d}{2f} \quad (2)$$

Where f, d and θ are, respectively, the focal length, the diameter and the incidence angle of the Fresnel lens.

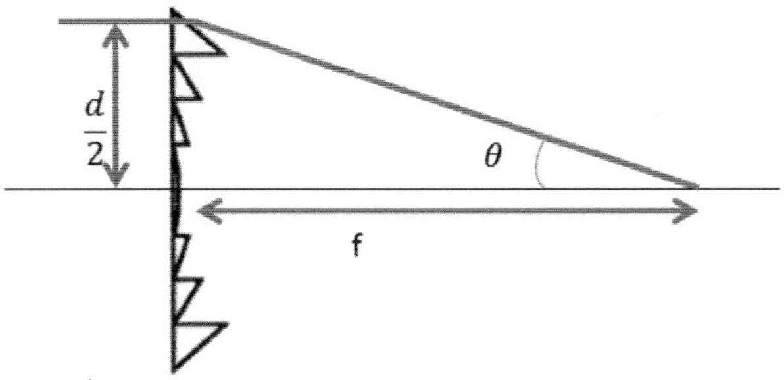

Figure 1. Illustration of the f_number for the Fresnel lens

A CPC is defined by its geometrical parameters (see fig.2): the exit and the entrance radiuses r_{cpc} and R_{cpc} respectively, the incidence angle, θ, the length L, and the focal length F_{cpc}. All these parameters are dependent and related by the following equations: [2]

$$R_{cpc} = \frac{r_{cpc}}{\sin\theta} \quad (3)$$

$$F_{cpc} = r_{cpc}(1 + \sin\theta) \quad (4)$$

$$L_{cpc} = \frac{R_{cpc} + r_{cpc}}{\tan\theta} \quad (5)$$

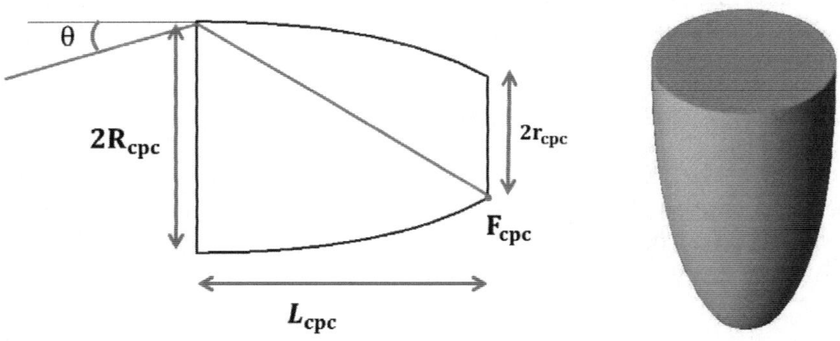

Figure 2. Geometrical parameters of the Compound Parabolic Concentrator (CPC)

The cone is defined byits output and input radiusesr_{cone}, R_{cone},the output and input anglesθ_f, θ, the angle of the cone α, its length $L_{cone}$and a_n is difference between the input and the output radius of the cone, [3]. The length of the cone depends on the inside number of ray reflections, n. All these parameters are given by the following equations [3, 4]:

$$\theta_f = \theta + 2n\alpha \tag{6}$$

$$R_{cone} = r_{cone} + \sum_{n=1}^{nmax} a_n \tag{7}$$

$$L_{cone} = \frac{\sum_{n=1}^{nmax} a_n}{\tan\alpha} = \frac{R_{cone} - r_{cone}}{\tan\alpha} \tag{8}$$

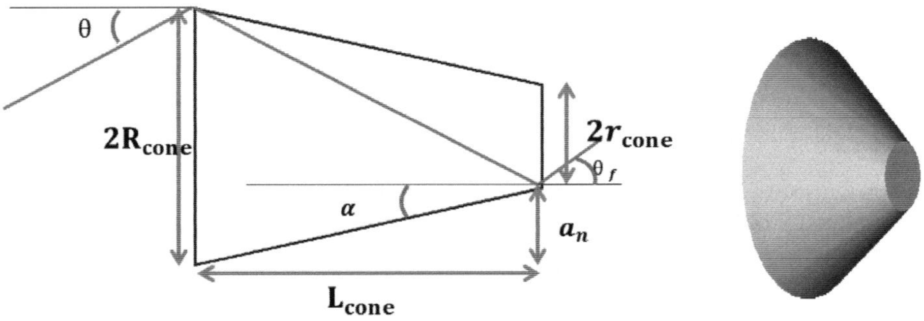

Figure 3. Geometrical parameters of the Conical Concentrator

3. RESULTS AND DISCUSSION

3.1. Influence of the Fresnel lens diameter on the sizes of the secondary elements

For the first comparison, we varied the Fresnel lens diameter from 50 mm to 500 mm to figure out its influence on the size, using the above equations (3-8). The number of reflections was firstly fixed to 4(n=4).

As shown on fig. 4 and fig. 5, both the length and the input radius of each optical element decrease when the Fresnel lens diameter increases but the input radius of the CPC is the larger.Concerning the length, we can notice that the CPC is longer than the cone as long as the Fresnel lens diameter is less than 200 mm otherwise the two lengths are approximately equal. It is important to note that we choose number of reflections to 4 because this number allows to a cone size almost near the size of the CPC.

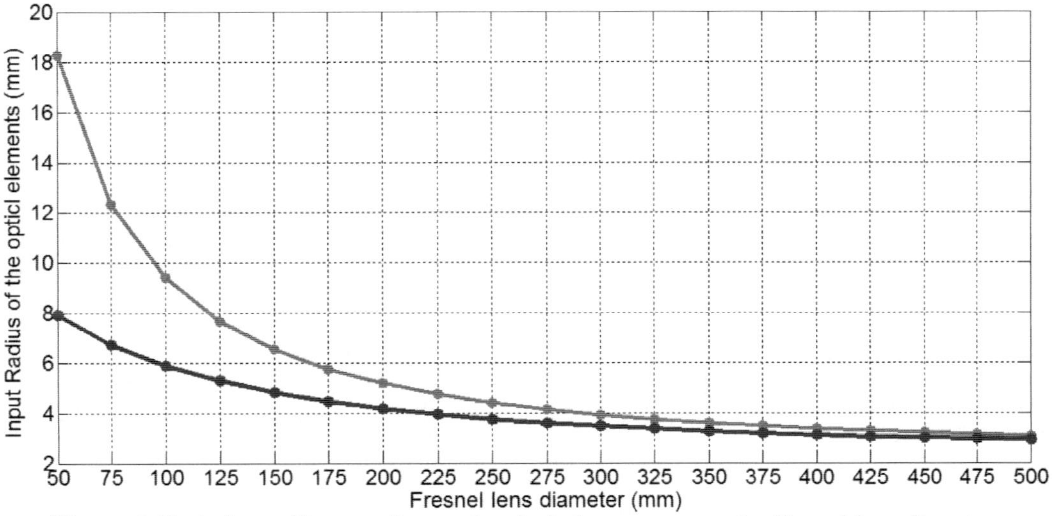

Figure 4. Variations of input radiuses of optical elements versus the Fresnel lens diameter

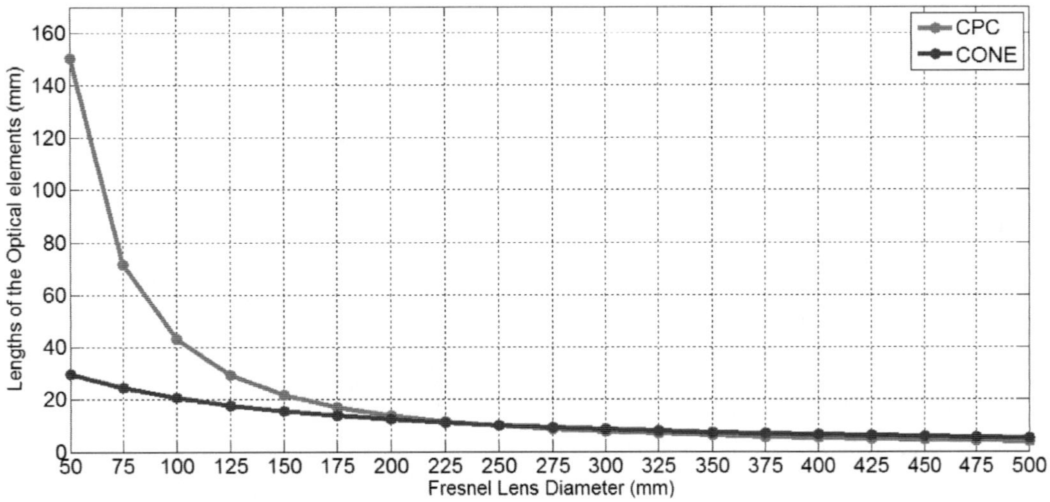

Figure 5. Variations on lengths of optical elements versus the Fresnel lens diameter

3.2. Design considerations

As a design example, we target a geometrical concentration ratio of 2000×. The exit radius of the two secondary optical elements is fixed to 5 mm, which is the classical size for commercial solar cells for CPV. The Fresnel lens diameter, calculated using the geometrical concentration ratio, is 223.6 mm and the acceptance angle of the two elements is fixed to 31.7°. Considering these parameters, we obtain for our secondary optical elements the sizes reported in table 1.

IOP Conf. Series: Materials Science and Engineering **186** (2017) 012033 doi:10.1088/1757-899X/186/1/012033

Table 1. Size of secondary optical elements (R, r and L) for the considered design example

	R (mm)	r *(mm)*	L (mm)
Cone	3.9469	2.5	11.3142
CPC	4.7656	2.5	11.7912

Finally, we studied optical efficiencies versus the distance of the secondary optical element from the Fresnel lens. The optical axe is set as a z-axis whose origin z = 0 coincides with the position of the Fresnel lens.
Three possible position of the secondary optical element are a priori possible without any optical losses. The first one, z, is defined by the focal length of the lens (fig. 6,a), the second and the third positions z_{max} and z_{min}, are respectively defined by the place where the incoming rays size fits with the input diameter of each element after and before the focal position of the lens (fig. 6, b and fig 6, c). These positions are given by:

$$z_{max} = f + \frac{R}{\tan\theta} \quad (9)$$

$$z_{min} = f - \frac{R}{\tan\theta} \quad (10)$$

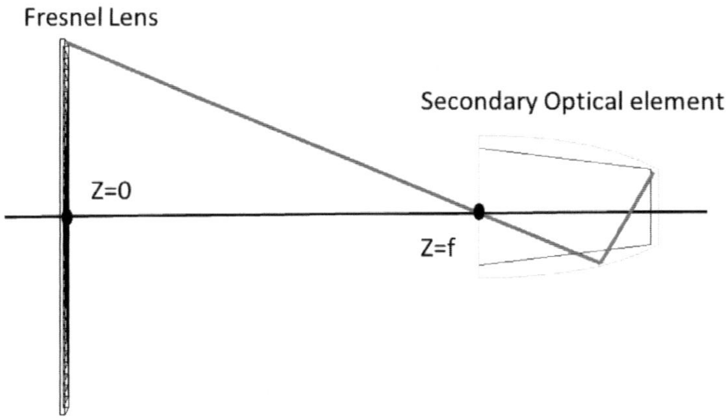

(a)

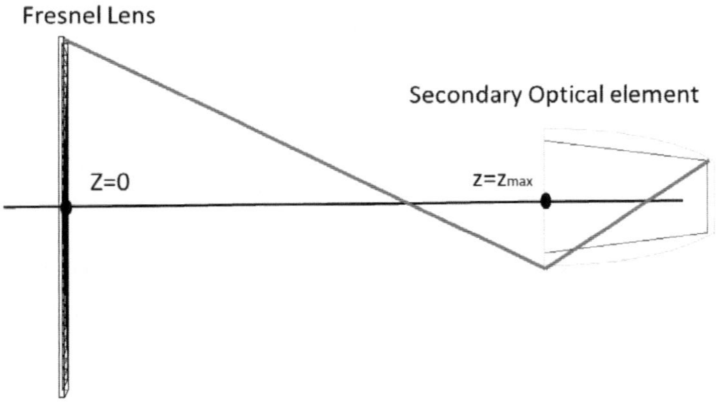

(b)

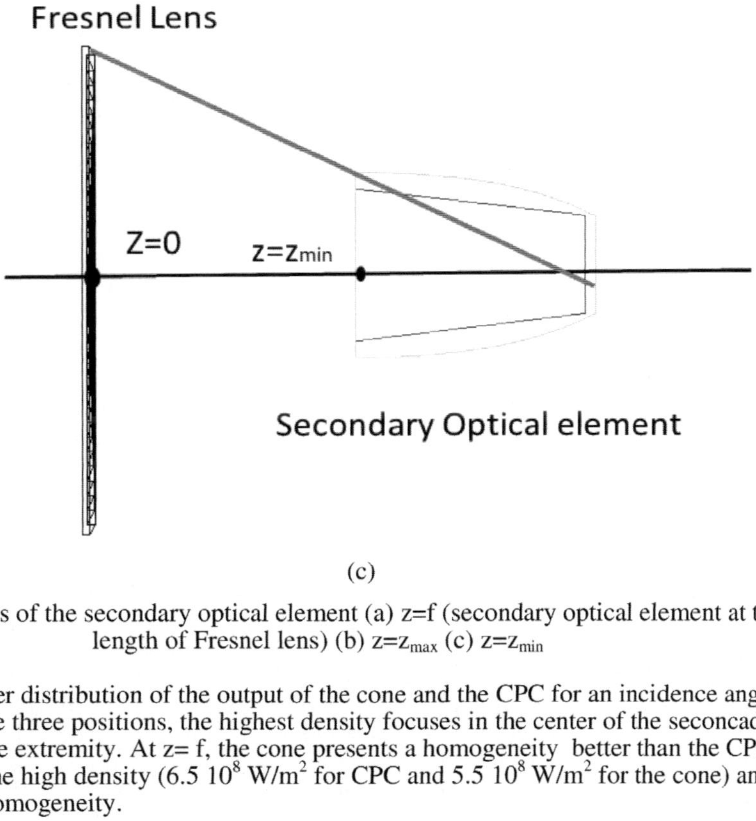

(c)

Figure 6. z-positions of the secondary optical element (a) z=f (secondary optical element at the focal length of Fresnel lens) (b) $z=z_{max}$ (c) $z=z_{min}$

Fig 7 shows the power distribution of the output of the cone and the CPC for an incidence angle of 0°. We notice that for the three positions, the highest density focuses in the center of the seconcadries exit, and it decreases in the extremity. At z= f, the cone presents a homogeneity better than the CPC. At $z= z_{max}$ presents the high density (6.5 10^8 W/m^2 for CPC and 5.5 10^8 W/m^2 for the cone) and the CPC has the worst homogeneity.

Table 2 resumes the values of the optical efficiency for the two concentrators with an incidence angle of 0°.

Table 2. The optical efficiency of secondary optical elements at 0°.

	z=f	$z=z_{max)}$	$z=z_{min}$
Cone	84.62%	85.64%	51.06%
CPC	85.033%	85.65%	49.59%

We note that the optical efficiency is higher when the secondary is in the position of $z=z_{max}$.

To calculate the optical concentration C, we use the following equation [5]:

$$\mathbf{C} = \mathbf{C_{opt}} \times \mathbf{C_{geo}} \qquad (11)$$

Where $\mathbf{C_{geo}} = \frac{\textbf{input aperture}}{\textbf{output aperture}}$is the geometrical concentration ratio and$\mathbf{C_{opt}} = \frac{\textbf{input rays}}{\textbf{output rays}}$.

From optical simulation and using the concentration equation (11), we calculate achieved concentrations ratios with several incidences angle from 0 to 2°.Results are presented on fig 8 andshow that the concentration decreases with the increment of incidence angle when the concentrator is placed at z or $z_{ma}x$ with a more pronounced decrease for the two studied elements. In this case, the concentration ratio at 0° is about 1700x. When the concentrator is placed atz_{min}, the concentration ratio increases to a maximum around 0.4° and then decreases. In this position ($z= z_{min}$)the achieved concentration ratio stills around 1000x indicating that this position is not optically efficient.

Z=f

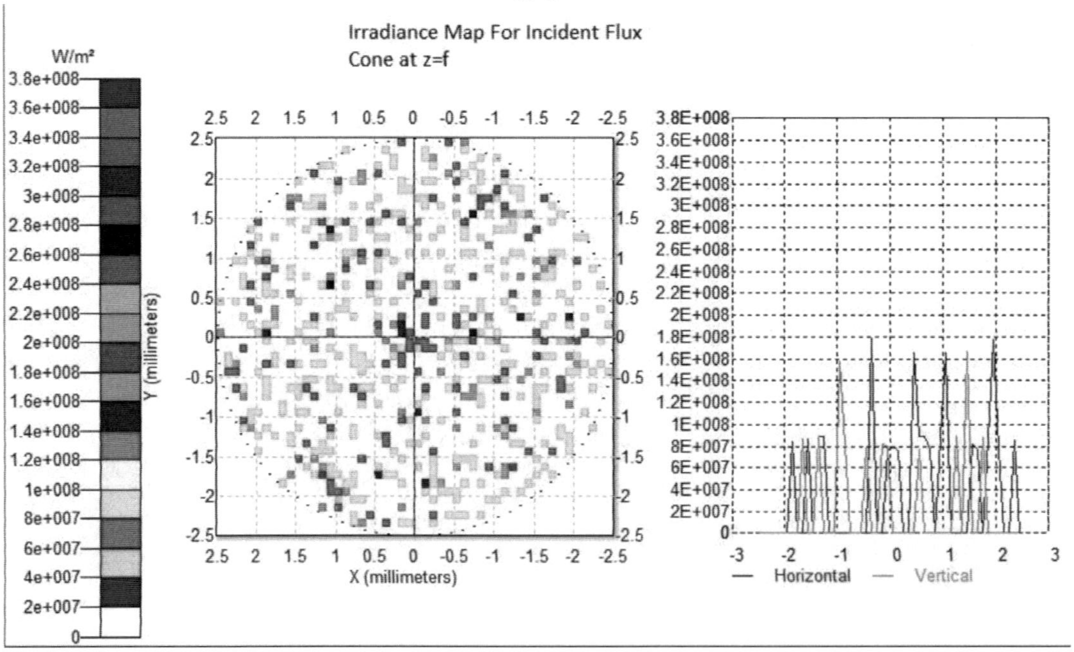

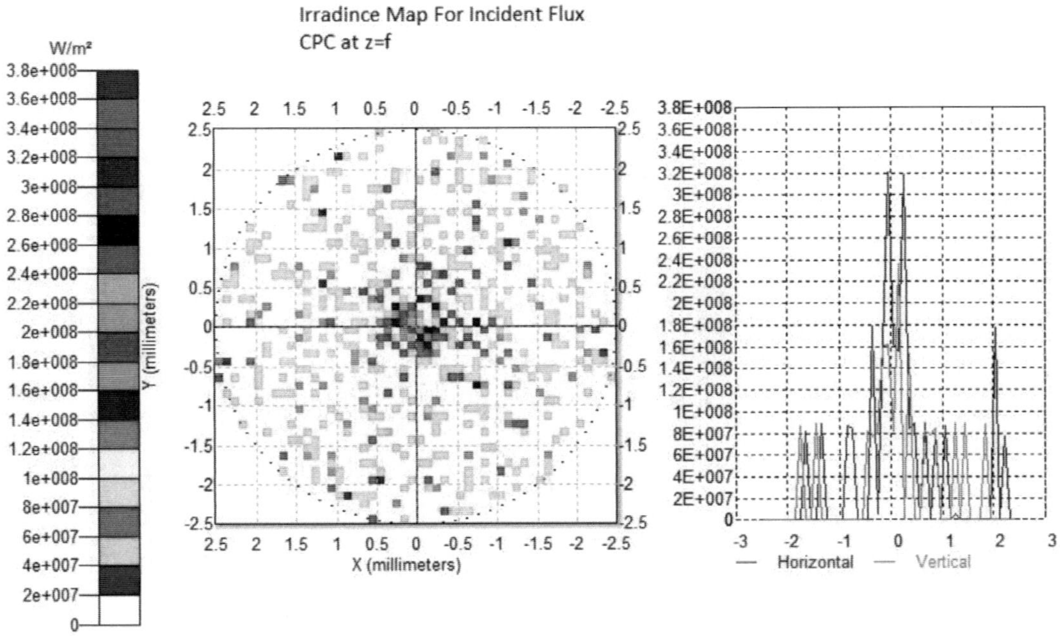
Irradince Map For Incident Flux
CPC at z=f
W/m²
Y (millimeters)
X (millimeters)
Horizontal
Vertical

(a)

$Z=z_{max}$

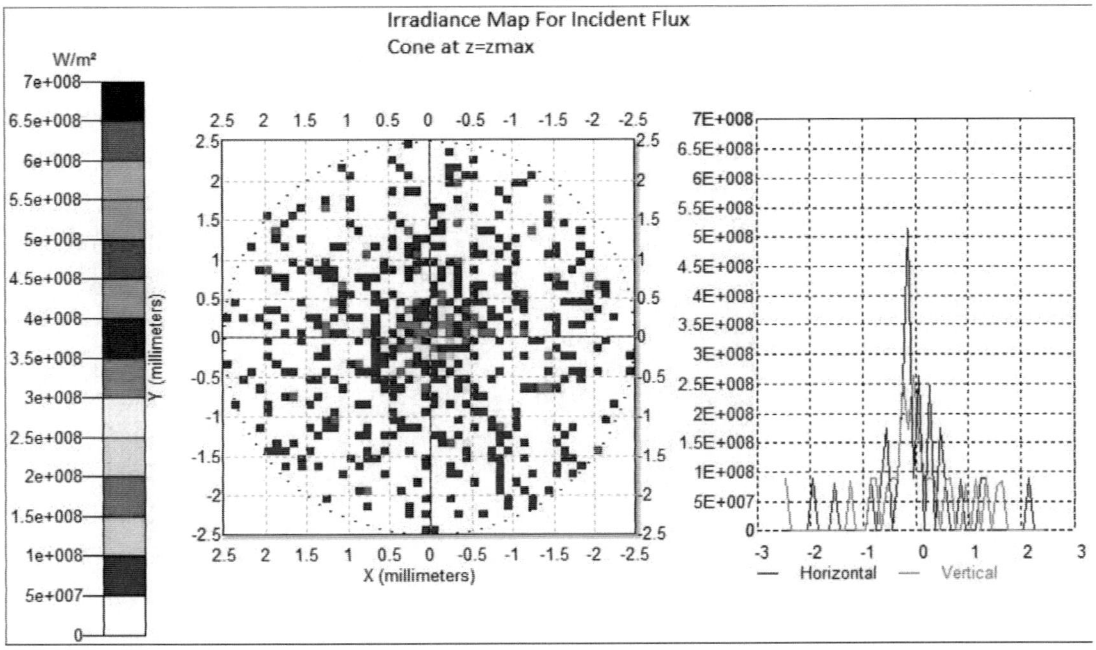
Irradiance Map For Incident Flux
Cone at z=zmax
W/m²
Y (millimeters)
X (millimeters)
Horizontal
Vertical

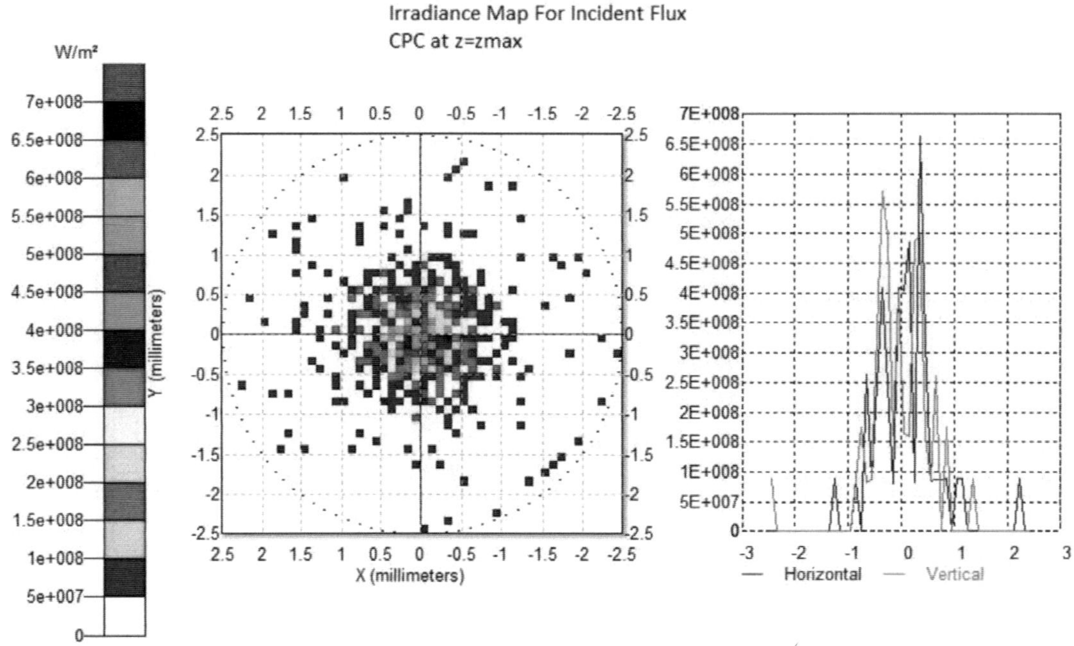
Irradiance Map For Incident Flux
CPC at z=zmax
W/m²
Y (millimeters)
X (millimeters)
Horizontal
Vertical

(b)

$Z{=}z_{min}$

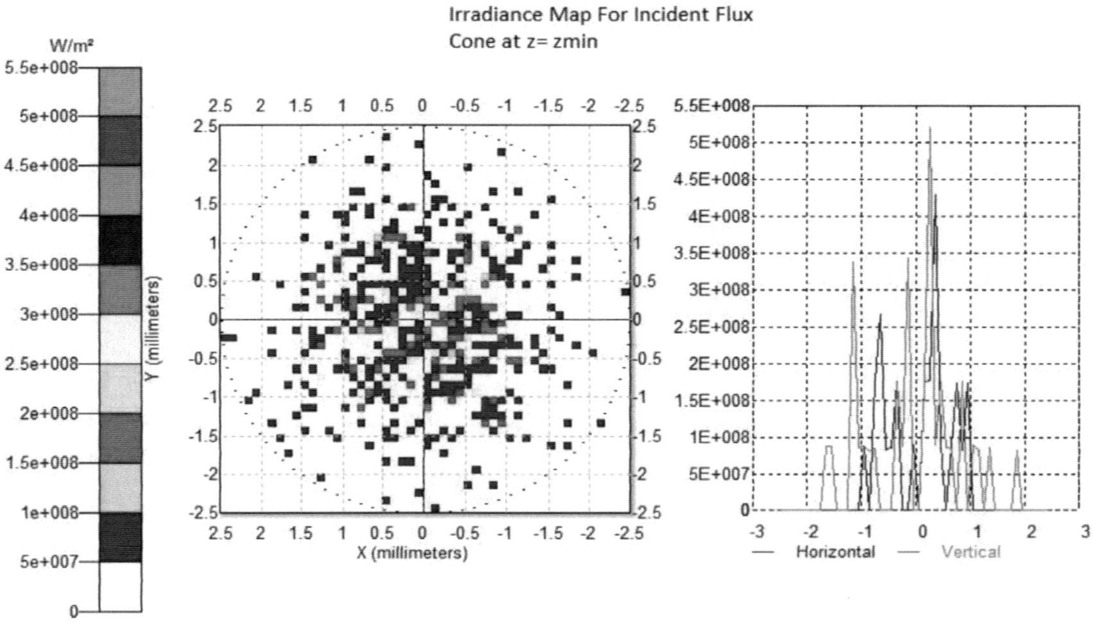
Irradiance Map For Incident Flux
Cone at z= zmin
W/m²
Y (millimeters)
X (millimeters)
Horizontal
Vertical

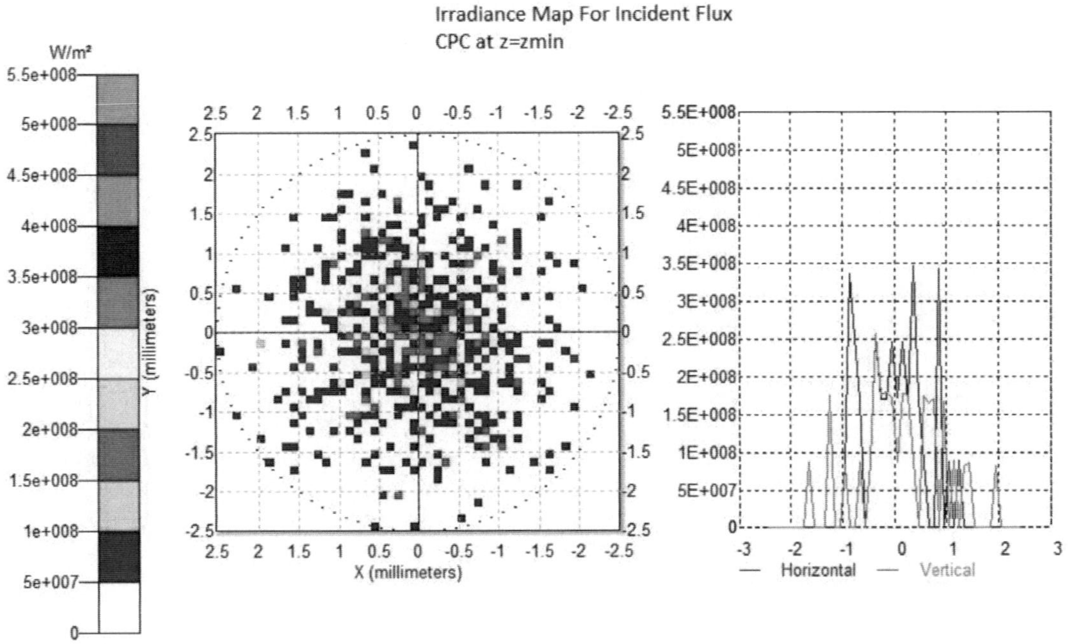

Figure 7. Example of the output energy flux mapping at exit of the cone and the CPC at(a) z=f (secondary optical element at the focal length of Fresnel lens) (b) $z=z_{max}$ (c) $z=z_{min}$

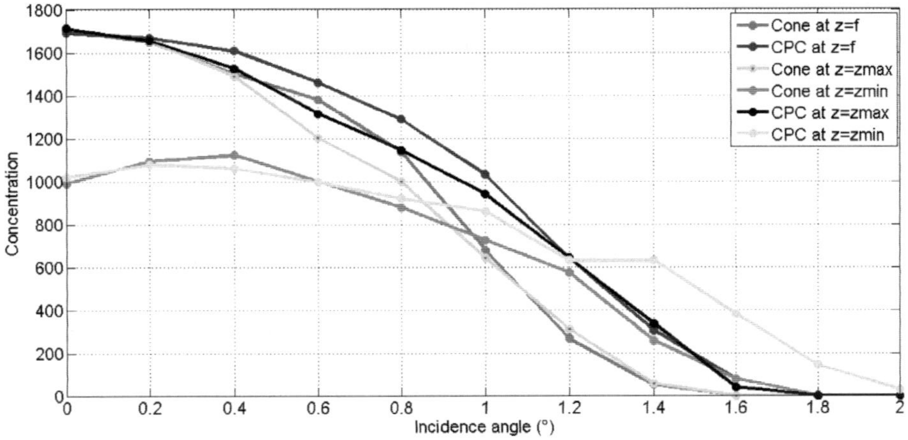

Figure 8. Evolution of the concentration versus incidence angle

4. Conclusion

In this work, we have presented a comparison of performances of two CPV optical concentrators, using a Fresnel lens as primary optical element. These elements are the CPC and the cone. Our results show that the length and the input radius of each optical element decrease when the Fresnel lens diameter increases but the input radius of the CPC is the larger. Concerning the length, we obtained that the CPC is longer than the cone as long as the Fresnel lens diameter is less than 200 mm.

Otherwise; the two lengths are approximately equal. Regarding concentration, and after studying the best placement of the secondary element, we noticed that it is enhanced in the case when the secondary optical element is a cone but when the angle of incidence increases, the concentration is higher in the case of the CPC placed at z_{max}.

Acknowledgments
This work is supported by the Research Institute for Solar Energy and Renewable Energies (IRESEN) in the framework of the LOuCOuM project

References

[1] Kiatgamolchai S. 2008, E.Chamni, *Theory and experiment of a two-dimensional cone concentrator for sunlight*, (*Solar Energy 82*).
[2] Languy F., C. Lenaerts , G. Loicq , T. Thibert , S. B. M. Coughenour, T. Stalcup, B. Wheelwright, A. Geary, K. Hammer, and R. Angel 2014, *Dish-based high concentration PV system with Köhler optics*, (*Opt. Express 22 S2 Suppl 2),* A211–A224
[3] Pan J., J. Huang , C. Wang b, H. Hong , Y. Liang 2011, *High concentration and homogenized Fresnel lens without secondary optics element*, (*Optics Communications)* 284 4283–4288.
[4] Williamson E.1952 , *Cone Channel Condenser Optics (the optical society of America*), 42,10.
[5] Andreev, V., V.D. Rumyantsev, and V.A. Grilikhes 1997, *Photovoltaic Conversion of Concentrated Sunlight*.: (*Wiley & Sons, Incorporated, John).* 308

IOP Conf. Series: Materials Science and Engineering **186** (2017) 012034 doi:10.1088/1757-899X/186/1/012034

Energy efficiency and recovery of heat lost in the Industrial Systems

S Mounkid and A Loukili

laboratory processes Renewable Energy and Environment, University Sidi Mohammed Ben Abdellah, Ecole Superieure de Technologie Fes, Morocco

Email: soukaina.mounkid@outlook.fr , aloukili203@gmail.com

Abstract. the economic importance of energy is manifested at all levels of farms, the demand for energy is today one of the major challenges of societies, it constitutes an indispensable element to any activity of production, it is for this, the industry has an interest to anticipate and invest in energy efficiency in order to gain competitiveness, this last represents a tremendous lever for performance and economy. The Energy diagnosis allows unveils the potential energy sinks and the discovery of the various sources of energy losses in a manufacturing process or in all system user of energy. Use with the effectiveness of the energy help the industry to meet the challenges of competitiveness.

1. Introduction

The mastery of the future energy demand and conditions of supply is a major political issue for the whole of the planet. It is made all the more difficult that humanity must face a double challenge. The satisfaction of the energy needs of a world population that will continue to grow during the next few decades [1], and with a large share aspires to a social and economic development based on an increased demand for energy. The one posed by the necessary reduction of emissions of greenhouse gases, inherent in the use of fossil fuels and emissions of other gaseous pollutants, which threaten the climatic balances environmental and of the earth. It is the issue of sustainable development, a concept in which the word development is not less important than the idea of responsibility vis-a-vis the future generations that involved the concept of sustainability, in its three dimensions economic, social and environmental.

An energy audit can lead to the discovery of various sources of energy lost in an installation. The consumption of this energy in the industrial sector is extremely concentrated and specific to each branch.

During the operation of a process of production or processing, the heat produced thanks to the energy provided is not used in full. A part of the heat is inevitably lost. It is because of this inevitability that we speak of "heat lost" [2] which can now be recovered. The recovery of the lost heat can respond to the needs of specific heat to the business or to the needs of heat from other companies, or more broadly, a territory, via a network of heat. It can also be converted into electricity for the internal or external use.

In different areas in particular the industrialists, the loss of energy is a major shortcoming in the optimization of its use. Its recovery is a very complex task; significant breakthroughs have been achieved in the development of new technologies, the purpose of which is to reduce its consumption and losses that is derived from it.

IOP Conf. Series: Materials Science and Engineering **186** (2017) 012034 doi:10.1088/1757-899X/186/1/012034

Knowing that such a loss cannot be avoided, it goes without saying that the innovation in its recovery is needed and that new technologies or techniques are to develop in this direction. The various possibilities of recovery of energy and the reduction of its consumption, lead to the reduction of its cost and the improvement of energy efficiency and to the improvement of the profitability of industries.
The energy optimizations on the processes represent a significant deposit that must be implemented early in the life of a project, in order to reduce in a visible manner the energy consumption and the associated costs, it is appropriate to practice for the management of the energy. These actions provide a profit on the Environmental Plan and on the economic plan. The system of management of the energy is a tool of pilotage allowing putting in place these good practices in an effective and sustainable manner.
Energy efficiency in the industry relies already on a number of clusters of excellence. The development and refinement of technologies allow exploiting a deposit of energy saving high in the industry .For optimal effectiveness, this search for energy performance must be based on a dual approach: global optimization of the performance of the systems and improving the performance of the equipment.

2. Energy consumption in the industry

As soon as the first oil shock in 1973 and 1982 [3], the world was re- accustomed to an energy little expensive, often wasted and not sufficiently valued. Its recent increase, the awareness of its scarcity inescapable and the climate impact point again the need of the mastery of the consumption of energy. With the increase in the long term the price of energies and its volatility, the energy performance of enterprises becomes a factor of competitiveness key: energy efficiency programs allow you to reduce the costs of production, the impact on the environment and identify new capabilities of investment for businesses.
The future of energy and economic development through the improvement of technologies in all areas of the use of energy. In particular, the energy consumption of the utilities and the transverse processes to all industrial sectors represents a significant share. The industry, because its motivation is before any economic, has been the "good pupil" in energy efficiency and important results have been achieved during the last thirty years [3]. However there is still the potential of energy efficiency very significant in this sector consuming 1/3 of energy. In effect, if the energy intensity in the industry has decreased from 2 to 3 per cent per year before 1990, she returned then to decay more moderate 1 per cent per year. The potential of economy of energy with the best existing technologies that is to say, outside technological breakthroughs, is evaluated in 15 to 20% of the consumption of the industry.
The energy consumption is divided into four areas: the transport, the building (residential and tertiary), agriculture, and industry. At the national level, the energy consumption in Morocco is divided in the following manner: 21% of the energy is consumed in the building, 44% in transport, 21% in industry, and 13% in agriculture, after ADEREE (National Agency for the development of renewable energy and energy efficiency).
Although the period of analysis is relatively short (10 years) of the consumption of energy by sector, we can certainly consider that the share of transport increases and that of the industry decreases on the long term.

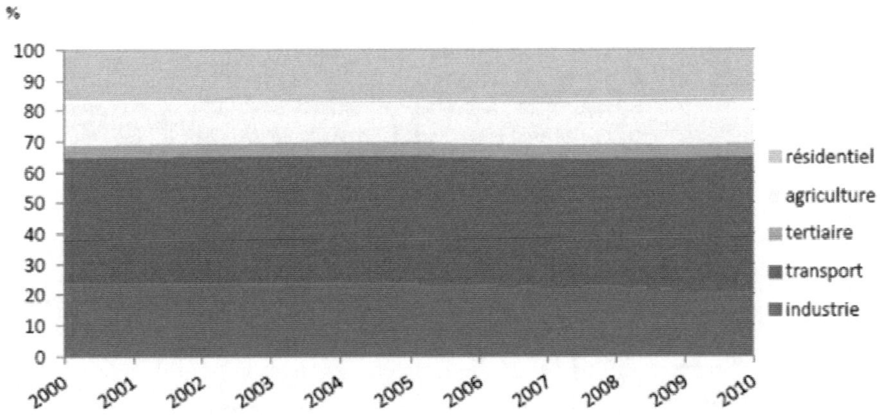

Figure 1. Energy consumption Moroccan 2000 -2010 in % (ADEREE)

In 2010, the energy consumption of industrial processes represented Moroccan 11.6 Mteq (mega tone of oil equivalent) [4], corresponding to 21% of energy consumption Moroccan, furnaces and kilns are processes with a high energy demand. Fossil fuels are the main source of energy consumed by the industry to produce heat.

The importance of energy efficiency has been underlined by many economic actors and international policies. The economies and the depletion of fossil resources are one of the driving forces in the development of processes more effective.

3. Energy analysis of the systems that produce heat

Regardless of the system studied, the objective is to provide the products a quantity of heat determined by:

- Its initial temperature;
- The temperature desired final;
- The mass of the products to heat;
- The specific heat of the products and, if there is place, the heats of transformation during the rise in temperature.

Either Qu this quantity of useful heat.

The establishment of the thermal balance of a system producing heat has for purpose to determine its performance and its specific consumption, i.e. the consumption reported to the unit produced useful. This implies the most precise knowledge as possible of all inflows and outflows and all the parameters of influence that are related to it.

The performance of a system producing the heat is the report of the useful energy to the energy Q that it must provide the system [3]:

With Qp the various thermal losses, it comes

$$Q = Qu + Qp \tag{1}$$

$$\eta = \frac{Q_u}{Q} = \frac{Q - Q_p}{Q} \tag{2}$$

The term Qp includes:

- The losses by the fumes Q_f ;
- The losses by the walls Qw ;
- The energy required to heat the walls and handling systems of the Qmv product;
- The losses by the openings of the oven Qe.

There is place to determine these losses to calculate the performance of the system.

This representation (Fig.1) gives a general idea on the system as well as the necessary parameters for the calculation of the losses and its performance.

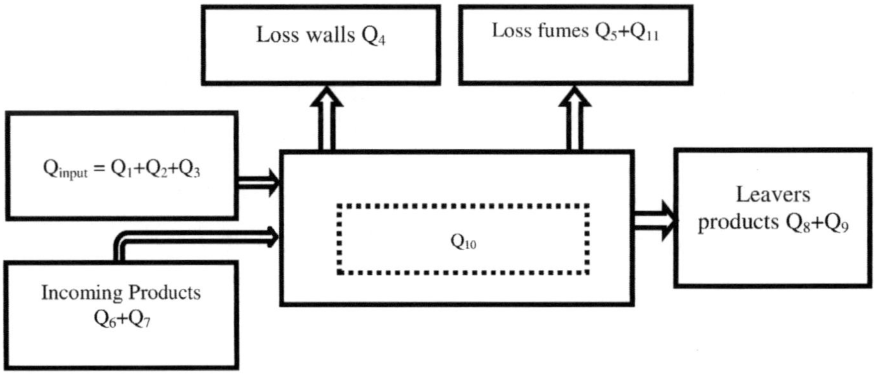

Figure 2. simplified diagram of a system for thermal balance sheet

Qinput = Quseful + Qoutput (3)
Qinput = Q1+Q2+Q3+ Q6+Q7 (4)
Qlost =Qfumes+Qwalls (5)
Qoutput= Qlost +Q8+Q9 (6)

NOMENCLATURE

- Q1 = m PCS : Heat of Combustion provided by the fuel (m=Mass of fuel, PCS (Kwh/kg), its gross calorific value);, sensitive heat provided by the fuel, (CG (Kwh/kg.k): thermal capacity of the mass fuel, TG its temperature);
- Q2 = m. Cg. Tg,sensible heat generated by the fuel (Cg (kwh / kg.K): specific heat capacity of the fuel, Tg its temperature.
- Q3 = Ma. Ca. Ta, Sensitive heat provided by the incoming air (includes the combustion air and gas introduced), (Ma: mass of air and the gas entering, Ca: their thermal capacity average mass and Ta their average temperature). heat is lost by the walls and the openings;
- Q_4: heat lost through the walls and the apertures;
- $Q_5 = M_f. C_f. T_f$: sensitive heat lost with the fumes, (M_f : mass of fumes, C_f their heat capacity, T_f their temperature);
- $Q_6 = M. C_c. T_1$, heat provided by the load, (M: its mass, C_c its Thermal capacity and mass T1: inlet temperature of the load and accessories);
- $Q_7 = M_d. C_d. T_1$: heat provided by the accessories or media handling, or accompanying products the load, (MD: mass of accessories and Cd their capacity mass thermal);
- $Q_8 = M. C_c. T_2$: heat is lost by the load, (T2: The temperature of the outgoing load and accessories);
- $Q_9 = M_d. C_d. T_2$: heat is lost by the accessories;
- $Q_{10} = M. C_r$: heat consumed during the treatment (useful heat), (C_r: enthalpy of KWH treatment /kg).
- Q11=latent heat lost by the fumes = M_f C_{11} (C_{11}: latent enthalpy of smoke increased the power fuel of some constituents rejected if applicable.

These equations allow us to calculate the losses that are related in part to the performance with which heat is produced and on the other hand to the amount of useful heat.
The energy analysis of the processes is based on the principle of conservation of energy, or more commonly designated by the first principle of thermodynamics. It indicates that the energy can neither

be created nor destroyed but retained when a transformation or a transfer (Lechatellier, (1884)). The two forms of energy found are the work Wand heat Q .

$$\Delta U + \Delta Ecin + \Delta Epot = W1{\rightarrow}2reçu + Q1{\rightarrow}2\ reçu \qquad (3)$$

The energy balance of the processes is based on the first principle of thermodynamics. It allows you to perform a quantitative analysis of the energy flows in a system thermodynamically open, but it is impossible to distinguish the quality of the different forms of energy.

4. Energy simulation of rotary kilns

The study of the thermal transfers occurring in a process can be performed from an energy modeling an open system. The rotary kilns are processes complex to model including the transportation of materials, fluid dynamics, and the heat transfer.

In the literature, a three-dimensional approach of the simulation of a rotary kiln has already been proposed for the heating of granular materials [5]. The study a torque A Modeling Computational Fluid Dynamics (CFD) with a DEM modeling (discrete element method). The method allows CFD modeling of the combustion reaction as well as the phenomena of turbulence is establishing in the gas phase of the oven. One of DEM allows a resolution of the solid phase including the transportation and the dynamic between each grain. The exploitation of the results shows that the average temperatures of the materials and gases are significantly overestimated by the model. The temperature of the gas in output on the process is normally close to 500K, but the temperature estimated by the model is closer to the 2300K. This result was expected because of the complexity of the process to simulate and shows the difficulty of putting in place a three-dimensional simulation of physical phenomena involved in the modeling of rotary kilns industrial.

The use of energy audits one-dimensional for the description of the phenomena of transfer is establishing in rotary furnaces allows you to simplify the modeling of rotary kilns and is already widely proposed in the literature. The models 1-D proposed allow studying the thermal transfers in rotary kilns of cement plant ([6] and [7]) or of the pyrolysis ([8], [9], [10]). Once the 1-D model is validated, it is easily possible to change some parameters to imagine changes in link with the optimization of energy efficiency.

Thus, the employment of energy balances allows one-dimensional considerably reduce the power and the computation time required to the modeling of rotary kilns. However, as any simulation, this methodology for modeling of the heat transfer is limited by the initial conditions, and limitations of the model.

The authors [6], present a model one-dimensional energy to simulate the key process is establishing in the bed of materials of cement kilns. In this model, an energy balance including transfers by conduction, convection and radiation between the different phases gas and solid has been resolved following the progress of the materials in the oven. This modeling 1-D has been validated by a comparison of temperature profiles obtained numerically to the experimental results. Then, the authors have used this model to study the influence of the operational parameters and design on the energy consumption of the oven. In addition to providing an effective tool in the simulation of performance of cement kilns, the model allows to provide a useful basis for the development of a full model in three dimensions.

Today, in the industry, the heat recovery is considered as a value added of the manufacturing process. The recovered heat can be used by the process, or in the production of additional services, such as the electrical energy or heat to a network of heat.

5. Recovery of energy at the level of industrial furnaces

Our study concerns particularly the recovery of energy in the ovens and industrial dryers or any other similar equipment, with a view to produce electrical energy. The solution envisaged remains specific

to each type of installation with the maximum care to respond to the call as long of news of the improvement of energy efficiency and the economy of energy.
The principle diagram of this recovery can be shown in Figure 3.

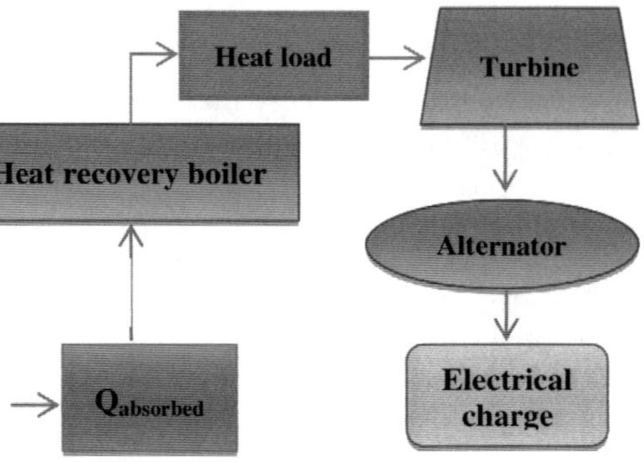

Figure 3. Presentation of the system for recovery of heat lost

The medium used for the recovery of the said energy lost, constitutes an innovative solution is optimal for respond to the problem. This solution was chosen from among several, in the course of the simulation study and the establishment of mathematical correlations linking the parameters of influence of the system and can describe the various systems in which it can be applied as well as the prediction of their thermal behavior.
As an example model, we considered a cement kiln on which a follow-up has been carried out in situ and in which we will have to confront our model.
The fundamental principles of the transfer of material and heat and the principles of thermodynamics applied to the ovens and industrial dryers are used to describe the operation of the system considered and the technical solution adopted as well as the means of conversion of the recovered energy into electricity.

6. Conclusion

The energy used in the industrial sector is considerable. This share can represent up to one quarter of the total energy consumption, the recovery or the upgrading of this energy allows increasing energy efficiency and reducing operating costs.
Energy efficiency is one of the major routes of improvement of the tools of production. By increasing the energy efficiency, we use less energy and we thereby reducing the emissions of greenhouse gases, thereby protecting the environment through the recovery of heat sources lost.

In fact, to recover and exploit the heat sources today lost on industrial processes requires a good knowledge of the industrial processes themselves, but also of technology upgrading to implement and know-how to select and dimension the best solutions, on the basis of technical criteria, economic and environmental. However a large part of these knowledge and know-how are today present in the academic world but broadcast little toward the industry, which, therefore, has a lot of difficulties to engage on the tracks of the heat recovery.

References

[1] Energie, National Center of Scientific Research.
[2] ADEME (Agency for the Environment and of the mastery of the ENERGIE), fatal heat industrial.
[3] Laurent LEVACHER1, Denis CLODIC, François MARECHAL, energy efficiency for the industry: existing technologies to innovative solutions
[4] The Energy Sector Key Figures Year 2012 provisional data (Department of Energy Mines of the water and the environment).

[5] Hobbs. A., 2009. Simulation of year aggregate dryer using coupled CFD DEM and methods. International Journal of Computational Fluid Dynamics 23(2), 199-207.
[6] Mujumdar, K.S., Ranade, V.V., 2006. Simulation of rotary cement kiln using a onedimensional model. Chemical Engineering and Design 83(A3), 165-177
[7] Chen. H, 2014. Modeling and Simulation of cement clinkering process with compact internal burning of carbon. Applied Thermal Energy 27, 1345-1351.
[8] Perron. J., Nguyen. H.T., Bui. R.T., 1992. Modeling of a calcination furnace of petroleum coke: i model. Canada Journal of Chemical Engineering 70, 1108-1119.
[9] Pattison, F., Lebas, E., Hanrot, F., Ablitzer, D., Houzelot, J.L. 2000. Coal pyrolysis in a Rotary Kiln: Part II overall model of the furnace. Metallurgical and Materials Transactions 31 (b), 391-403
[10] Li, S-Q., MA, The-B., WAN, W., Yao, Q., 2005. A mathematical model of heat transfer a rotary kiln thermo-reactor. Chem. Eng. Technol 28 (12) , 1480-1489

IOP Conf. Series: Materials Science and Engineering **186** (2017) 012035 doi:10.1088/1757-899X/186/1/012035

Thermophysical and mechanical characterization of clay bricks reinforced by alfa or straw fibers

Y Elhamdouni[1], A Khabbazi[1], C Benayad[1], S Mounir[1] and A Dadi[2]

[1]Université Med V-Agdal, Ecole Mohamedia d'Ingénieurs, Laboratoire Énergétique, Matériaux et Environnement, 227 Avenue Prince héritier EST- Salé, Maroc.
[2]Institut universitaire des sciences et techniques, Abeche, Tchad .

E-mail: yassineelhamdouni@gmail.com

Abstract. This work is part of the valuation of local materials such as clay (earth), alfa fiber and straw fiber. The goal is to use these materials as bricks in rural construction. These materials are abundant, natural, and renewable. The objective of this work is to study the thermal and mechanical behavior of a new material by mixing clay (chosen as the binder) with different mass percentages of alfa fiber (0.5%, 1%, 2%, 3%, 4%), and to compare these results with those of materials often used in the construction of individual houses in rural Morocco (clay + straw). The results obtained prove to us that using straw fibers can reduce the thermal conductivity compared to alfa fiber, which allows to have energy savings of 2% to 7%. By against, alfa fibers can improve the mechanical behavior of clay-based materials when compared to the clay + straw material (an increase of 8% to 17% in the tractive resistance by bending and 6% to 18% for compression resistance). These results also specify the optimal usage conditions of these fibers (alfa and straw) in the clay bricks.

1. Introduction

Because of their abundance and thermal insulation, clay [1] and straw fibers [2] are the two most used materials in the construction of rural houses in Morocco. However, there are other local materials, such as alfa fiber[3] , that are abundant in the northern region of Morocco and are unused in construction, despite their benefits and the in prices which are much less expensive than straw fibers.

The manual and visual comparison between the Straw fibers and the fibers alfa has led us to point out that the rod of alfa is more hard than that of straw and that the surface of the stems of alfa is rough [4] (Figure1) compared to the surface of the stems of straw which is thin , This will allow to create a very good connection between clay and fiber alfa. This adhesion will reduce the voids and consequently improve mechanical behavior.

By against, straw fibers contain more voids inside [2] than alfa fibers (Figure 2). This which will favor the thermal behavior of material clay + straw relative to alfa + clay material.

Based on these remark of departure, we began to work on both (clay + esparto and clay + straw) materials.

To characterize composite materials , we based primarily on the method of the plan hot in permanent regime [5] [6] to determine the thermal conductivity and the compression and tensile tests [7] [8] to get an idea about the mechanical behavior of these two materials.

The final objective is twofold: firstly valorize local materials and also lower energy expenses (cooling and heating), thus contributing to a sustainable environment.

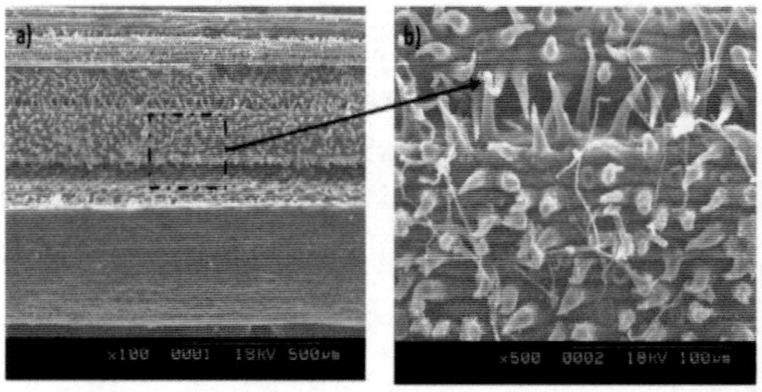

Figure1. Image MEB of an alfa fiber surface and enlargement

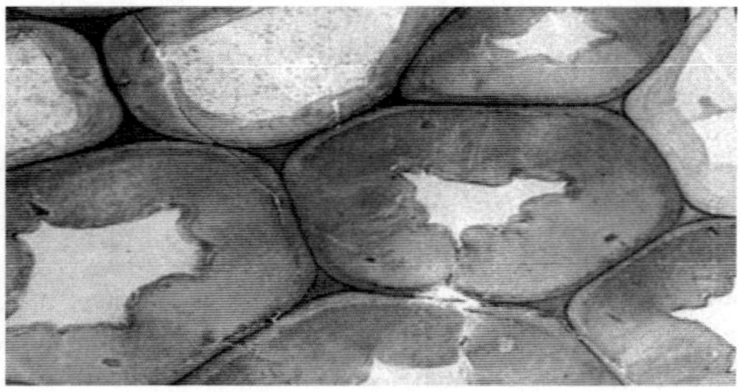

Figure 2. Straw bundle , cross section, transmission electron microscopy(TEM)

2. Materials used

2.1 Clay (earth)

In this study, we are using a clay soil from the Ksar-Elkebir area (northern Morocco), whose chemical analyses are given in Table 1.

Table1. Chemical analysis of used clay.

	Constituents											
	SiO2	Al2O3	Fe2O3	MnO	MgO	CaO	Na2O	K2O	TiO2	P2O5	P.F	Total
Percentages(%)	45,79	15,68	12,83	0,17	6,26	9,6	3,54	1,39	2,84	0,6	0,31	99,01

2.2 Alfa and Straw

The Straw fibers and alfa(Figure 3a,3b) are retained for reasons of availability and economy .the physical characteristics of these two fibers are shown in table 2:

Table 2. Physical characteristics of the fiber used physical

Characteristics	Unit	Alfa fibers	Straw fibers
apparent density	Kg/m^3	140	215
Diameter	mm	0.9-1.2	1-4
Length	mm	10-50	10-50

2.3 Mixing water

After several tries of estimate of the quantity of necessary water to thin the pate of clay and have a normal consistency, we found that the quantity of water must prove following relation:

$$\frac{W}{C}=0.20 \quad (1)$$

With W and C are respectively the quantity of water (g) and the quantity of clay dryness (g)

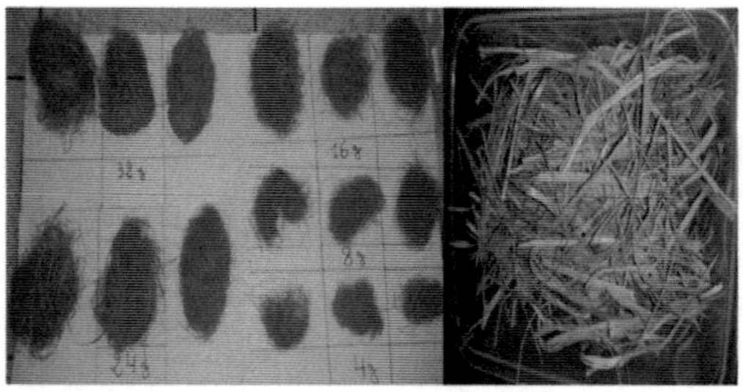

Figure 2a. Pictures of alfa used

Figure 2b. Pictures of straw used

2.4 Preparation of samples

The mixtures were done manually in the salt LEME laboratory. The amount of water added to the mixture was selected in order to get a normal consistency. We took a quantity of clay and varied the fiber mass percentage (0% – 0.5% – 1% – 2% – 3% – 4%).

There are three types of samples used in this study:

- Parallelepiped dimensions (11cmx11cmx3cm) to perform the thermal tests (Figure 3).
- Cylindrical dimensions (11cmx22cm) to perform compression tests according to NF P18-406 (Figure 4).
- Prismatic dimensions (4cmx4cmx16cm) to perform the Tensile Test by bending according to NF P18-40 (Figure 5).

Samples are then kept in a etuve in a temperature of 60°C. They is dried up to a constant mass. Then the mass dried being measured and then they were wrapped in bags of plastic has at the end that they support a rate of humidity close to zero.

Figure 3. Example of samples parallelepipeds.

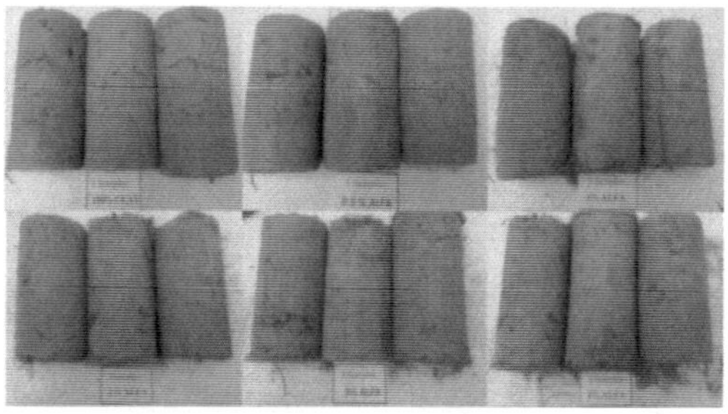

Figure 4. Examples of cylindrical samples

Figure 5. Examples of prismatic samples.

3. Results and interpretation

3.1 Thermal Conductivity by asymmetrical method in permanent regime

Figure 6 shows the evolution of the two thermal conductivities depending on the fiber percentages.

The thermal conductivities tend to decrease as a function of the percentage of fibers, this is perfectly logical, more than the percentage of fibers is high, more than the material contains pores which are the cause of this decrease.
We also point out that the thermal conductivity of our material (clay+paille) is always lower than that of clay with alfa (figure 6). This is due to the shape of the stems of the straw (figure 1) which allows to store a large quantity of empty has the inside . The presence of straw fibers improves so the thermal insulation of the samples compared to alfa fiber.
The measurement error was calculated by using the method three times for each sample, followed by taking the average of these three trials (it does not exceed 1%).

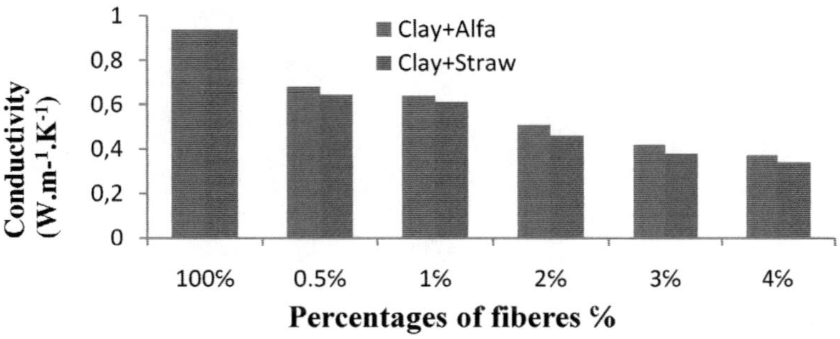

Figure 6. Evolution of thermal conductivity based on the percentages of fibers

To determine the economy of energy which can bring the straw fiber compared to fiber alfa. We will consider 2 exterior walls formed by (the material clay + alfa and the other in clay + straw) having the same thickness and are subjected to the same temperature gradient, We can be deduced from the report of the flow of heat which passes through two of these walls:

$$\frac{\phi_{clay+straw}}{\phi_{clay+alfa}} = \frac{\lambda_{clay+straw}}{\lambda_{clay+alfa}} \quad (2)$$

This allows us to calculate the energy savings that are gained by the (clay + straw) material compared to the material (clay + alfa):

$$Energie_{saving} = 100x\left(1 - \frac{\emptyset_{clay+straw}}{\emptyset_{clay+alfa}}\right) \quad (3)$$

We apply the relations (2) and (3) to the different samples, we get the energy savings based on the fiber percentages (Table 3).

Table 3. Energy savings of clay + straw material compared to material clay + alfa

Samples	Energy savings (%)
EP1: 0.5% Straw	5.71
EP2 : 1% Straw	4.37
EP3: 2% Straw	9.43
EP4 : 3% Straw	9.78
EP5 : 4% Straw	8.60

Based on Table 3, we can conclude that the use of material (clay+ straw) like bricks can give an energy saving of 4% to 9% compared to material (clay+ alfa).

3.2 Tensile test by bending

The results of the bending tension tests are shown in Table 4 and Figure 7. By breaking the specimen, the resistance to tension is determined. The resistance to tension is the average resistances obtained from at least three specimens.

Figure 7 shows two areas. In 0% to 2% fiber mass, we observed an increase in the tensile strength by bending, this increase is due to the role played by the fibers (alfa fiber resists well tractive effort and also a good adhesion between fibers and clay). From 2% fiber mass, the tensile strength by bending starts to decrease, because of the adhesion between the fibers and the clay paste, which becomes weak because of the increased fiber percentage and the reduced binding to the clay

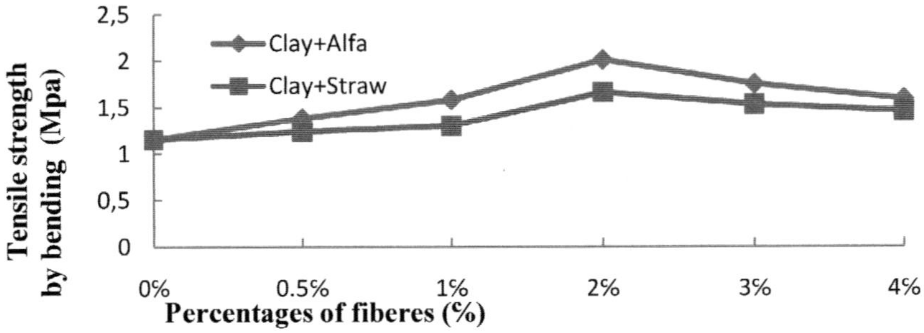

Figure 7. Evolution of the tensile strength by bending of samples in function of the mass percentages of fibers.

In addition, the clay + alfa material has a tension resistance that is greater than that of the straw +clay (Table 4). This is due to the hardness of the alfa fiber rod compared to that of the straw fiber.

Alfa fiber is better than straw in terms of mechanical behavior (the difference between the two resistances can reach 17.72%).

Table 4 . The values of the tensile strength by bending of the various samples.

Samples	Average tensile Strength of clay+ alfa(Mpa)	Average tensile Strength of clay+ straw (Mpa)	Gap to tensile strength of clay+alfa compared to that of clay +straw(%)
ER :100%clay	1.15	1.15	-
1% fibre	1.58	1.3	17.72
2% fibre	2.01	1.66	17.41
3% fibre	1.75	1.53	12.57
4% fibre	1.59	1.46	8.17

3.3 Compression test

The results obtained from compression tests on different mixes are shown in Table 5. The compressive strength is an average of resistances obtained from three samples. Overall, these resistances, in comparison to concrete, remain low and in the order from 2.35 Mpa to 4.63 Mpa.

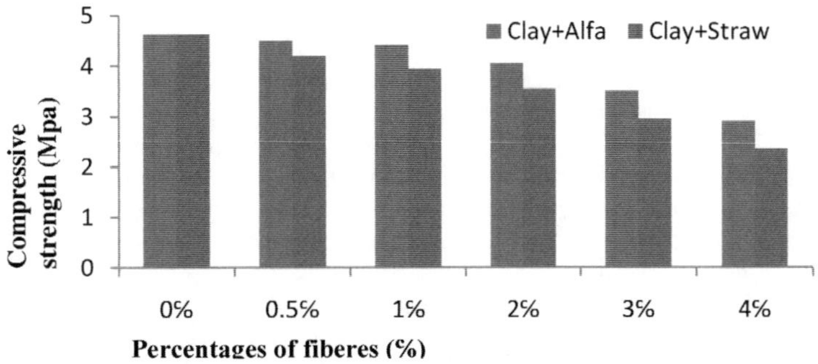

Figure 8. Progression of the resistance to compression based on fiber mass percentages.

According to Figure 8, the increase of fiber percentages in clay samples has a negative influence on the compressive strength. Also, the samples containing alfa fibers have higher compressive strength compared to clay with straw (Table 5). This is due to the rough surface of the alfa fiber which guarantees a better adherence to the clay, and which reduces the voids between the alfa fiber and the clay.

Table 5. The compressive strength of the two materials: clay + alfa and clay +straw

Samples	Average value of the compressive strength of clay+ straw (Mpa)	Average value of the compressive strength of clay + straw (Mpa)	Gap of the compressive strength of clay + esparto compared to that of clay +straw (%)
100% clay	4,63	4,63	-
0.5% fibre	4,5	4,2	6.66
1% fibre	4,42	3,95	10.63
2% fibre	4,05	3,54	12.59
3% fibre	3,5	2,95	15.71
4% fibre	2,9	2,35	18.96

4. Conclusions

The article is a comparative study of the thermo-mechanical properties of two materials: clay + alfa and clay + straw. We have shown that increasing the percentages of fibers (alfa or straw) improves the thermal and mechanical properties of both materials (a decrease in thermal conductivity with an increase in the tensile strength by bending), as long as the percentage of fibers does not exceed 2%.
The fiber percentage of 2% remains the optimal condition for using these fibers in clay bricks.
In others, we found that straw fiber and clay materials can bring energy savings of 4% to 9% compared to that of the material (clay + alfa). However, the material alfa+ clay can increase the mechanical resistance by 8% to 17% for the tractive resistance by bending and 6% to 18% for the compressive strength compared to that of straw fiber.
This study provides indices for the selection of local materials in rural construction in northern Morocco. However, other criteria such as thermal effusivity, thermal diffusivity, and mass heat are needed for the final choice of materials, and also the impact of rain water on the thermal and mechanical properties of the two composites to control durability.

References

[1] Soumia Mounir, Youssef Maaloufa, Abou bakr Cherki, Abdelhamid Khabbazi 2014 Thermal properties of the composite material clay/granular cork *Construction and Building Materials* **70** 183-190.

[2] Ahmat-Charfadine Mahamat, Mahamat Barka, Abakar Mahamat Tahir, Malloum Soultan ,Salif Gaye, Aboubakar Cheikh Beye 2015 Thermo-Physical Properties of Local Materials Used in the Construction of Chad *American Journal of Civil Engineering* **3(5)**: 140-149

IOP Conf. Series: Materials Science and Engineering **186** (2017) 012035 doi:10.1088/1757-899X/186/1/012035

[3] Mohamed DALLEL, thése de doctorat : Caractérisation physico-chimique de la fibre alfa,2012.
[4] Mohamed Dallel 2012 thése de doctorat Caractérisation physico-chimique de la fibre alfa , page 70,Université de Haute Alsace,Laboratoire de Physique et Mécanique Textiles (LPMT).
[5] Soumia Mounir, Abdelhamid Khabbazi, Asmae Khaldoun, Youssef Maaloufa,Yassine El Hamdouni 2015 Thermal inertia and thermal properties of the composite material Clay–wool *Sustainable Cities and Society* **19** 191-199.
[6] Abdallah Dadi Mahamat, Oumar Idriss Hamid, Malloum Soultan, Mahamoud Youssouf Khayal, Yassine Elhamdouni, Mohammed Garoum and Salif Gaye 2015 *Research Journal of Applied Sciences, Engineering and Technology* **10(4)**: 464-470,.
[7] S.Gaye 2001 Caractérisation des propriétés mécanique, acoustiques et thermiques de matériaux locaux de construction, Thèse d'Etat Energétique, FST/UCAD-Sénégal, 41-79.
[8] Emmanuel Ouedraogo and al 2015 Mechanical and Thermophysical Properties of Cement and/or Paper (Cellulose) Stabilized Compressed Clay Bricks *Journal of Materials and* Engineering Structures 2 68–76.

Mesoscopic Study of a Square Cavity Ventilated and heated by the bottom

Z E Bouayed [1] and S Houat [2]

[1, 2] Laboratoire MNEPM, Equipe MSMPT, Département de Génie Mécanique, Université Abdelhamid Ibn Badis, Mostaganem, Algérie

E-mail : [1] bouayedzino@gmail.com, [2] sa_houat@yahoo.fr

Abstract. In this paper, we present a numerical study of the laminar mixed convection in a ventilated cavity square; one of the walls is subjected to a constant temperature, while the other walls are considered adiabatic. The thermal lattice Boltzmann method with the double population model was used. A computer code was developed with the D_2Q_9 model for the velocity field and D_2Q_5 for the temperature field to determine the whole structure of the flow. The results are presented in the form of thermal fields for the Richardson number Ri = 10. Those show that each time the Richardson number increases, the number of Nusselt increases too what justifies an increase in the heat transfer.

1. Introduction

The transfer of heat by mixed convection aroused the considerable interest of several researches for technological applications such as: the ventilation of the buildings, the chemical plating of the thin layers, the cooling of the electronic parts, squanderers of heat in the solar collectors and nuclear reactors. Several simple and complex geometrical configurations were examined by report to a theoretical, digital end experimental approach.

A large number of numerical studies were interested on fixing of only one entry and exit [1-3] with an isothermal wall. Others research considered the wall heated with a heat flow [4-5], where [6] studied various configurations of the entry and exit position in order to detect the best possible position of the entry and exit opening, and to obtain a more effective cooling. Other numerical and experimental [7-9] studies treated the effect of the geometry of an obstacle like the source of heat inside the cavity in order to maximize the total conductance. The position of the air entry and the exit has a great effect of the hydrodynamic and thermal structures [10]. One cavity that has several entries [11] improves its ventilation.

The lattice Boltzmann method (LBM) being a digital method relatively recent and original which came out at the beginning of the Nineties. It is interested, either with the macroscopic quantities (celerity, pressure and density), but directly the distribution of the various particles constituting a fluid. We speak then about mesoscopic representation. What makes it competing with the other conventional methods such as finite volume, finite elements and the finite differences. It is initially resulting from the lattice gas method [12], from the automata cellular theory [13] and while being based on the physic statistical formalism [14]. It is important to be able to locate its performances compared to those of the classical digital methods used to simulate and reproduce the flows with the heat transfer in the ventilated enclosures. The purpose of the study presented in this work is the analysis of the phenomenon of the heat transfer with mixed flow of the convection in the laminar mode, in a square cavity provided with two openings. The interior walls are supposed to be adiabatic except for the located on the low side, it is considered isothermal. The thermal model of the lattice Boltzmann method with nine celebrities (D_2Q_9) is used to reproduce the dynamic field and that simplified at five celebrities (D_2Q_5) is used for the temperature field.

Thus, a thermal analysis will have to be carried out with LBM. That will enable us to determine the performances of this new digital method in this field.

2. Physical description of the problem

The model chosen is square cavity of coast H filled with two ventilation openings, the first located in the lower left corner of L_1 side and the second located in the upper right side $L_2 = L_1 = 20\%$ H. The walls of this cavity are adiabatic except the lower wall.

The walls of this cavity are adiabatic except the low wall which is maintained by a source of heat at a constant temperature T_h. Air entering through the left opening of the wall with a temperature T_0 and a uniform velocity U_0 as fig.1 shows it. The assumptions used are summarized in the case of a incompressible fluid, Newtonian in two dimensions, laminar, satisfying the assumption with Boussinesq, stationary with a transfer of heat by radiation and a dissipation of heat by effect of viscosity negligible.

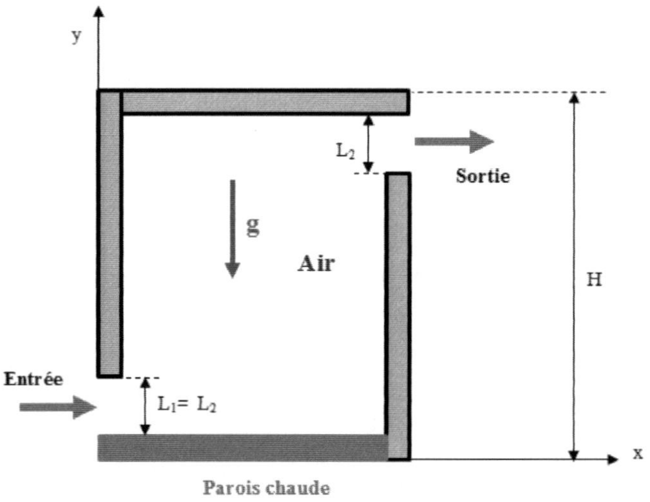

Figure 1. Geometry configuration of physic problem.

3. Lattice Boltzmann method

In statistical physics, a gas is described by a cloud of particles obeying the Boltzmann equation. This continuous equation as it was proposed by Boltzmann in 1872 can be written as follows:

$$\frac{\partial f}{\partial t} + c.\frac{\partial f}{\partial r} + F.\frac{\partial f}{\partial c} = \Omega(f). \quad (1)$$

This equation is the basic equation of the kinetic theory of gases. It describes gas molecules distribution function which is a function of time, space and velocity of the molecule.

With the approximation Bhatnagar, Gross and Krook (BGK) [15] The collision operator is written as follows :

$$\Omega = \omega(f^{eq} - f) = \frac{1}{\tau}.(f^{eq} - f). \quad (2)$$

The collision frequency is given by: $\omega = \frac{1}{\tau}$. (3)

The local equilibrium distribution function is denoted by f^{eq}; which is the distribution function of Maxwell-Boltzmann. In general the equilibrium distribution function Maxwell Boltzmann [16] is expressed by :

$$f_i^{eq}(x,t) = \rho.w_i.\left[1 + \frac{e_i.u}{c_s^2} + \frac{(e_i.u)^2}{2.c_s^4} - \frac{u^2}{2.c_s^2}\right]. \quad (4)$$

Discretizing equation (10) in space, time and compared with speeds gives the Boltzmann equation system given by the following formula:

$$f_i(r + e_i\delta t, t + \delta t) - f_i(r,t) = -\frac{1}{\tau}\left(f_i(r,t) - f_i^{eq}(r,t)\right). \quad (5)$$

The Local conservation laws: the Density, The velocity and Energy are expressed as follows [17]:

$$\rho(x,t) = \sum_{i=0}^{i=N} f_i(x,t) = \sum_{i=0}^{i=N} f_i^{eq}(x,t). \quad (6)$$

$$u = u(x,t) = \frac{1}{\rho(x,t)}\sum_{i=0}^{i=N} e_i.f_i(x,t) = \frac{1}{\rho(x,t)}\sum_{i=0}^{i=N} e_i.f_i^{eq}(x,t) \quad (7)$$

$$E = E(x,t) = \frac{1}{\rho(x,t)}\sum_{i=0}^{i=N}\frac{1}{2}(e_i - u)^2.f_i(x,t) = \frac{1}{\rho(x,t)}\sum_{i=0}^{i=N}\frac{1}{2}.(e_i - u)^2.f_i^{eq}(x,t) \quad (8)$$

3.1. D_2Q_9 model

For the discretization of speeds, the model of the square lattice nine speeds (Fig. 2) with a unit spacing. Each node of the network is connected to eight by eight other neighbouring links. The particles can only reside on a node and move to their earnest neighbours along these links for a unit time. There are three types of particles on each node with nine different speeds.

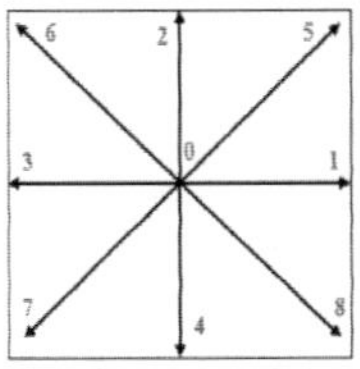

Figure 2. D_2Q_9 lattice

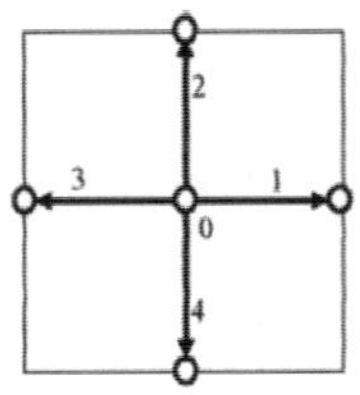

Figure 3. D_2Q_5 model

The lattice velocities e_i are: $e_i = 0$ for $i = 0$

$$e_i = c\left(\cos((i-1)\pi), \sin((i-1)\pi)\right) , \quad i = 1-4 \tag{9}$$

$$e_i = \sqrt{2}c\left(\cos\left((i-5)\frac{\pi}{2}+\frac{\pi}{4}\right), \sin\left((i-5)\frac{\pi}{2}+\frac{\pi}{4}\right)\right), \quad i = 5-8$$

Where $c = \frac{\Delta x}{\Delta t}$, and The corresponding weight factors are : $w_0 = 4/9, w_1 = w_2 = w_3 = w_4 = 1/9, w_5 = w_6 = w_7 = w_8 = 1/36$

The dynamic viscosity is given by the formula: $\mu = \left(\tau - \frac{1}{2}\right).\frac{\rho.c^2}{3}$ (10)

3.2. D_2Q_5 model :

For the discretization of the temperature, the square lattice pattern of five-speed (Figure 3) with a spacing unit. Each node of the network is connected to four other neighbours by four links. The particles can only reside on a node and move to their nearest neighbours along these links for a unit time.

Since there is no coupling between the celerity equation and the energy, two different distribution functions have to be solved, for example for speed f and g for energy. The effort provided for the establishment of a stable model for the lattice Boltzmann method thermal (TLBM) has paid off recently. The celerity function equation is the same as that of D_2Q_9. For the energy function, the equation is :

$$g_i(x+\Delta x, t+\Delta t) = g_i(x,t).[1-\omega_s] + \omega_s.g_k^{eq}(x,t). \tag{11}$$

The summation of the distribution functions in each D_2Q_5 lattice node gives us the temperature:

$$T = \sum_{i=1}^{5} g_i. \tag{12}$$

The distribution equilibrium function is:

$$g_i^{eq} = w_i.\emptyset(x,t).\left[1 + e_i.\frac{\vec{u}}{c_s^2}\right]. \tag{13}$$

$\mathbf{e_i}$ are the lattice celerities defined by :

$$e_i = c.\left(\cos\frac{2\pi(i-1)}{4}, \sin\frac{2\pi(i-1)}{4}\right) \quad i = 1, \dots 5. \tag{14}$$

The corresponding weight factors are: $w_0 = 1/3,\ w_1 = w_2 = w_3 = w_4 = 1/6.$

The equation of the thermal diffusion coefficient is given by

$$\alpha = \frac{\Delta x^2}{3.\Delta t}.\left(\frac{1}{\omega} - \frac{1}{2}\right). \tag{17}$$

4. Results an discussion

4.1. Numerical code validation

We studied the natural convection in a square cavity closed differentially heated, where two horizontal walls of the top and bottom are considered adiabatic. It emphasizes that the right side wall of the cavity is

subjected to a cold temperature and the left at a hot temperature. We compare the results of the value of Nusselt using the two methods with those of the literature.
In this study, we perform simulations for moderate Rayleigh numbers $10^3 \leq Ra \leq 10^5$ and Prandtl number Pr = 0.71. The mesh study based on lattice Boltzmann method with dual population (TLBM) was performed in the laboratory. Resolution mesh was satisfied between 71x71 and 151x151 for different Rayleigh numbers.

Table 1. Presentation of the average Nusselt results for different Rayleigh number values with their comparison with the results of the literature

Ra	Present work *TLBM*	Literature results	
		De Vahl Davis [17]	*Markatos & Pericleous [18]*
10^3	**1,115**	1,118	1,108
10^4	**2,221**	2,243	2,201
10^5	**4.461**	4,519	4,430

The Table 1 presents the numerical results obtains of the average Nusselt number with the TLBM code. These are compared with the results of the literature using different numerical methods. It is clear that the results of our TLBM model are agreed with the results of the literature in the range of Rayleigh numbers tested. The error calculated for the average Nusselt number Nu, does not exceed 1.2% compared to the reference results [17], 1.9% compared to [18].

4.2. *Results for the ventilated cavity*

The geometrical configuration treated in this study is the ventilated square cavity presented in (fig. 1). The selected fluid inside is the air with (Pr = 0.71). We use the grid of 121*121 for the lattice Boltzmann method in all our calculations. We study the influence of the Richardson number on the total structure of the flow. For this case, the Grashof number is fixed to Gr = 105 and the Richardson number is varied by

Ri = 3 Ri = 5 Ri = 10 Ri = 15

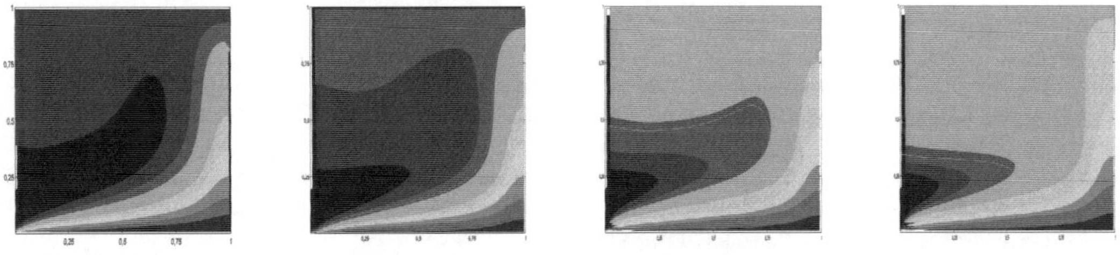

Ri = 20 Ri = 30 Ri = 50 Ri = 100

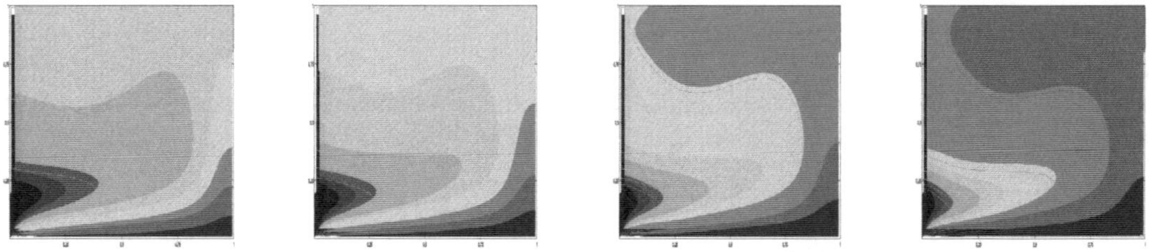

Figure 4. Isothermal values of Ri = 3, 5, 10, 15, 20, 30, 50 et 100 ($G_r = 10^5$ and Pr = 0,71)

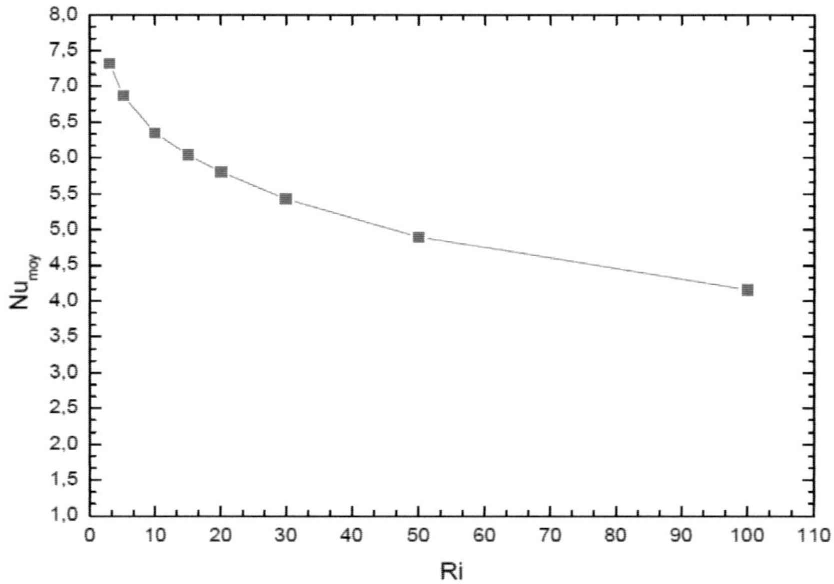

Figure 5. Variation of the average Nusselt values with Richardson number and ($G_r = 10^5$ and Pr = 0,71)

taking the following values: Ri = 3, 5, 10, 15, 20, 30, 50 and 100. The results are presented in the isotherm form (figure 4).The geometrical configuration treated in this study is the ventilated square cavity presented in (figure 1). The selected fluid inside is the air with (Pr = 0.71). We use the grid of 121*121 for the lattice Boltzmann method in all our calculations. We study the influence of the Richardson number on the total structure of the flow. For this case, the Grashof number is fixed to Gr = 10^5 and the Richardson number is varied by taking the following values: Ri = 3, 5, 10, 15, 20, 30, 50 and 100. The results are presented in the form of the isotherms (figure 4).

We note that the distribution heat in the cavity is in conformity with the circulation of the fluid. Indeed, we note a heating of the fluid starting from the entry (low left corner) throughout the low wall to the exit (right high corner) for the eight values of the Richardson number Ri. For a Richardson number = 3, we note a thermal absence of the stratification, which coincides with the beginning of the mixed convection. Starting from a Richardson number = 5, we note the appearance of a thermal stratification. From values of

Ri $\geq$ 10, we note a widening of temperature occupying a considerable portion of the cavity, also we observe a horizontal stratification of the layers of an air in which the natural convection is prevalent.
The position of the hot wall has an influence on the heat transfer. We notice that the high temperatures are localised in narrow spaces in the vicinity of the hot wall which correspond to the thermal thickness of the boundary layer and which is largely influenced by the Richardson number. Far from the hot wall, the gradients of the temperature are weak.

In figure 5, the variation of the average Nusselt number depending on the Richardson number is presented. We note that the Nusselt number decreases with the increase of the Richardson number what justifies a decrease in the heat transfer.

5. Conclusion

This work focuses on the numerical study of laminar mixed convection in a ventilated cavity with two openings. The results obtained in this work were used to evaluate the performance and LBM's ability to reproduce the mixed convection in a ventilated cavity. The TLBM model with a double population model used has all the advantages, including good numerical stability and the ability to generally manage the overall heat transfer by convection problems.

References

[1] Zermane S, Boudebous S and Boulkroune N 2005. Etude numérique de la convection mixte laminaire dans des cavités ventilées. *Sciences &Technologie B–* N°23, pp. 34-44.

[2] Radhi Z K 2011. Numerical study of mixed convection heat transfer through double square cavity connected with each other. *Al-Qadisiya Journal For Engineering Sciences* Vol. 4 No.4.

[3] Hamini A, Bouabdallah S and Benchatti A 2012. Numerical study of Mixed Convection in a Square Cavity Containing Inlet and Outlet air. JM'EMP08 EMP, (Bordj El Bahri, Algeria).

[4] Rahman M M, Alim M A, Mamun M A H, Chowdhury M K and Islam A K M S 2007. Numerical study of opposing mixed convection in a vented enclosure. *Asian Research Publishing Network (ARPN).* Vol. 2, N°2.

[5] Mahmoudi A H, Shahi M and Talebi F 2010. Effect of inlet and outlet location on the mixed convective cooling inside the ventilated cavity subjected to an external nanofluid. *International Communications in Heat and Mass Transfer* 37 - 1158–1173.

[6] Raji A, Hasnaoui M and Bahlaoui A 2008, Numerical study of natural convection dominated heat transfer in a ventilated cavity: Case of forced flow playing simultaneous assisting and opposing roles. *International Journal of Heat and Fluid Flow* 29 - 1174–1181.

[7] Radhakrishnan T V, Verma A K, Balaji C and Venkateshan S P 2007. An experimental and numerical investigation of mixed convection from a heat generating element in a ventilated cavity. *Experimental Thermal and Fluid Science* 32 - 502–520.

[8] Rahman M M, Alim M A, Saha S and Chowdhury M K 2008. Mixed convection in a vented square cavity with a heat conducting horizontal solid circular cylinder. *Journal of naval architecture and marine engineering* DOI: 10.3329/jname.v5i2.2504.

[9] Benachour E, Draoui B, Rahmani L, Mebarki B, Asnoune K and Imine B 2010. Simulation Numérique de la Convection Mixte d'une pièce (type d'habitat) avec la présence d'un Corps de Chauffe. 10thInternational Meeting on Energetical Physics SIPE 10 (Bechar, Algeria). *Journal of Scientific Research* N° 0 Vol. 1.

[10] Oztop F H 2010. Influence of exit opening location on mixed convection in a channel with volumetric heat sources. *International Communications in Heat and Mass Transfer* 37 - 410–415.

[11] Belhi M, Boudebous S 2007. Etude numérique de la convection mixte dans une cavité carrée munie de plusieurs entrées. *Congrès Français de Thermique, SFT* (Île des Embiez, France), 29 mai - 1 juin.

[12] Dieter A. Wolf-Gladrow 2005. *Lattice-Gas Cellular Automata and Lattice Boltzmann Models - An Introduction. Springer edition.*

[13] Biggs M J and Humby S J 1998. Lattice-gas automata methods for engineering. *Chemical Engineering Research & Design*, 76 (A2):162-174.

[14] Lifshitz E M, Pitaevskii L P 1980. *Statistical Physics, Part 1, Pergamon Press, Oxford.*

[15] Bhatnagar P, Gross E, and Krook M 1954. A model for collision process in gases. I. Small amplitude process in charged and neutral one-component systems, *Phys. Rev.* 94(3), 511-525.

[16] Han K , Feng Y T, Owen D R J 2008. Modeling of thermal contact resistance within the framework of the thermal lattice Boltzmann method. *International Journal of Thermal Sciences 47 - 1276–128.*

[17] De Vahl Davis G 1983. Natural convection of air in a square cavity: a benchmark numerical solution. Int. J. Numer. Meth, Fluids3, pp249.

[18] Markatos N C, Pericleous K A 1984. Laminar and turbulent natural convection in an enclosed cavity. *Int. J. Heat Mass Transfer*, 27, 755-772.

Optimal design of inverted truncated pyramid with Fresnel lens for concentrated photovoltaic Units

S El-Yahyaoui, S El Himer, A Mechaqrane and A Ahaitouf

Laboratory of Renewable Energies and Intelligent Systems, Faculty of Sciences and Technology, Sidi Mohamed Ben Abdellah University, PO. Box 2202 Fez Morocco

Email: sara.elyahyaoui1@usmba.ac.ma

Abstract. The aim of the presented work was to determine the optimum parameters of inverter truncated rectangular pyramid with Fresnel lenses. The use of secondary optical element (SOE) in a concentrated photovoltaic system can be effective in redirecting the sun light into the solar cell, increasing the concentration as well as improving the energy uniformity on the solar cell Ray tracing technique was used to simulate the optical characteristics of the CPV unit with various design parameters of the component. Finally, a typical concentrator was designed by using three possible materials, the Fused Silica, the BK7 and the PMMA.

1. Introduction

Concentrating photovoltaic (CPV) is becoming a promising option for clean energy use due to the reached high efficiencies in these photovoltaic systems. A CPV system is based on the combination of three components : receiver (solar cell), focusing optics and solar tracker. The focusing optics concentrate solar energy into a small surface solar cell, generally multijunction III-V based materials cells. This concentration can be done in one step: the sunbeam are directly routed to the cell via a primary optics, or in two steps, the rays are first focused on a secondary optical placed above the cell. Using a secondary optics has several objectives: increase concentration, homogenize the irradiance uniformity on the cell, and increase the acceptance angle of the module.

In terms of the primary optic, Fresnel lens and parabolic mirror are the best-known examples. But the availability of lenses on the market and their cost tend to impose them as the most used in current technologies [1]. Two methods are used to manufacture lenses, one using a type of plastic polymethyl methacrylate or PMMA [2] and the other the silicone-on-glass[3]. In the case of a system using a Fresnel lens, the chromatic aberration phenomenon causes nonuniformity of the concentrated distribution on the solar cell surface[4]. Localized hot spots and nonuniformity increase the cell temperature, cell resistance, and lower the efficiency. Secondary optics helps to solve this problem. Today there is some variety in design of secondary optics, which can be grouped into reflective and refractive optical elements of various forms. Refractive secondaries are more widespread, they consist of a transparent dielectric (glass or plastic) that redirects the beam at the cell. We can mention among other the compound parabolic concentrator (CPC), the refractive truncated pyramid (RTP), the dome, or the dielectric total internal reflector (DTIR) [5].

To improve the performance of CPV module, several work has been done. Ning et al. investigated the two-stage photovoltaic concentrators with Fresnel lenses as primaries and DTIR as secondaries. It was found that the two-stage concentrator offered increased acceptance angle, higher concentration and also a more uniform flux distribution than the Fresnel lens alone[6]. Andree et al. presented optimization of SOE for CPV unit with a flat Fresnel lens [7]. Furthermore Miñano et al. designed and manufactured a CPV system, which contained two optical elements a Fresnel lens and free form optics as secondaries) [8].

In this work we present a design of a secondary concentrator, the inverted truncated rectangular pyramid. The 3-D model of this design is established by TracePro and optimal parameters such as inclination angle and component height are determined with Fresnel lens. Optical efficiency, acceptance angle and irradiance distribution over the cell are also studied for three materials.

2. Design Process

A two-stage system consists of two optical components: a PMMA Fresnel lens as a primary optical element (POE) and a truncated rectangular pyramid as a secondary optical element (SOE). Figure 1 illustrates such a two-stage design. The POE has a diameter d of 350mm, a thickness of 3mm and a focal length f of 370mm. The taken size of the solar cell is 10 mm × 10 mm, and the theoretical geometric concentration ratio is 1225.

The Fresnel lens is characterized by its f_number, FN/# defined by: [9]

$$FN/\# = \frac{f}{d} \tag{1}$$

$$\tan\theta = \frac{d}{2f} \tag{2}$$

θ is the semi-angle of the outco-mingbeam from the Fresnel lens

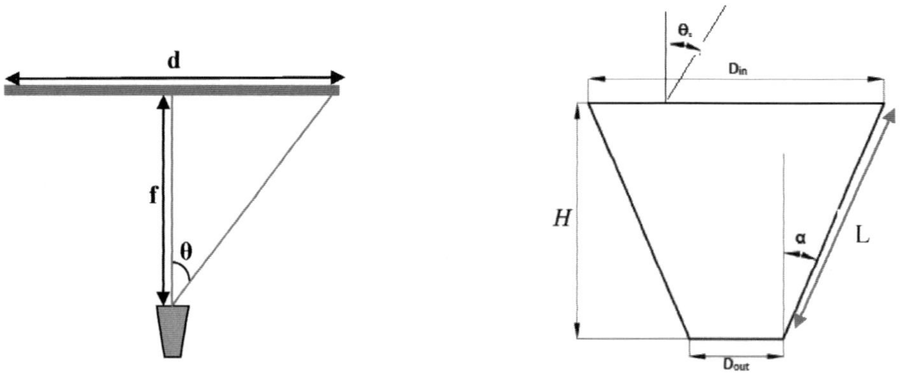

a b

Figure 1. (a)two-stage concentrator (b)Design parameters of pyramid

The design of a concentrator involves some geometrical parameters. For instance the pyramid, shown in Fig.1 is characterized by the inclination angle α, the entrance aperture D_{in}, the exit aperture D_{out}, the slant length L , the height H and the angle between the vertical (normal incidence) and the incident light rayθ_z.

$$H = \frac{D_{in} - D_{out}}{2 \tan \alpha} \quad (3)$$

$$L = \frac{D_{in} - D_{out}}{2 \sin \alpha} \quad (4)$$

The required condition on the inclination angle and n the number of reflections for which the ray light reaches the exit aperture is as the following [10]:

$2(n+1)\alpha + \theta_z \leq 90$(5)

To minimize the losses, the incident light is allowed to experience no more than one total reflection in the SOE before reaching the cell. θ_z should be less than or equal to θ the incidence angle of Fresnel lens to effectively collect all the rays.

$\theta_z \leq \theta$(6)

3. Simulation and results

3.1. Preliminary Design of the SOE

We suppose the initial value of inverted rectangular pyramid height H is 20mm. Since the exit aperture D_{out} is 10mm, the entry aperture D_{in} can be varied by changing the value of inclination angle α. considering the conditions cited before (equation (5) and equation (6)) α can be changed from 1° to 16°.The Fresnel lens is made by PMMA and The material used for SOE was BK7 glass. The incident light intensity on the Fresnel lens is set as 1000W/m^2. We simulated the system in TracePro software. Different optical properties and design parameters of the POE and SOE are mentioned in Table.1.

Table 1. Different parameters of CPV module

Title	value
Fresnel lens diameter	**350mm**
Focal length of Fresnel lens	**370mm**
Thickness of Fresnel lens	**3mm**
Material of Fresnel lens	**PMMA**
Material of the pyramid	**BK7**
Index of Refraction of BK7	**1.518**
Size of solar cell	**10mm x 10mm**
Light source	**1000W/m^2**

Figure 2 gives the evolution of the optical efficiency of the concentrator versus the inclination angle α. As we can notice, the optical efficiency increases with α, we reach86.4 % at 16°.

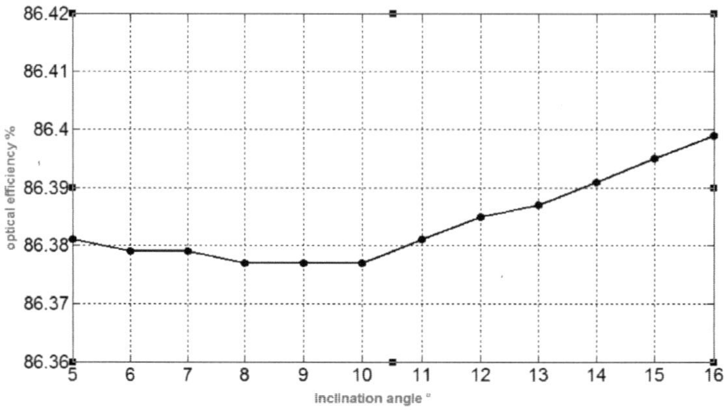

Figure 2. Optical efficiency of the Fresnel lens with pyramid

3.2. Optimization of the size of the pyramid, method 1

From simulation results we notice that the optical efficiency maximizes when α is set to 16°.So we fix D_{in}at 21,47mmthe corresponding value to α=16° and we vary the height value H from 15mm to 28mm and simulate the obtained system.
As shown in Figure 3 the optical efficiency decreases with an increase of the pyramid height. It is maximal when H is 15mm (86,50%).

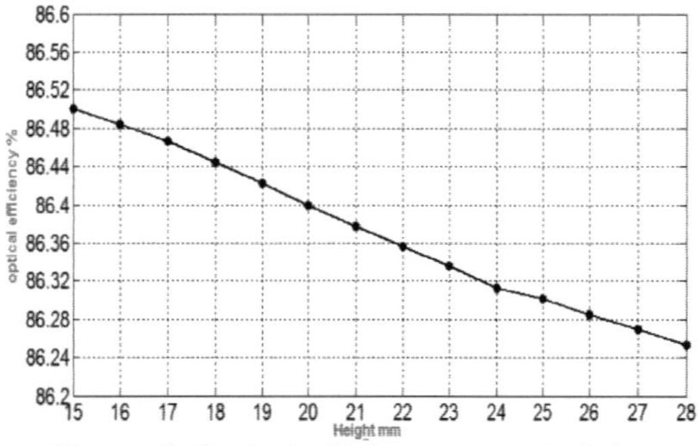

Figure 3. Optical efficiency versus height

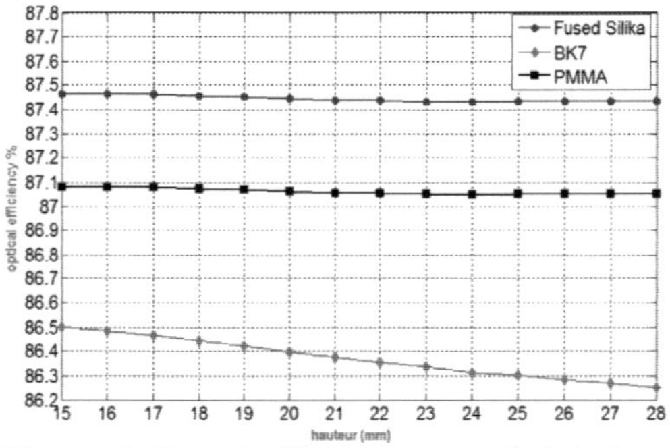

Figure 4. Optical efficiency versus height for three materials

We keep the same conditions and we simulate for two other materials: the PMMA and the Fused Silica. The refractive indexes of these materials are 1.493 and 1.46 respectively.
Figure 4 illustrates the effect of material on the optical efficiency. We can see that Fused Silica achieves a higher optical efficiency (87,462 %) than the PMMA(87,08 %) and BK7 (86,50 %).

3.3. Optimization of the size of the pyramid, method 2

As mentioned before two of the basic design parameters of the inverted truncated pyramid are the height H and the inclination angle α. To determine the optimum size of the pyramid a parametric design process with two values of H and various vales of α were performed [11].

IOP Conf. Series: Materials Science and Engineering **186** (2017) 012037 doi:10.1088/1757-899X/186/1/012037

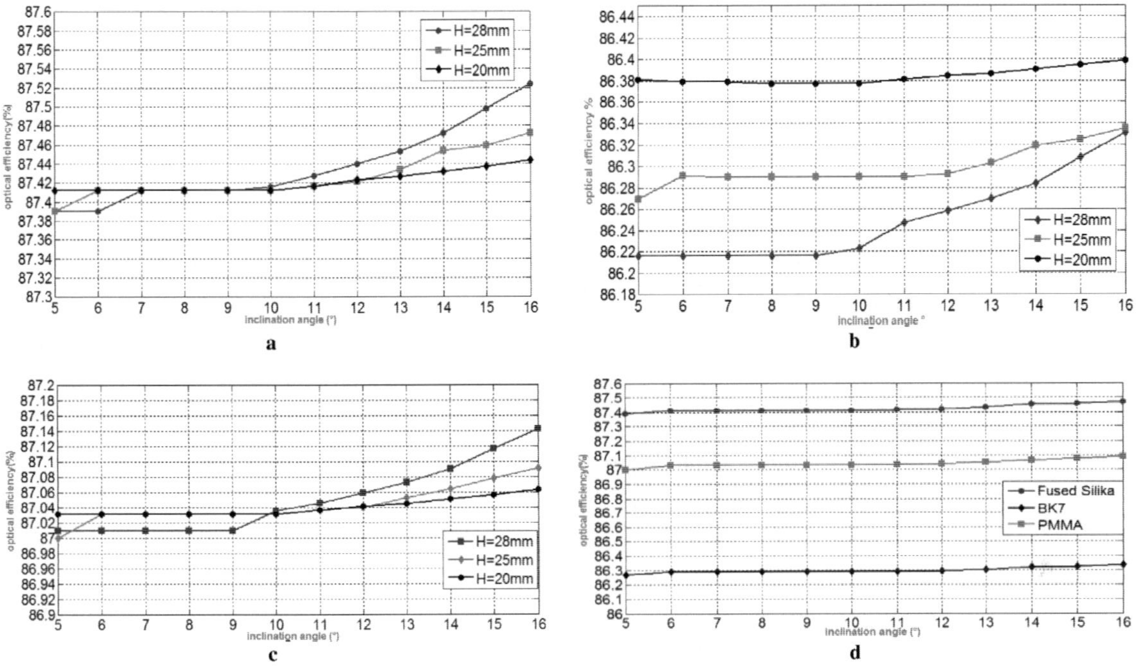

Figure 4. optical efficiency versus inclination angle:
(a): Fused Silika, (b): BK7, (c): PMMA, (d)H=25mm

Figure 4 revealed that the optimum value of inclination angle α was 16° to achieve a highest optical efficiency for the three values of H. Moreover when H is fixed to 28mm the optical efficiencies were maximal for Fused Silika (87,524%), and PMMA (87,124), while it is at maximum for BK7 at H=20mm (86,4%).

3.4. Optical tolerance and irradiance distribution

Figure 5 shows and compares the optical efficiencies of the three values of H (20mm, 25mm and 28mm) of the inverted truncated pyramid for the fused silica under different incidence angles. It can be observed that the pyramid with height equal to 28mm is sensitive to the deviation of the incidence angle.

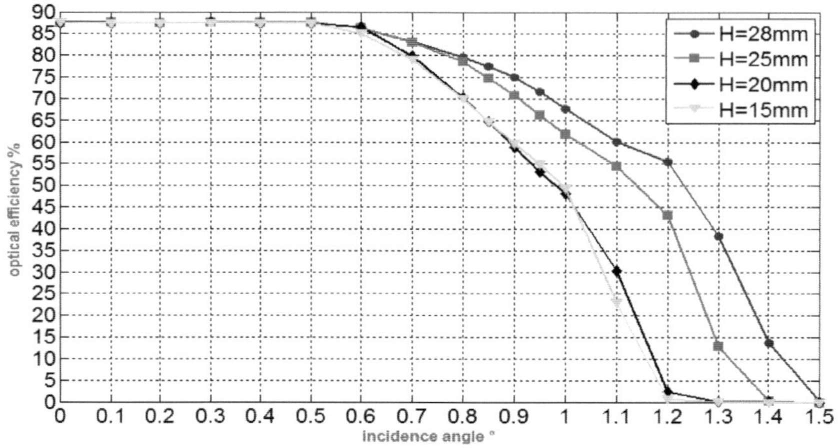

Figure 5. Optical efficiency versus incidence angle

Table 2. Comparison of the optical performance of the Fresnel lens with pyramids

Height (mm)	28	25	20	15
$\theta_{90\%}$	0,82	0,81	0,72	0,71
$\theta_{80\%}$	0,96	0,91	0,82	0,8
Optical efficiency under normal incidence (%)	87,524	87,472	87,444	87,462
Optical efficiency at $\theta_{90\%}$(%)	78,771	78,724	78,699	78,715
Optical efficiency at $\theta_{80\%}$ (%)	70,019	69,977	69,955	69,97

Table 2 summarize the optical characteristics and comparison of CPV unit with pyramids with various heights as shown pyramid with H=28mm exhibits a better acceptance angle $\theta_{90\%}$=0,82° and $\theta_{80\%}$=0,96°($\theta_{90\%}$and$\theta_{80\%}$are the incidence angle corresponding respectively to 90% and 80% of the maximum optical efficiency at normal incidence)than other pyramids and the corresponding optical efficiency at the acceptance angles are respectively 78,77% and 70,02%.

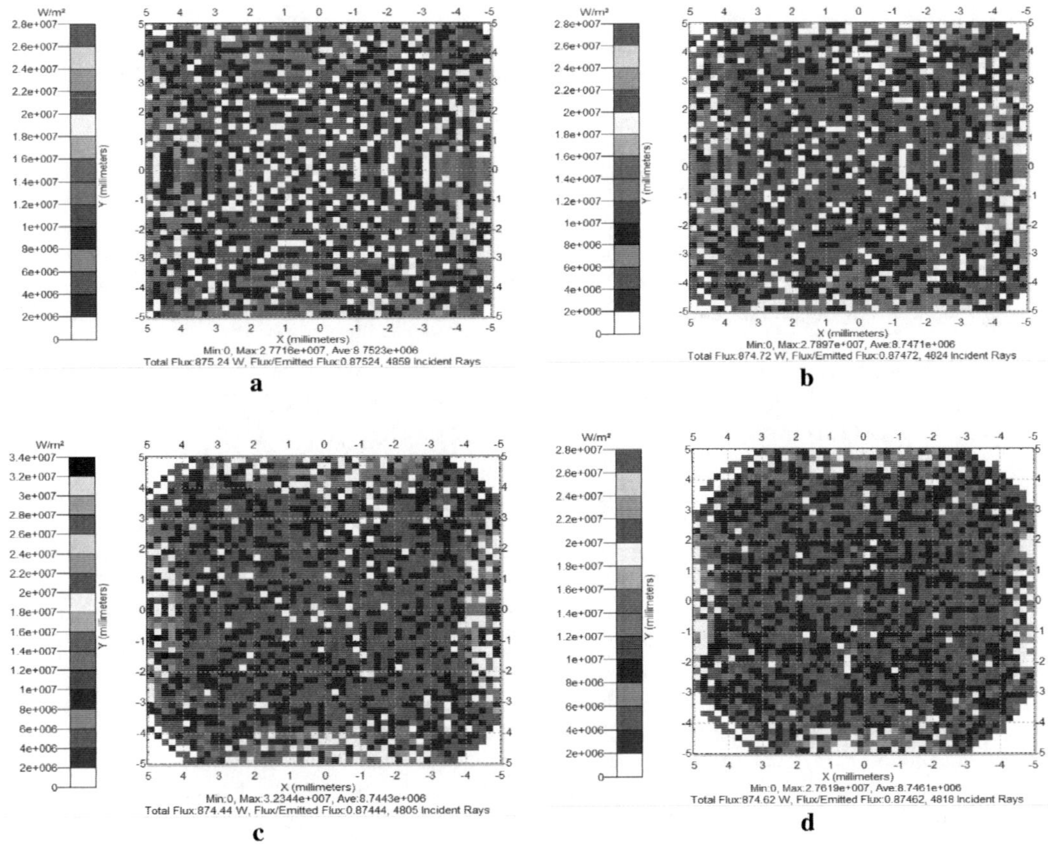

Figure 6. Maps of output flux at the exit of the pyramid:
(a) H=28mm, (b) H=25mm, (c) H=20mm, (d)H=15mm

Figure 6 displays the irradiance distribution under normal incidence for different values of H. It can be seen that all pyramids give uniform irradiance distribution over cell and that pyramid with H=28mm gives the best homogeneity.

4. Conclusion

This study presents a design of an inverted truncated pyramid to improve the performance of CPV module. Inclination angle and height are optimized to achieve high values of optical efficiency and acceptance angles. Different sizes of pyramids were evaluated by ray-tracing simulation. We found that when concentrator height and inclination angle are set to be 28mm and 16° respectively and the used material is fused silica, we achieve a highest optical efficiency 87,524% as well as a good acceptance angle $\theta_{80\%}$=0,96°

Acknowledgments

We are very grateful to the Research Institute for Solar Energy and Renewable Energies (IRESEN), the Ministry of Higher Education, Scientific Research and Management Training, the CNRST, the French Embassy in Morocco and the French Institute of Morocco to support this work.

References

[1] Xie W T, Dai Y J , R Z Wanga, Sumathy K, Concentrated solar energy applications using Fresnel lenses: A review. Renewable & Sustainable Energy Reviews, 2011. 15(6): p. 2588-2606.

[2] Luce T and Cohen J, "The path to volume production for cpv optics," in Proceedings of the 35th IEEE Photovoltaic Specialists Conference, 2010.

[3] Lorenzo E and Sala G, "Hybrid silicone-glass fresnel lens as concentrator for photovoltaic applications," in Proceeding 2nd European Photovoltaic Solar Energy Conference, 1979.

[4] Kurtz S and O'Neill M, "Estimating and controlling chromatic aberration losses for two-junction, two-terminal devices in refractive concentrator systems," in Conference Record of the Twenty Fifth IEEE PVSC, pp. 361–364, 1996.

[5] M. Victoria, New concepts and techniques for the development of high-efficiency concentrating photovoltaic modules. PhD thesis, Universidad politecnica de Madrid,2014

[6] Ning XH, O'Gallagher J, Winston R. Optics of two-stage photovoltaic concentrators with dielectric second stages. Applied Optics 1987;26(7):1207–12

[7] Andreev, V.M.; Grilikhes, V.A.; Soluyanov, A.A.; Vlasova, E.V.; Shvarts, M.Z. Optimization of the secondary optics for photovoltaic units with Fresnel lenses. In Proceedings of the 23th European Photovoltaic Solar Energy Conference, Valencia, Spain, 1–5 September 2008; pp. 126–131.

[8] Miñano JC, Benítez P, Zamora P, Buljan M, Mohedano R, and Santamaría A, Free-form optics for Fresnel-lens-based photovoltaic concentrators, Optics Express 21 (2013) A494–A502.

[9] Kiatgamolchai .,Chamnii E,Theory and experiment of a two dimensional coneconcentratorfor sunlight, Solar Energy 82 (2008) 111–117

[10] Burkhard DG, StrobelGL and Shealy DL, Solar concentrating Properties of truncated hexagonal, pyramidal circular cones, Applied Optics. Vol. 17, No. 15, (1978), p. 2431-2440

[11] Yi-Cheng Chen and Hung-Wei Chiang Design of the Secondary Optical Elements for Concentrated Photovoltaic Units with Fresnel Lenses applied sciences *Appl. Sci.* 2015, *5*, 770-786; doi:10.3390/app5040770

IOP Conf. Series: Materials Science and Engineering **186** (2017) 012038 doi:10.1088/1757-899X/186/1/012038

Contribution of electromagnetic and tomographic technique to the study of the impact of salinity in soils of the experimental station of Al Ain Atti (Errachidia, Morocco)

A Benamara[1*], M EL Harnafi[2] and A Ammar[3]

1*Faculty of Sciences and Techniques of ERRACHIDIA
Geology Department
[2]Institut Scientifique de Rabat
[3]Faculté des Sciences UMV, Rabat

Email: a.benamara@fste.umi.ac.ma

Abstract. Soil salinity is widespread particularly in arid areas. Much work has been done to remedy this natural contamination and make them more favorable to receive experimental soil of vegetation adapted to grow in these contrasting environments of scarce water resources. Over the 80 years, Morocco has experienced excessive dryness whose effects have been severely felt in the region of Errachidia. The depletion of the Quaternary alluvial aquifer has required the exploitation of deep reservoirs. In order to optimize the use of salt water in arid, the Institute of Agricultural Research tested soil experimenting three plant species; the Triplex, Acacia and Cypress. This experiment was not only economic objectives; increasing agricultural yields and production, but also reduce the impact of desertification in this region. So these are environmental reasons which have led researchers to realize this experiment whose purpose is to examine the ability of these plants to grow and adapt to irrigation by saline water from the deep aquifer in place. The experimental site is located on the left bank of the national road to Erfoud (Errachidia). In its entirety, the redesigned covering about 10 hectares. The studies that have been conducted mainly concerned pedogenetic analyzes and observations the development height of the plantation tested, regardless of the experimental ground heterogeneity. This is why we undertook a geophysical survey which aims to provide information on the quality of the tested soils and accumulation of salinity at different depths of the experimental ground. To achieve our goal, we used the technique of electrical and electromagnetic investigation in order to identify the main lines of a possible spatial heterogeneity. Recall that the work of OKAY (2010) on an experiment coring by examining the evolution of the newly formed fracturing have proved that the result of resistivity measurement is correlated with that of the chargeability. The electrical resistivity is closely related to the concentration of mineral particles. Characterization studies at the La Ronde tailings mine (Agnico-Eagle Ltd, in Abitibi), showed quite encouraging results (Campos, 2004; Anterrieu, 2006). Since it is a salt-bearing areas (highly conductive) and given that the induced polarization is very sensitive to the ground conductivity; by measurement of residual potential, so we took measures chargeability tomography, the result is compared to the resistivity measured at different soils (control and irrigated), which will better clarify the impact of salinity on the quality of the disturbed soil. Indeed, the chargeability response it possible to characterize the different irrigated soils and its variation is attributed not only to the effect of saline irrigation but also the existence of an heterogeneity of the original ground. Prospecting in electrical tomography has displayed vertically and horizontally anomalies existing within the experimental site of the station of Ain Al Atti, it showed that:

- accumulation of salinity becomes wider going from the control ground to that of Acacias.
- surveyed the ground at least appear more homogeneous in surface, but their conductivity varies in depth.
- salinity origin and the concretion formed on the surface greatly influenced chargeability and resistivity measured on the different experimental soils.
Keywords. Drought, depletion of groundwater, Al Ain Atti experimental site, saline irrigation, technique electromagnetic, electrical tomography, induced polarization.

1. Introduction

The region-Tafilalet Errachidia is located in the southern Atlas region of Morocco. It faces enormous challenges including scarcity and poor water quality, following long periods of severe drought affected who in recent years and whose ecological consequences are considerable, as evidenced by the decline in irrigated area under the action of desertification and rising saline soils.
Agriculture represents 90% of the economic activity of the population. Agricultural activity is, however, directly related to the availability of wadis, springs, and khettaras that have been used since ancient times by the people in place. In order to optimize the use of salt water in arid areas, the National Institute for Agricultural Research (INRA) tested soil experimenting four plant species; Atriplex, Acacia, Eucalyptus and Palm. This experiment was not only economic goals; increasing agricultural yields and production, but also reduce the impact of desertification in this region. So these are environmental reasons which have led researchers to realize this experiment whose purpose is to examine the ability of these plants to grow and adapt to irrigation by saline water from the deep aquifer up.
It should be recalled that the previous work has borrowed several methods including soil analysis technique. The aim of our work is the introduction of the electrical tomography technique to better visualize vertically and horizontally the impact of salt water on soil amendment, firstly, and secondly, to highlight the heterogeneity that crosses the prospected ground. The result then requested provide useful lessons that will enrich the aforementioned work. The working procedure involves the construction of six profiles in pseudo-section, three concern the measure induced polarization and three in the measurement of electrical resistivity. Which will compare the results of the two respective techniques.

2. Presentation of the study area

2.1 Geographic Location and geological setting

The experimental Site Ain Al Atti is situated on the left bank of the road leading from Errachidia to Erfoud. It is part of the Cretaceous basin-Boudnib Errachidia (Morocco Southeast), this one is recognizable by its morphological characters cliff; it is regular, continuous, uniform and includes all land Cretaceous outcrop between Tinghir in west and Boudnib in east.
The basin-Boudnib Errachidia is limited: in the north by the South Atlas accident, the south by the mountain chain of the Anti-Atlas, to the east by outcrops of tertiary Hamada Guir. It occupies the western part of the Cretaceous basin "Ziz-Guir" [1] and to the west, the basin tapers into a narrow channel that reaches its minimum width in Tinghir, at which it connects to the Ouarzazate basin (Figure 1).

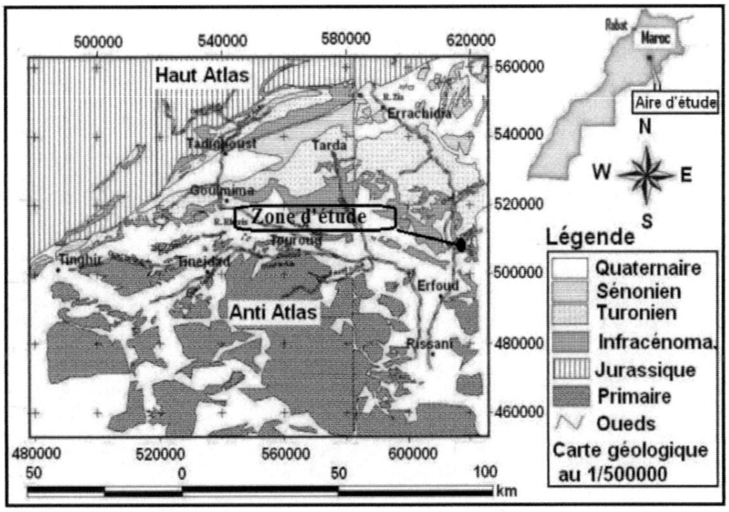

Figure 1. Geographical location of the study area [2]

Regarding its lithostratigraphy, the Cretaceous basin Errachidia-Boudnib is characterized by a stratigraphic series that runs from Paleozoic to Quaternary.
The Paleozoic outcrops south of the basin at Erfoud. It is formed mainly by limestone, conglomerates, shale, marl and quartzite sandstone.
For the Mesozoic Triassic outcrops Foum Zaabel in Tunnel located north of Errachidia, it unconformable overlies Paleozoic series. It is represented by a lagoon series of Anhydrous red marl and red marl, separated by diabase basalts.
Jurassic outcrops on the northern edge of the basin, it generally has layers of red sandstone and red or green marl and then alternating marl and limestone, red clay and conglomerates.
The Cretaceous is characterized by Cenomanian-Turonian transgression on the Middle Jurassic which invaded the Moroccan Southeast. Sedimentation is performed in epicontinental regime and has significant variations in thicknesses and facies. The Infra-Cenomanian corresponds to glyptogenic continental formations, resulting from the destruction of anti-Cretaceous reliefs [3].
The Infra-Cenomanian consists of continental deposits (coarse sandstones and conglomerates) and lagoon (sand, clay and marl). After the regression Albian and early Cenomanian, seas (Mediterranean and Atlantic) meet in the South Atlas furrow where the Cenomanian-Turonian platform develops [4].
In the outcrop, the Cenomanian Turonian is formed by limestone benches resting directly on the red clay to gypsum. At the top of the Cretaceous series, detrital deposits for overcoming the Turonian limestone. Senonian is composed of clay-gréso continental formations very heterogeneous, with gypsum and anhydrite, varying in thickness from 70m to the east and 500m north.
Quaternary present various facies (poudinges, gravel, silt and alluvium). It forms varying thickness layers between 5 and 40 m.

2.2 Climatological, hydrological and hydrogeological context

The climate that reigns in the Errachidia Boudnib-basin is an arid climate, strong continental influence. It is characterized by large variations in temperature and seasonal distribution of rains are scarce and very irregular.
The hydrographic network is represented by Oued Ziz that starts in the High Atlas and crosses the North region in the South. Groundwater resources play a key role in meeting water needs. These resources consist of groundwater, located along the valleys, and deep aquifers that are divided from

north to south in three well separated hydro-geological units [2]: The High Atlas, the Cretaceous basin Errachidia-Boudnib and the Anti-Atlas (fig.2).

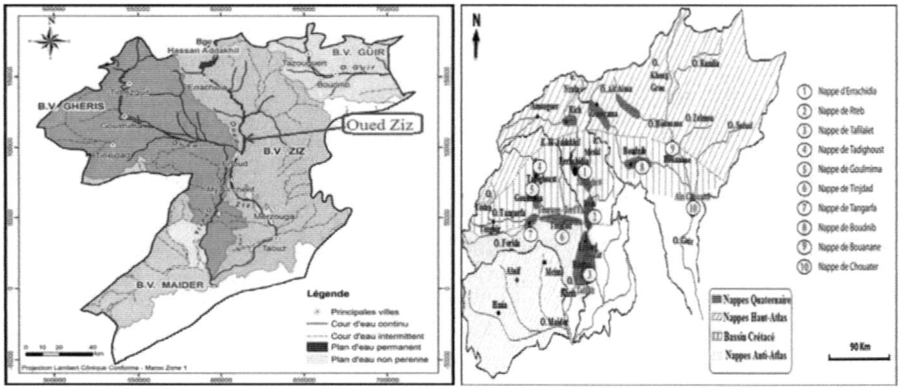

Figure 2. Water system in Guir, Ziz, Rheris and Maider basin and hydro-geological map showing the various layers of Errachidia Basin.[5]

The aquifer Infra-Cenomanian consists of continental deposits (coarse sandstones and conglomerates) and lagoon (sand, clay and marl).
The Turonian aquifer consists of fractured limestone and marine origin dolomite as karst facies.
South of the basin (tablecloths of Tafilalet), quaternary aquifer is formed at the base more or less cemented conglomerate and sandstone limestone lake, which contain the essential reserves of the tablecloth. [6]

3. Geophysical measurements in the experimental station of Ain AL Atti

3.1 Recall of measures result electrical prospecting in station of Ain AL Atti

The test site Ain Al Atti covers about 10 hectares. The Institute for Agricultural Research, which tested the development of four plant species: Atriplex, Acacia, Eucalyptus and Palm, has stopped halt saline irrigation there are already several years. We recall also that the interpretations of geophysical measurements were obtained taking into account the drilling data (Fig. 3) dug in the purpose of this experiment.

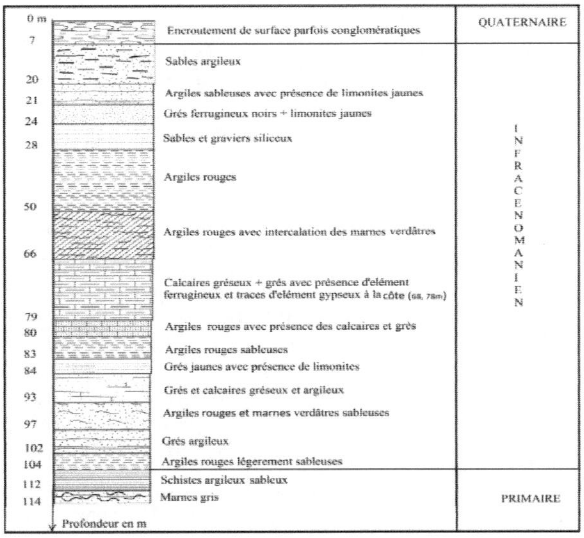

Figure 3. Stratigraphic log of drill Ain AL Atti (No. 4042/57).

Electrical methods to better understand the subsurface structure through the study of electrical resistivity formations that compose it. The resistivity was measured by injecting into the soil a continuous current whose intensity I is measured by means of two metal electrodes denoted A and B planted in the ground and connected to the two terminals of a DC generator. the potential difference resulting from the current flow is measured by means of two other electrodes denoted M and N. The electrical resistivity (ρ) is obtained from the potential difference (ΔV), the current (I) and the geometry of the electrodes (K) according to the following relationship:

$$\rho a = \frac{K\Delta V}{I}$$

Several devices can be used depending on the objective sought exploration. For vertical survey, the transmitting electrodes are spaced apart progressively. Horizontal exploration, measurement is obtained by a constant dimensions device and the base moves with constant field prospecting. These two techniques are often used in hydro-geological exploration.

With the aim of characterizing the heterogeneity within experimental soils, more oriented profiles (NE-SW) were carried out to map the resistivity in the control plot and in the Atriplex. [7]

The control plot (Fig.4) shows a well individualized conductive anomaly in the middle of the edge of the profile, its resistivity value is considerably high (155 ohm-m), where the field conglomeratic resistant comes into contact with a conductive zone (30 ohm -m) to the SW, which reflects the location of an anomaly of fracturing ground prospected. Another conductive zone appears to the Northeast part to be related to a discontinuity of conglomeratic field. In the rest of the profile, the area surveyed appears more homogeneous and least conductive.

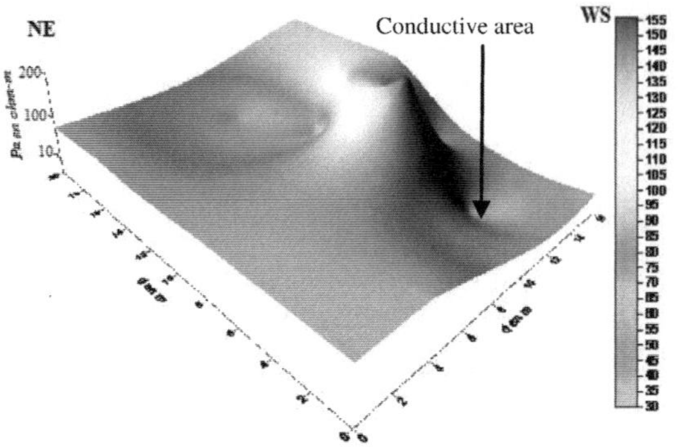

Figure 4. Variation of apparent resistivity according to the distance, measured in dipole profile at the experimental station in control plot (at SW edge, with XC = 3 and 6 = XP, step=2m, line spacing = 4m)

The result of the diagram (Fig.5) shows a heterogeneous field likely reflecting the real quality of the field, strongly affected by the accumulation of soil salinity prospected surface. The apparent resistivity has a large variation along the profiles, the accumulation of this salinity occurs therefore not uniformly during the irrigation periods. Note that it must not neglect the heterogeneity related to the natural terrain.

The resistivity response obtained in the irrigated parcel and planted by the triplex is very different from the control field.

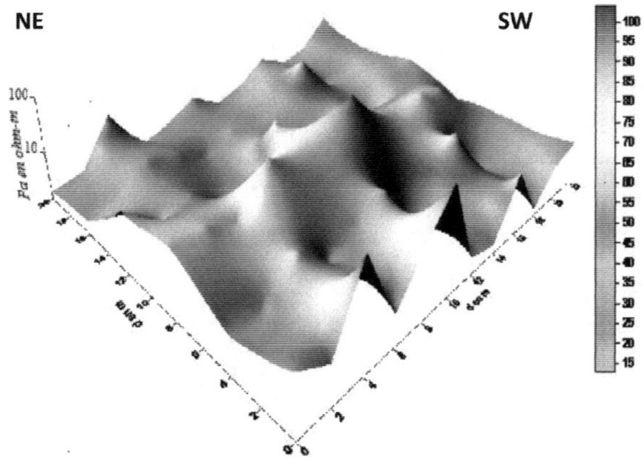

Figure 5. Variation of the apparent electrical resistivity according to the distance, measured in dipole profile at the irrigated soil (Atriplex, with XC = 3 and 6 = XP, step = 2m, line spacing=4m)

3.2 Result of measures in electromagnetic prospecting

With the aim to enrich the results of the electrical prospecting, we brought upon the VLF electromagnetic measurements technical (very low frequency method) using a fixed transmitter. Parameters measured ellipticity (b/a) is the ratio between the smallest and largest polarization axis; equivalent to the imaginary component (Im = secondary field) and the inclination or tilt (θ) the major axis of polarization with respect to the horizontal; equivalent to the real component (Re= secondary field).

$$\frac{b}{a} \cong \frac{H_s * \sin\alpha * \sin\varphi}{H_p} \cong \frac{I_m}{H_p}$$

$$Tan\theta \cong \frac{H_s * \sin\alpha * \cos\varphi}{H_p} \cong \frac{\mathrm{Re}}{H_p}$$

with α angle between Hp and Hs and φ the phase shift between Hp and Hs.

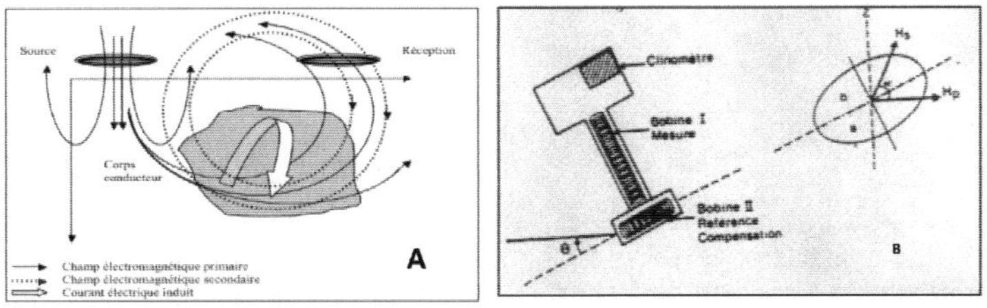

Figure 6. Schematic representation of the electromagnetic prospecting. A-production fields and induction phenomenon, B -setting out of a VLF receiver with two orthogonal coils.

This instrument can also be used for measuring the electrical resistivity. Indeed, the primary field radiates as two components in the direction of the issuer; a vertical electric component and the other horizontal magnetic Hy perpendicular to the first. After induction in the ground, another horizontal electric component Ex occurs along the direction of propagation. The electrical resistivity is given by the ratio Ex / Hy in Form magneto-telluric (also referred to as MT-VLF):

$$\rho a = \frac{1}{\varpi\mu}\left|\frac{E_X}{H_Y}\right|^2$$

With $\omega=2\pi f$ and μ is the magnetic permeability ($\mu=4\pi.10^{-7}$)

The investigation by this technique is limited to a very shallow (about 10 m), that is why we do not often use to explore the deeper formations.
Five profiles spaced of 40 m and oriented NE-SW were realized by this technique in Ain AL Atti station.
Profile (Fig. 7) is carried out at the West edge experimental field. The resulting response suggests the existence of an anomaly of contact with a relatively strong field located at the beginning of the profile (d = 20m) where the inclination has a maximum of 2%, while the ellipticity is at minimum. For the rest of the profile, tilt opposes the ellipticity. The first presents negative values, while the other is positive and the curve which shows no substantial changes with distance, it is therefore a zone is relatively homogeneous and conductive.
The second profile made within the experimental field relates to the plot planted with the palms. The graph curves (Fig.7) show a variation comparable to that observed in the first profile (control field). But after the strong anomaly (in beginning of the profile), the tilt drop maintains a relatively lower value than that observed at the natural terrain (Profile 1). This is attributed to the effect of saline irrigation has received this experimental ground.

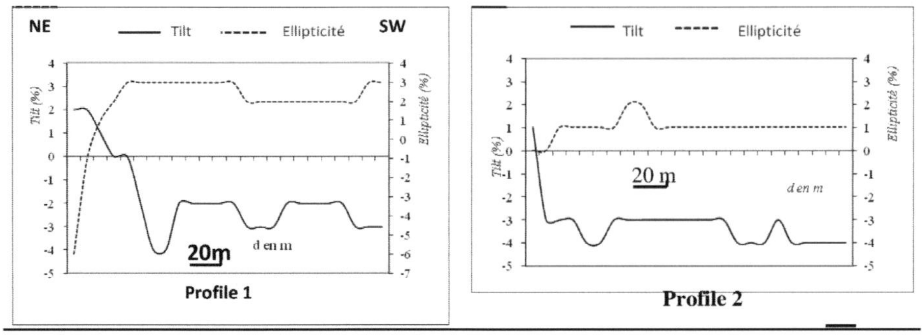

Figure 7. Variation of inclination and ellipticity according to the distance (Profile 1 and 2, step= 20 m spacing = 40m, NE-SW)

Regarding profile 3 (Figure 8, Eucalyptus plots and Atriplex), the response suggests that there is no opposition between the two components, according to the two curves, this response must be attributed to a homogeneous and conductor ground.

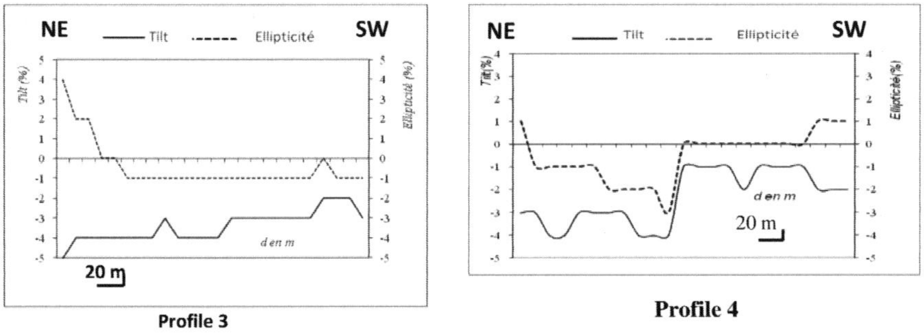

Figure 8. Variation of inclination and ellipticity according to the distance (Profile 3 and 4, step= 20 m spacing = 40m, NE-SW)

Profile 4 (Fig.8) begins first by a conductor field which stretches over a distance approximately 100 m and represents the border of the land concretionary following the discharge of salt water from the irrigation canal. In the middle of the profile, the signal shows the area indicated by the heterogeneity electrical profile (fig. 3), which marks the transition to a relatively resistant ground for which the tilt and the ellipticity significantly increase in natural area with no undergoes significant transfer of salinity.

Profile 5 (Fig. 9) is made at the unirrigated natural ground. The first part of the profile seems more affected by the spill salt water, this is the reason why the field components show negative values at the beginning of the profile. At about 280 m, the two parameters become almost constant for the rest of the profile, so they mark the transition to a homogeneous and slightly resistant field.

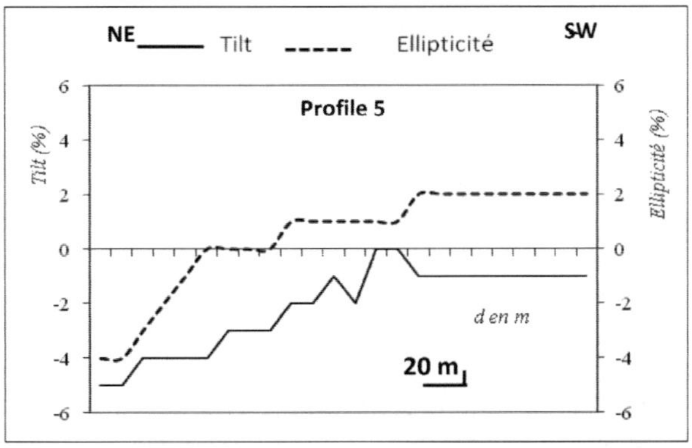

Figure 9. Variation of inclination and ellipticity according to the distance (Profile 5, step= 20 m spacing = 40m, NE-SW)

Electromagnetic profiles made at the experimental station Ain Al Atti allowed to characterize the heterogeneity of the original land they have also allowed to highlight the impact of spraying salt water on experimental soils.
Comparing the first four profiles, it is noted that the inclination has a negative maxima (-4%) for the profiles 1, 2 and 4, while the maximum is especially lower (-5%) for the profile 3 (forest plots; Eucalyptus and Atriplex), following the accumulation of salts in irrigated soil.
At the control plot, inclination has a negative maximum of -5% related to the concretion of the ground surface which becomes very compact but driver. The anomaly resistant is individualized for the profile 4.
From the electrical and electromagnetic profiles, we were able to identify two types of anomalies:
-Abnormal fracturing, in the middle of profiles and marked by resistivity measurements.
-An abnormality of contact with a compact ground, located at the northeastern edge of the terrain and marked by electromagnetic surveys.
The importance of accumulation of salinity differs depending on whether the plot concerned is irrigated or not. Thus for irrigated soils, salt deposits accumulate progressively irrigations (decrease of surface dilution). While for the non-irrigated soil, the contamination occurred by side chemical transfer.

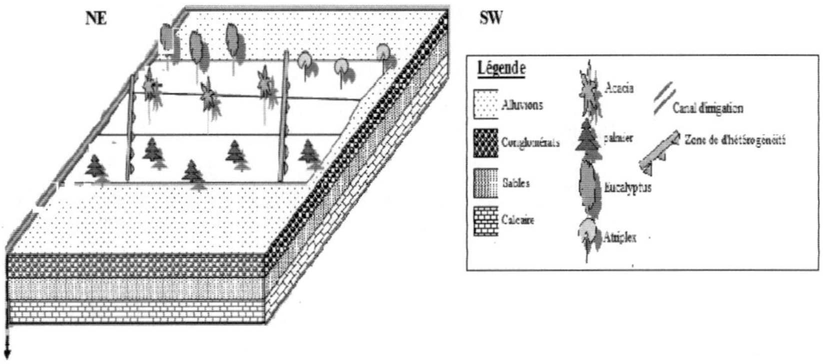

Figure 10. Schematic Figure illustrating the location of the anomalies identified in the experimental station Al Ain Atti

3.3 Result electrical tomography measurements realized in the experimental station Ain Al Atti

In order to obtain better recognition site studied, we have made the electrical tomography, which allows to obtain an image of the basement, that is to say a cutting resistivity or chargeability depending depth. This electrical imaging inform in detail on the impact of this salty irrigation on soil quality, and on the heterogeneity of the land surveyed.
The induced polarization is the origin of the electrochemical processes that occur when the current passes from one medium at ionic conductivity (water) in a medium at electronic conductivity or from ionic conductivity medium in somewhat conductive medium, or in contact with a different ionic conductivity medium.
The polarization measurement technique caused also called electric chargeability, consists of measuring the delay of the response of a medium subjected to an external electric field. This delay is linked to a non-zero time required for a medium to return to the equilibrium when the injected current is cut (phase shift, chargeability).
It can be measured in the time domain (PPT) or frequency (PPS). In the same way as for resistivity, induced polarization measurements are carried out using a quadruple and the results are in the form of profiles or of pseudo-sections. The interpretation can be improved by combining the profiles in the form of a pseudo-sectional three-dimensional image. [8]
In the time domain, applying a direct current in the ground, followed by its stoppage will produce a voltage-time curve (Figure 11). The voltage Vo is one that is observed and measured and is due to the application of the current and polarization effects. The continuous voltage Vs is that which would be observed immediately after the power was cut. This current was applied long enough for V_0 reaches its maximum. It is easily seen that:

$$V_s = V_0 - V$$

The real chargeability of a medium can be measured only in places where there are heterogeneities. What we really measure is the apparent chargeability **(ma)** which depends on real resistivity and chargeability of surveyed materials.
Measuring the voltage-time curve is very complicated (lack of fitted equipment), then another measurement voltage which can be expressed as:

$$V_t = F(m_a, V_0, t)$$

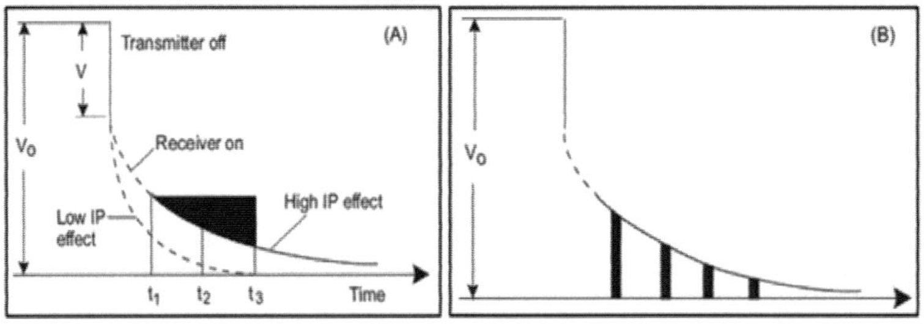

Figure 11. discharge curve

Since Vo is known, the function F (ma, Vo, t) depends only on **ma**. If the measurements are made at set times so you can get an idea of Vt relative values for different m.

$$m = \frac{1}{V_0} \int_a^b V_t(t) dt$$

For significant figures, there will be measures in ms or mV / V.

IOP Conf. Series: Materials Science and Engineering **186** (2017) 012038 doi:10.1088/1757-899X/186/1/012038

Electrical tomography surface allows to obtain a resistivity of cutting (electric image of the basement) according to the depth by measuring the resistivity profile for different combinations of current electrodes and potential.

To achieve our goal, we realized three electrical tomography profiles (type Wenner) in the test site Al Ain Atti (Figure 12). To interpret the various pseudo-sections, we made use of NIRZ 4042/57 drilling data realized in the study area (Figure 3).

Figure 12. location of electrical profiles made at the test site of Al Ain Atti (taken from Google Earth)

The procedure involves the use of an electrical panel based on an acquisition system (ABEM, fig.13) for controlling the measurements, ABEM connected to a resistivity measuring the apparent resistivity with a base electrodes 64 made of stainless steel and 4 cables to connect the electrodes to the acquisition system. With a 12 V battery, the device injects the electric current between the electrodes A and B and measuring the electrical potential between the electrodes M and N. The data-logger can choose from the following devices (Schlumberger, Wenner, dipole - dipole ...).

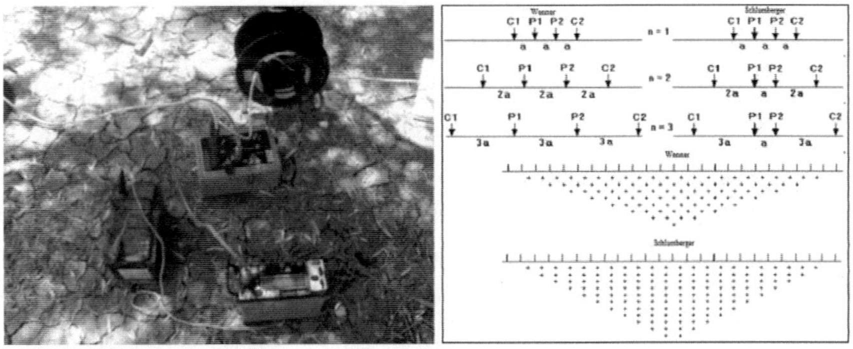

Figure 13. Investigation by the device of tomography and principle of building a pseudo-section

Both resistivity and chargeability parameters are closely related to the mineralized structures. Recently, in an experiment performed on a coring by combining the results of resistivity measurements and the chargeability measures OKAY (2010) showed that newly formed fractures are sub-vertical structures moving in depth and without solid fill (open fractures). In this sector, as pFG1 drilling (Figure 14), the depth of the new fracture crossing is 75 cm. This result confirms the vertical extension of the EDZ characterized by resistivity of inverted sections (Figure 15).

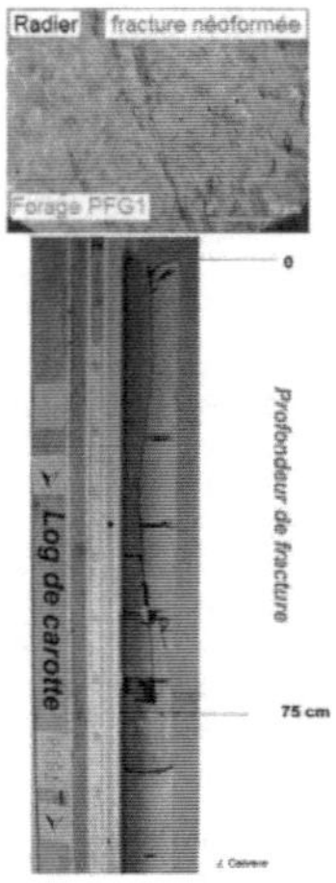

Figure 14. Drilling pFG1 (depth 75 cm). The purple dots represent the location of the electrodes. Small points concern the location of sensors for seismic LCPC (OKAY, 2010)

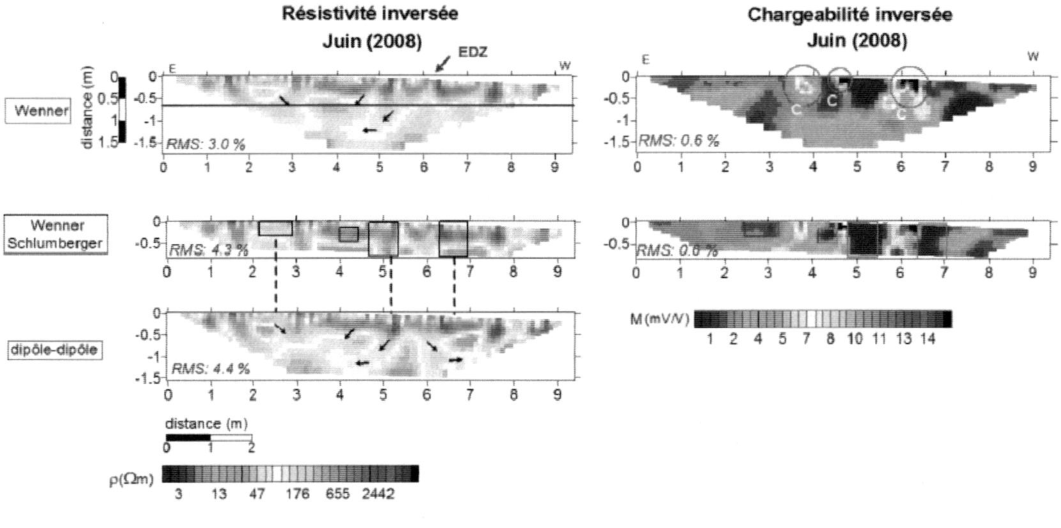

Figure 15. Sections resistivity and chargeability paid for the Wenner-α devices, Wenner-Schlumberger and dipole-dipole on the longitudinal profile to strike in the eastern gallery of the 96-Tournemire experimental station in June 2008 (OKAY, 2011). (The dipole-dipole device is acquired only in resistivity mode C: fracture filled with calcite, arrows indicate a probable path of the desideration process, injection slot: 1 s, the number of iterations used for inversion: 4)

The electrical resistivity is closely related to the water content, the concentration of mineral particles and the nature of dissolved solids, that is why we have combined the chargeability. Characterization studies at the LaRonde tailings mine (Agnico-Eagle Ltd), located in Abitibi, showed fairly conclusive results (Campos, 2004; Anterrieu, 2006). The geophysical measurements on the dump from a height of about 25 m were used to map several anomalies on the slopes and in the heart of the dump with strong resistivity contrasts. These anomalies are associated with particular particle size segregation and compaction level (resulting from construction mode). Anterrieu (2006) coupled the results of electrical measurements with data from geochemistry and hydrogeology to propose a preliminary model of the internal structure of the dump (Figure 16).

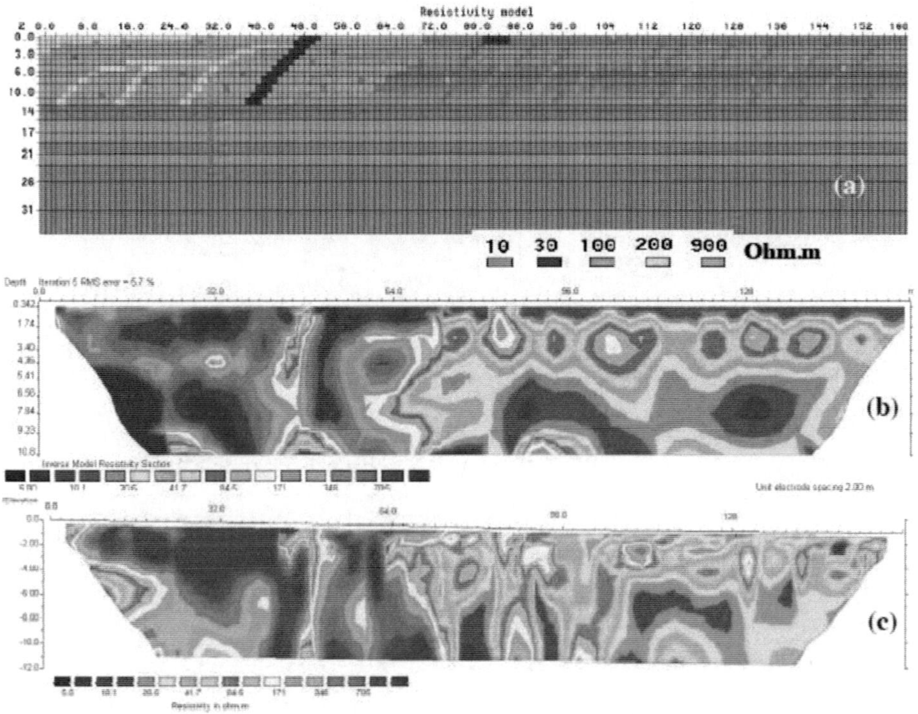

Figure 16. (a) Electrical model of the internal structure of halde- (b) reversed synthetic section (following the proposed model (a)) - (c) reversed section (actual measurements) (adapted Anterrieu, 2006)

This model shows that the upper bed (h <14 m) is formed in part of an alternation of oblique resistive and conductive layers. This alternation is caused by the successive discharge of these variable size materials.
In the station Ain Al Atti,The first profile analysis (Figure 17) suggested a large vertical variation and horizontal discontinuities in chargeability. It is low for level surface formations and higher for those located deep. This is related to atmospheric precipitation inputs (after stopping irrigation salt) that led to the dilution and the mobilization of the mineralization, its lateral transfer and its vertical infiltration. This is why the formation (10 to 40m) seem more conductive (chargeability between 500 and 800 mV / V) in depth of forest plots (Eucalyptus and Acacia). The presence of clay in depth (about 50m) also amplified accumulation of the salinity level of the overlying formations.
In the Atriplex, the chargeability is relatively low (30-100 mV / V), the soil of this land has not been overhauled and has not also suffered significant salty irrigation.
The end of the profile (SW side) is marked by the presence of strong chargeability values over a relatively large depth (20 to 40m). This is likely related to the transition of a fractured area indicated by electrical dragged (Figure 4) [7].

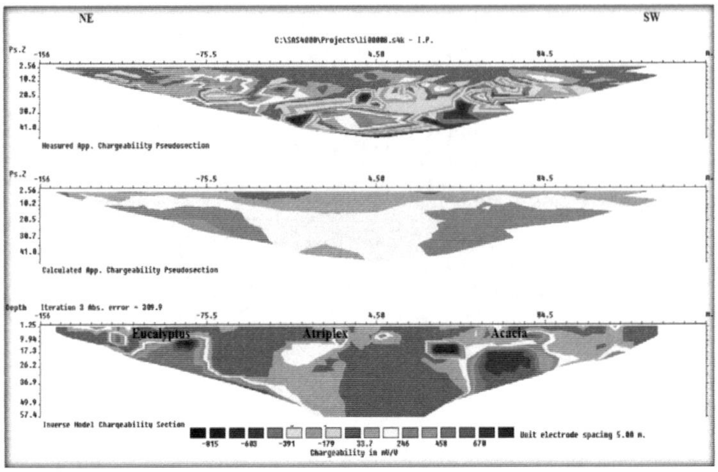

Figure 17. Pseudo-section of the chargeability obtained in experimental plots Eucalyptus, Acacia and Atriplex, (c.l: X = 616428,44m, Y = 507028,81m and Z =854m)

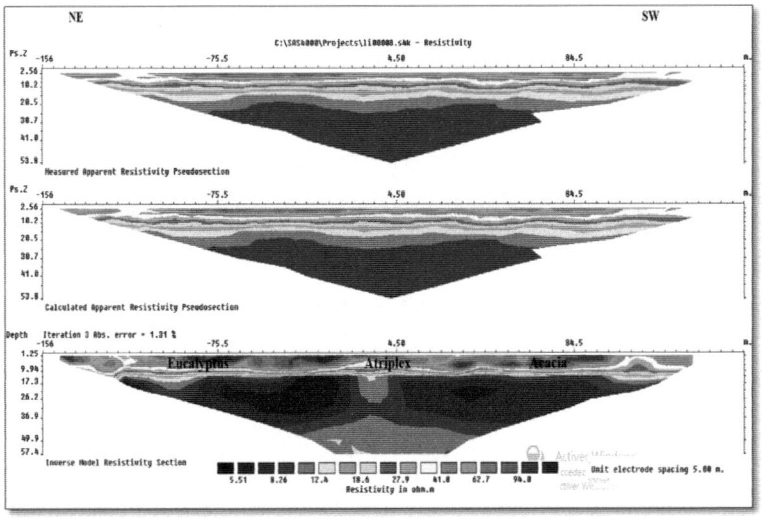

Figure 18. Pseudo-section of the resistivity obtained in experimental plots Eucalyptus, Acacia and Atriplex)

Regarding the resistivity measurements, the pseudo-section (Figure 18) suggests that it is high (> 50 ohm.m) in surface formations and small in depth, this is related to both the presence sometimes conglomerates and also the concretion which is taking place after the iron oxide from the salt water. This concretion, which increases the hydraulic resistance is less pronounced in the plot of Atriplex.

In depth, the ground appears slightly conductive for the entire profile with lower resistivity values at Eucalyptus plots and Acacia (5 to 7 ohm.m), relating to the accumulation of salinity in a transition to marl areas.
These results confirm those obtained by the result of the chargeability. The observed response to the passage of the fractured area seems less pronounced compared to that obtained by the chargeability.
The second profile of chargeability (Figure 19) is realized out in the experimental field, at 30 meters south of the first profile and in the same direction (NE-SW). The variation in chargeability was made

in the same manner as in the first profile, it is strong in the areas where the lateral flow is from forest heavily irrigated areas (Acacia and Eucalyptus) and low in the areas related party of less irrigated (Atriplex vegetation). But note that the values obtained by this pseudo-section are proportionally lower than those observed in the first profile. Indeed, the phenomenon of lateral transfer salinity is amplified by the presence of a deep sand (20 to 28m), located by electrical surveys and mechanical boreholes drilled in this zone [7]. Contamination of land surrounding the experimental site is done especially during rainy periods. Note also that this profile met the fractured zone observed in SW flank where the chargeability is high (300 mV / V).

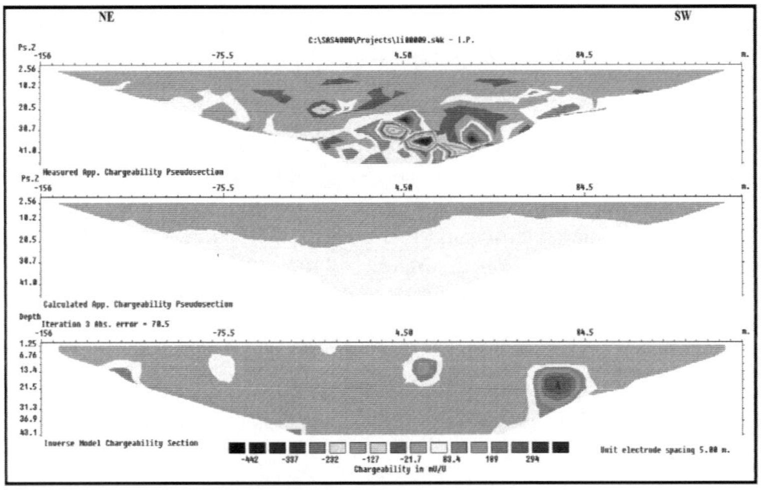

Figure 19. Pseudo-section of the chargeability obtained in ground control at the southern end of the experimental station (c.l: X = 616,428.44 m, Y = 507028,81m and Z = 854m).

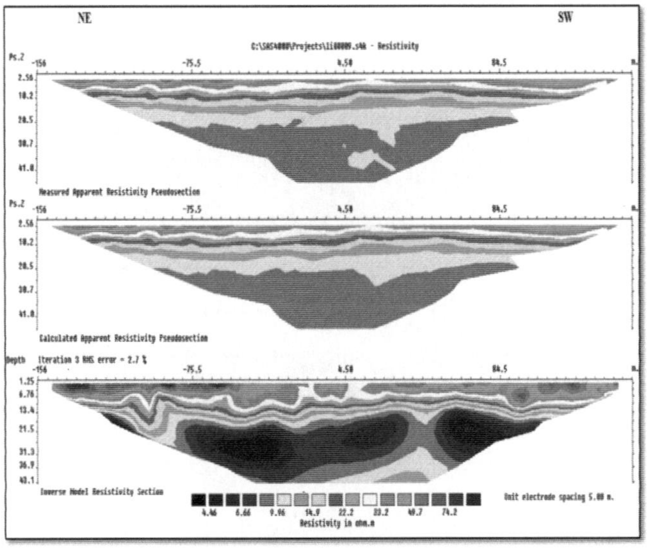

Figure 20. Pseudo-section resistivity obtained in ground control at the southern end of the experimental station.

According to the pseudo-section (Figure 20), the resistivity shows the same variation as chargeability. Subsurface formations are less conductive and correspond to dry silt and gravel. The deep formations (13-43 m) are conductive and licks whose resistivity varies between 6 and 13 ohm-m. The resistivity values are substantially the same as those encountered with the first profile. The conductivity of sand and clay formations in depth is important in terms of forest plantation (Eucalyptus and Acacia) and slightly lower in the Atriplex (middle profile).

The analysis of these two profiles provided important lessons for the Contrast met by the side of prospecting different experimental soils. Indeed, the contrast obtained by the analysis of chargeability profiles seems more better than that observed by the variations in resistivity. The result has confirmed the location of the heterogeneous natural terrain marked by electromagnetic profiles.

The third profile (Figure 21) is the other ground control (outside of the experimental station) parallel to the first, but its location is in the north and upstream of the natural flow of water that affect this area of experimentation.

The response shows no horizontal or vertical variation of chargeability and the recorded value is approximately 40 mV/V, it corresponds to a silty-sandy clays and conductive formation (Infra-Cenomanian). The prospected ground seems more homogeneous. The only observed heterogeneity is marked by the presence of a narrow area marked on the side NE at a depth of 8m (upper 183mV/V) which must correspond to the location of an area contaminated by the transfer from water spills drilling used for irrigation and located on the side of the land (approximately 100 m).

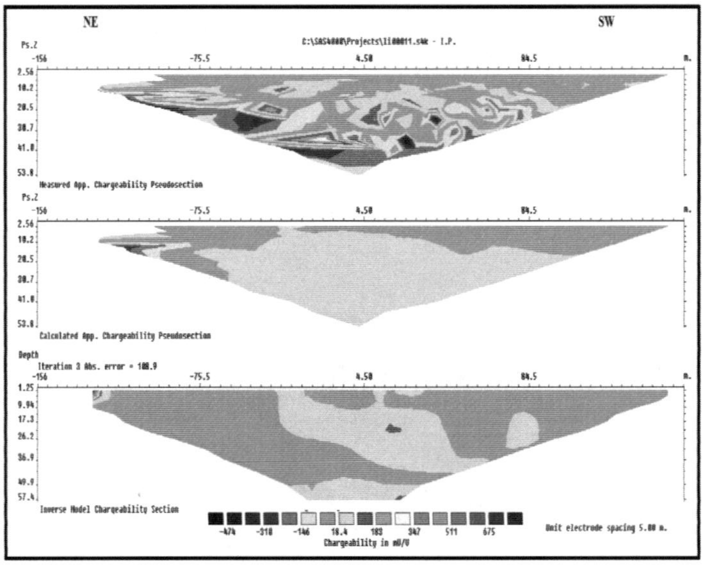

Figure 21. Pseudo-section of the chargeability obtained in ground control at northern part of the experimental station (c.l X = 506942,33m,Y = 616,435.95 m and Z = 851m).

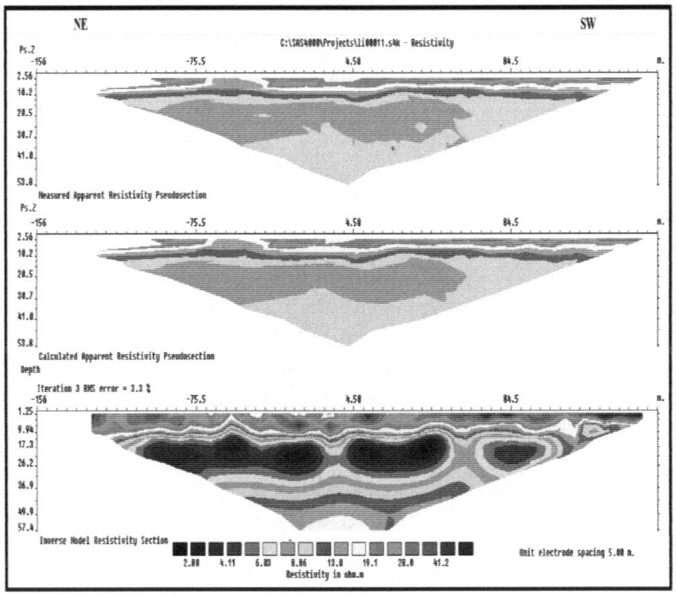

Figure 22. Pseudo-section of resistivity obtained in ground control at the northern part of the experimental station Ain Al Atti.

The formations identified by the profile (Figure 22) show a sub-table structure. The resistivity values obtained confirm the variations observed by the measures chargeability.
On the surface, dry and relatively resistant alluvium based on wet and conductive sand formations which overcome the sandy loams of Infracenomanien (around 6 ohm.m) and located at a depth 26m. The conductivity is related to the accumulation of the likely outcome salinity evaporate formations towards contaminating all smooth sheet of the region. In depth (50 m) appears the conductive sandy-clay formation (20 ohm-m).
On the surface, induced polarization measurements results show no significant variation between the experimental plots (around 40mV/V), this is mostly related to lateral transfer salinity and infiltration depth under the effect of precipitation atmospheric.
In depth, the response of the chargeability provides sufficient contrast to distinguish between the different plots. It is considerably high at the forest units (> 500 mV/V) than at Atriplex (<40 mv / v).
The response of the two pseudo-sections obtained in ground control shows that the north side was not affected by salt contamination, lateral transfer salinity is therefore carried from North to South in the direction of the natural flow groundwater.

4. Conclusion

The coupling of both electromagnetic and electrical tomography techniques has provided information significantly important and rewarding to study the impact of salinity in soil of Al Ain Atti station. He has displayed vertically and horizontally different heterogeneities that exist within the land surveyed.
The pseudo-sections obtained have characterized the impact of spraying salt water on the quality of soil amendment. They showed, in a causal manner, the location of the fracture conductive anomalous zone that crosses the field prospected according a alignment oriented in NNE-SSW and identified by the result of electrical and electromagnetic measurements.
In depth, the contrast of chargeability seems better than that obtained by the resistivity measurements.

It is also noted that saline irrigation has caused not only buildups of surface soils, but also an important clogging pedogenitic horizon amendment. Which resulted in a lateral anisotropy of the ground explored is why the signal obtained by chargeability seems discontinuous. It must be said that the result would have been better if the measures had been carried out during irrigation periods.

For non-irrigated ground, soil contamination occurred side by chemical transfer, without forgetting the part of salt that rise subsurface under the effect of evapo-transpirator phenomenon during drought period.

References

[1] Levy. R., (1949). Le bassin du Ziz-Guir. Rapport Géologique Inédit, Maroc, Rabat. 35 p.

[2] Ammary B., (2007). Etude géochimique et isotopique des principaux aquifères du bassin Crétacé d'Errachidia et de la plaine de Tafilalt. Thèse Doc. Uni. Med V Agdal, Rabat. 134p.

[3] Choubert G., (1920-1945). Essai sur la paléogéographie de Mésocrétacé marocain. Vol. Jubi.Soc. Sc. Natu. Maroc. PP.307-325.

[4] Choubert G. et Fauve-Muret A., (1960-62). Evolution de domaine atlasique marocain depuis les temps paléozoïques. In: Livre mémoire P.Fallot, Mém. h. ser. soc. Géol. france, 447-527.

[5] DRHGRZ, (2006). Rapport édit., novembre 2006 : débat national sur l'eau.

[6] Margat J., (1977). Etude hydrogéologique du bassin Quaternaire de Tafilalet. Ressources en Eau du Maroc, p.310-380.

[7] Benamara A., (2009). Application de la technique électrique et électromagnétique à l'étude de l'épandage des eaux salées au niveau de la station expérimentale de Ain AL Atti (région d'Errachidia). 2eme Colloque sur la création d'entreprises dans le domaine de l'écotourisme. 21 et 22 Mars 2009. FSTE d'Errachidia.

[8] Guérin R., (2004). Contriution à l'hydrogéophysique. Thèse d'Hbilitation présenté à l'Université de Pierre et Marie Curie.

IOP Conf. Series: Materials Science and Engineering **186** (2017) 012039 doi:10.1088/1757-899X/186/1/012039

Performances of nanofiltration and low pressure reverse osmosis membranes for desalination: characterization and modelling

Y A Boussouga and A Lhassani

Laboratory of Applied Chemistry, Faculty of Science and Technology-Fez,
Sidi Mohamed Ben Abdellah University, P.O. Box 2202, Fez, Morocco

Email: boussouga.youssef@gmail.com

Abstract. The nanofiltration and the reverse osmosis processes are the most common techniques for the desalination of water contaminated by an excess of salts. In this present study, we were interested in the characterization of commercial, composite and asymmetric membranes of nanofiltration (NF90, NF270) and low pressure reverse osmosis (BW30LE). The two types of characterization that we opted for our study: (i) characterization of electrical proprieties, in terms of the surface charge of various membranes studied by the measurement of the streaming potential, (ii) hydrodynamic characterization in terms of hydraulic permeability with pure water, mass transfer and phenomenological parameters for each system membrane/salt using hydrodynamic approaches. The irreversible thermodynamics allowed us to model the observed retention R_{obs} of salts (NaCl and Na_2SO_4) for the different membranes studied, to understand and to predict a good filtration with a membrane. A study was conducted on the type of mass transfer for each system membrane/salt: convection and diffusion. The results showed that all tested membranes are negatively charged for the solutions at neutral pH, this is explained by their material composition. The results also showed competitiveness between the different types of membranes. In view of that the NF remains effective in terms of selective retention with less energy consumption than LPRO.

Keywords: membrane, nanofiltration, reverse osmosis, streaming potential, mass transfer

1. Introduction

The nanofiltration and the reverse osmosis are the most used techniques for water desalination. The NF membrane is a type of pressure-driven membrane which properties are situated between reverse osmosis (RO) and ultrafiltration (UF) membranes. This positioning of this type of membrane offers advantages, in terms of, low operational pressure, high flux, high retention of multivalent anion salts, relatively low investment and low operation and maintenance costs [1-2].

In order to understand the functioning of membranes, and to make the choice of a membrane for a given application, we have developed two types of characterizations: first (i) characterization of electrical proprieties, in terms of streaming potential and secondly (ii) hydrodynamic characterization, in terms of, the hydraulic permeability, the salts retention, type of mass transfer (diffusion and/or convection).

In this present study, we were interested in the characterization of commercial, composite and asymmetric membranes of nanofiltration (NF270 and NF90) and low pressure reverse osmosis(BW30LE). The composition of the active layer for these membranes is in polyamide, and the support layer is made of polysulfone (Figure 1).

Figure 1. Structure of organic composite membranes studied

2. Theory

2.1. Streaming potential

The membrane charge into contact with an aqueous electrolyte solution can be determined by the measurement of the streaming potential SP which can be linked to the zeta potential by the equation [3-4-5-6-7]:

$$SP = \frac{\Delta\phi}{\Delta P} = \frac{\varepsilon\xi}{\mu\chi} \quad (1)$$

With $\Delta\phi$ is trans-membrane potential difference; ΔP is the trans-membrane pressure; ε is the permeability; $\zeta$is the zeta Potential; μ is the dynamic viscosity and is χ he ionic conductivity of the electrolyte solution.

2.2. Hydrodynamic approaches

This approach is based on the irreversible thermodynamics processes [8-9-10]. The equations describe the relationship between the solvent flux and the solute flux which cross through a RO/NF membrane:

$$J_v = L_p(\Delta P - \sigma\Delta\pi) \quad (2)$$

$$J_s = P_s(C_m - C_p) + (1-\sigma)J_v C_m \quad (3)$$

Where J_v and J_s are respectively the solvent flux and the solute flux, ΔP and $\Delta\pi$ define respectively the pressure and osmotic differences between each side of the membrane, L_p is the hydraulic permeability to pure water, σ is the reflection coefficient, P_s is the solute permeability, and the C_m with C_p the concentrations respectively at the surface of the membrane in the bulk side and in the permeate.
The observed rejection as reported by some authors may be expressed as [11-12-13-14]:

$$R_{obs} = \frac{1}{\dfrac{(1-\sigma)\exp\left(\dfrac{J_v}{k}\right)}{\sigma\left(1-\exp\left(-\dfrac{(1-\sigma)}{P_s}J_v\right)\right)} + 1} \quad (4)$$

Where, *k* is the mass transfer coefficient.

2.3. Diffusion and convection of salts
In Equation 3, the total flow of salts appears as the sum of diffusion and convection terms [15-16-17]. Thus, it's possible to write the following equation:

$$C_p = J_{diff}\frac{1}{J_v} + C_{conv} \quad (5)$$

3. Materials and methods

3.1. Streaming potential measurements
The cell is a dead-end cell which has a capacity of 50 ml and used with a flat-sheet membrane. The unit operates at pressure differences using pressurized nitrogen gas as a driving force. Figure 2 shows a schematic representation of the cell which is useful to support membrane samples for the streaming potential measurements.

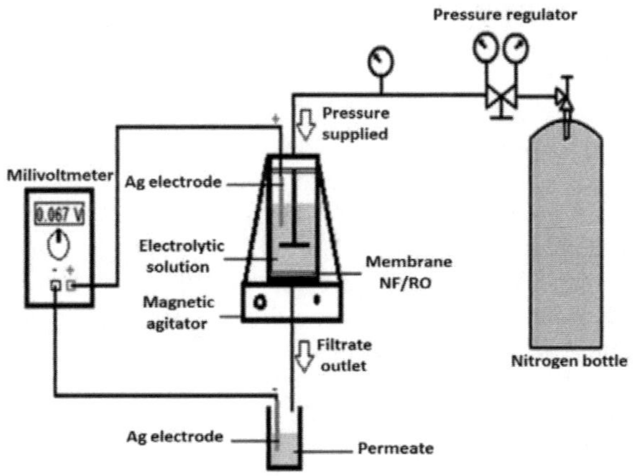

Figure 2. Schematic representation of the SP measurement device.

3.2. Laboratory scale filtration unit
A flat sheet laboratory scale NF/RO setup consisting of a planar cross flow module, with a membrane area of 138 cm^2.The total volume of the system was 5 L. A schematic representation of the equipment is illustrated in Figure 3.

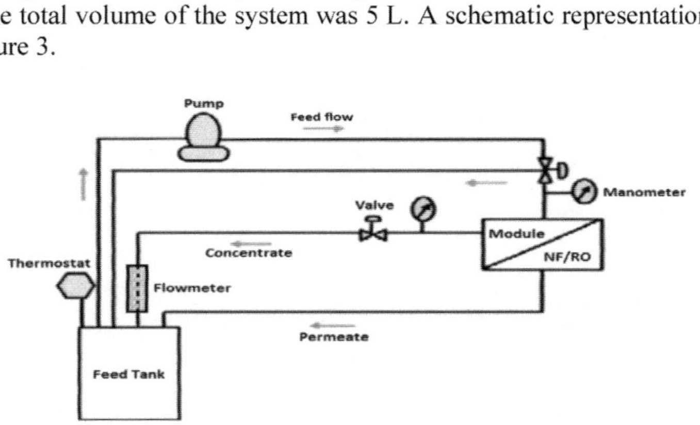

Figure 3. Schematic representation of NF/RO filtration assembly.

IOP Conf. Series: Materials Science and Engineering **186** (2017) 012039 doi:10.1088/1757-899X/186/1/012039

4. Results and discussion

4.1. Characterization of electrical proprieties

As shown in Figure 4, the experience consists to measure the coefficient of streaming potential SP in function of pH.

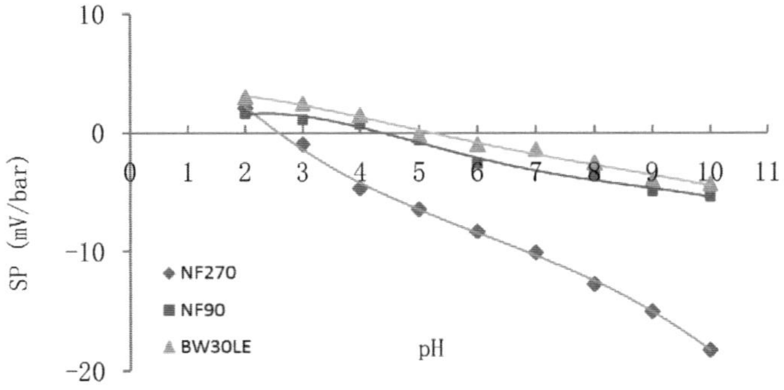

Figure 4. Streaming potential as a function of pH; KCl10^{-3} M; NF270, NF90 and BW30LE membranes.

In the case of the organic membrane made in polyamide membrane, the amphoteric character of the amino groups of the membrane surface reveals a positive value or a negative value of the streaming potential coefficient SP, according to the pH value [18]. The isoelectric point IEP is the pH value in which the streaming potential coefficient cancels.

Table 1. Streaming potential (SP) and isoelectric point (IEP) of the NF and LPRO membranes.

NF/LPRO	NF270	NF90	BW30LE
Charge (neutral pH)	Negative	Negative	Negative
SP (mV/bar)(KCl 10^{-3}M; pH =6.6)	-9.4	-6.2	-5.2
IEP (KCl 10^{-3}M)	2.6	4.4	4.9

4.2. Hydrodynamic characterization

4.2.1. The hydraulic permeability. The permeability with ultra-pure water (1µs/cm)L_p of investigated membranes is showed in Figure 5.

Table 2. Values of pure water permeability of the NF and LPRO membranes.

NF/LPRO	NF270	NF90	BW30LE
L_p (10^{-6} m.s^{-1}.bar^{-1})	1.52	1.23	1.14

The high permeability characteristic generally indicates a high porosity [12];this is what is observed in the results of the permeability in Table 2which means that NF270 has the highest porosity and provides a higher permeate flow than NF90 and BW30LE.

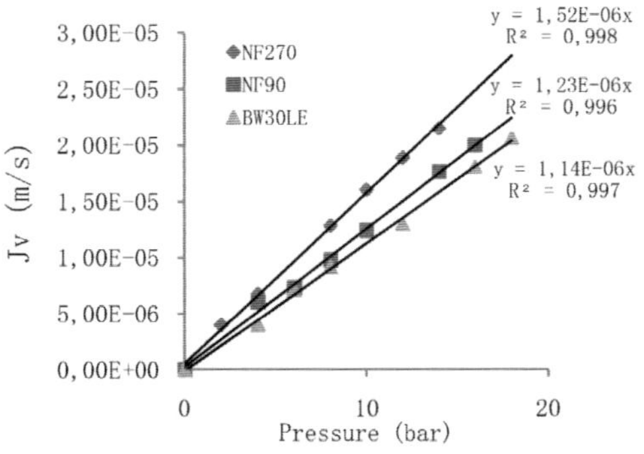

Figure 5. Pure water flux as a function of the applied pressure for the NF and LPRO membranes (T = 25°C, pH =6.6).

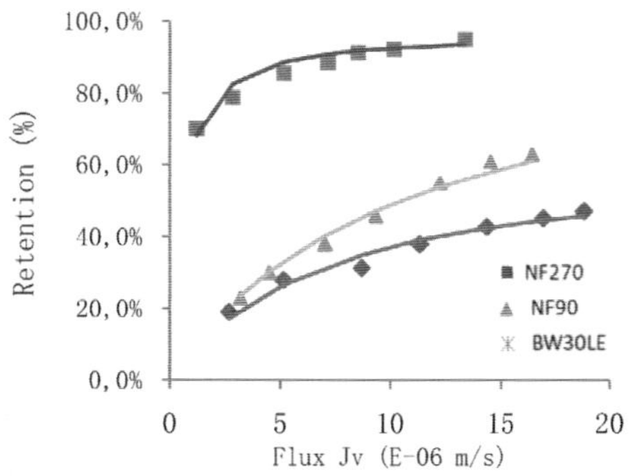

Figure 6. Evolution of NaCl salts retention as a function of the permeate flow for NF/LPRO.

4.2.2. Modelling of salts retention. As shown in Table 3, the retention of divalent salts Na_2SO_4 is higher than monovalent salts NaCl for both type of membrane, this phenomenon can be explained by the hydration energy of salts [6]. The retention of both salts is also dependent of the porosity for each membrane. The experimental data of rejection and flux for all investigated membranes and salts were fitted using the best fit values of P_s and σ obtained according to Equation 4.In Figure 6, the modelling showed us that there is a good agreement between experimental and theoretical results. The reflection coefficient σ is a characteristic of the convective transport of the solute [13]. For σ value of 100% (or 1), mass transfer by convection is totally hindered, it's the case of BW30LE with both salts and NF90 with Na_2SO_4 only (Table 4).The results of the permeability of salts P_s decreases initially with the decrease of the pore size of a membrane [19]. The decrease of P_s can also due to the increase of a salt size (or hydration energy of a salt).

Table 3. Retention of NaCl and Na_2SO_4 salts at three pressures and 10^{-3}M of concentration for the NF and LPRO membranes.

NF/LPRO	%Robs of Salts (10^{-3} M)					
	NaCl			Na_2SO_4		
	P = 4 bar	P = 8 bar	P = 12 bar	P = 4 bar	P = 8 bar	P = 12 bar
NF270	28.0	38	45.2	90.1	93.3	95.4
NF90	30	45.9	61	93.8	95.4	96.8
BW30LE	74	80.5	85.3	94.2	95.7	98

Table 4. Phenomenological parameters and mass transfer parameters estimated for NF and LPRO membranes for NaCl and Na_2SO_4.

NF/LPRO	Salts (10^{-3} M)							
	NaCl				Na_2SO_4			
	σ	P_s (10^{-6} m.s^{-1})	C_{conv} (g.L^{-1})	J_{diff} (10^{-6}m.s^{-1})	σ	P_s (10^{-6} m.s^{-1})	C_{conv}(g.L^{-1})	J_{diff} (10^{-6}m.s^{-1})
NF270	0.53	6.26	0.0252	0.1165	0.96	0.343	0.0054	0.0333
NF90	0.87	9,62	0.0182	0.0971	0.98	0.139	0.0018	0.0198
BW30LE	0.98	0.525	0.0023	0.0250	0.99	0.111	0.0019	0.0131

4.2.3. Type of mass transfer. In Figure 7 and Figure 8 the plot of C_p vs. $1/J_v$ is linear and the coefficient of determination R^2 is very near to 1which is confirming the relation of Equation 5.

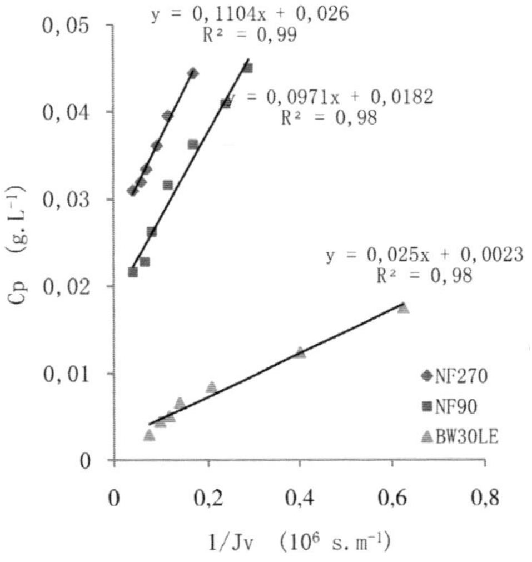

Figure 7. Permeate concentration C_p evolution as a function of the ratio $1/J_v$ for NF and LPRO membranes; ([NaCl]=10^{-3} M; pH = 6.8; *Y = 5%; T = 24°C).

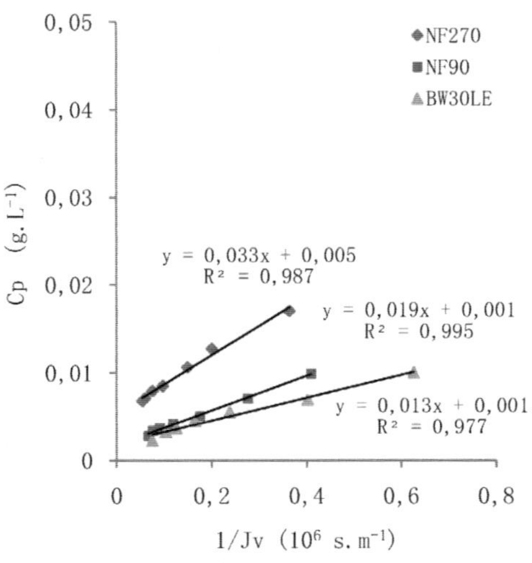

Figure 8. Permeate concentration C_p evolution as a function of the ratio $1/J_v$ for NF and LPRO membranes; ([Na_2SO_4]=10^{-3}M; pH = 6.8; *Y=5%; T=24°C).

*Y: Conversion rate (Y% = (permeate flux/ Total flux)*100)

In this experience, we observed that the mass transfer for reverse osmosis membranes is a diffusion type; it's the case of BW30LE for both salts. In nanofiltration membranes, we observe two types of solute transport mode [13], with NaCl is by convection which is a physical nature (Figure 7), and with Na_2SO_4 is by diffusion which is a chemical nature[20] (Figure 8).

5. Conclusion

The characteristic of electric and hydrodynamic properties which made for commercial and composite membranes showed a competitiveness of NF overlooked LPRO, in terms of, a selective separation of salts (selective desalination), an important hydraulic permeability and lower energy consumption in NF than in LPRO membranes. This study also showed us the functioning and the performance of each membrane studied for a given application.

References

[1] Hilal N,Al-Zoubi H, Darwish N A,Mohammad A W and AbuArabi M2004 *Desali***170** 281–308.
[2] Lhassani A, Dach H, Pontie M and Diawara C, 2008 *Journal des Sciences et Technologie***7** 32-8
[3] Kim K J, Fane A G, Nystrom M, Pihlajamaki A, Bowen W R and Mukhtar H 1996 *J. Memr. Sci***116** 149-59
[4] Afonso M, Hagmeyer G and Gimbel R 2001 *Sep. Purif. Technol* **22** 529-41
[5] Moritz T, Benfer S, Arki P and Tomandl G 2001 *Colloids Surf. A***47**25-33
[6] Fievet P and Szymczyk A 2002 *C. R. Chimie***5** 493–505
[7] Boussu K, Zhang Y, Cocquyt J, Van der Meeren P, Volodin A, Van Haesendonck C, Martens J A and Van der Bruggen B 2006 *J. Membr. Sci***278** 418-27
[8] Spiegler K S and Kedem O 1966 *Desali***1** 311–26
[9] Kedem O, Katchalsky A 1962 *Biophys.* **J2** 53-78
[10] Baranowski B 1991*J. Membr. Sci***57** 119
[11] Jain S and Gupta S K 2004*J. Membr. Sci***232** 45–61

IOP Conf. Series: Materials Science and Engineering **186** (2017) 012039 doi:10.1088/1757-899X/186/1/012039

[12] Pontié M, Dach H, Leparc J, Hafsi M and Lhassani A 2008 *Desali***221** 174–91
[13] Kelewou H, Lhassani A, Merzouki M, Drogui P and Sellamuthu B 2011*Desali***277** 106–12
[14] Wu F, Feng L and Zhang L2015*Desali* **362** 11-17
[15] Lhassani A, Benjelloun D and Rumeau M 2000 *Trib. Eau***53** 100–07
[16] Pontie M, Diawara C, Lhassani A, Dach H, Rumeau M, Buisson H and Schrotter J C 2006 *Adv. Fluor. Sci***2** 49-80
[17] Diawara C K, Sidy Lô M, Rumeau M, Pontie M and Sarr O 2003 *J. Membr. Sci***219** 103-12
[18] Brant J A, Johnson K M and Childress A E 2006*Colloids Surf. A***280** 45-57
[19] Hilal N, Al-Zoubi H, Darwish N A, Mohammad A W and Abu Arabi M, 2004 *Desali***170** 281-308
[20] Maurel A, 1993*Tech. Ing. Génie Proc***3J2790** J2790.1

Institute of Physics
Dirac House, Temple Back
Bristol BS1 6BE UK

ISSN: 1757-8981
ISBN 978-1-5108-3773-7

CURRAN ASSOCIATES INC.
proceedings.com